Israel Gohberg and Friends

On the Occasion of his 80th Birthday

Harm Bart
Thomas Hempfling
Marinus A. Kaashoek
Editors

Birkhäuser
Basel · Boston · Berlin

Editors:

Harm Bart
Econometrisch Instituut
Erasmus Universiteit Rotterdam
Postbus 1738
3000 DR Rotterdam
The Netherlands
e-mail: bart@few.eur.nl

Marinus A. Kaashoek
Department of Mathematics, FEW
Vrije Universiteit
De Boelelaan 1081A
1081 HV Amsterdam
The Netherlands
e-mail: m.a.kaashoek@few.vu.nl

Thomas Hempfling
Editorial Department Mathematics
Birkhäuser Publishing Ltd.
P.O. Box 133
4010 Basel
Switzerland
e-mail: thomas.hempfling@birkhauser.ch

Library of Congress Control Number: 2008927170

Bibliographic information published by Die Deutsche Bibliothek
Die Deutsche Bibliothek lists this publication in the Deutsche Nationalbibliografie;
detailed bibliographic data is available in the Internet at <http://dnb.ddb.de>.

ISBN 978-3-7643-8733-4 Birkhäuser Verlag, Basel – Boston – Berlin

© 2008 Birkhäuser Verlag AG
Basel · Boston · Berlin
P.O. Box 133, CH-4010 Basel, Switzerland
Part of Springer Science+Business Media
Printed on acid-free paper produced of chlorine-free pulp. TCF ∞
Printed in Germany

ISBN 978-3-7643-8733-4   e-ISBN 978-3-7643-8734-1

9 8 7 6 5 4 3 2 1   www.birkhauser.ch

Contents

Preface

Mathematicians do not work in isolation. They stand in a long and time honored tradition. They write papers and (sometimes) books, they read the publications of fellow workers in the field, and they meet other mathematicians at conferences all over the world. In this way, in contact with colleagues far away and nearby, from the past (via their writings) and from the present, scientific results are obtained which are recognized as valid. And that – remarkably enough – regardless of ethnic background, political inclination or religion.

In this process, some distinguished individuals play a special and striking role. They assume a position of leadership. They guide the people working with them through uncharted territory, thereby making a lasting imprint on the field. Something which can only be accomplished through a combination of rare talents: unusually broad knowledge, unfailing intuition and a certain kind of charisma that binds people together.

All of this is present in Israel Gohberg, the man to whom this book is dedicated, on the occasion of his 80th birthday. This comes to the foreground unmistakably from the contributions from those who worked with him or whose life was affected by him. Gohberg's exceptional qualities are also apparent from the articles written by himself, sometimes jointly with others, that are reproduced in this book. Among these are stories of his life, some dealing with mathematical aspects, others of a more general nature. Also included are reminiscences paying tribute to a close colleague who is not among us any more, speeches or reviews highlighting the work and personality of a friend or esteemed colleague, and responses to the laudatio's connected with the several honorary degrees that were bestowed upon him.

The documents collected here give a fascinating and sometimes moving insight in the human factors that influence the development of mathematics. The focus is not on formal mathematical results but instead on the personal relationships that constitute the underlying propelling-power in scientific cooperation. Centered around the remarkable figure of Israel Gohberg, a picture emerges of the development of operator theory and its applications during the last four or five decades.

The book is divided in seven parts which we will now briefly describe.

Part I. This part contains two presentations written by Gohberg himself in which he speaks about his mathematical development in former years and on some general, philosophically tinted aspects of mathematics. It also features a piece written by the two Gohberg daughters, Zvia and Yanina, giving an impression of what it meant to be the children of a loving father leading a life devoted to mathematics.

Part II. This part starts with some documents carrying factual information: Gohberg's curriculum vitae, a list of his publications and a list of his Ph.D students. Also included are a translation of a letter of reference written by M.G. Krein, when Gohberg was a master student, and translations of letters and telegrams supporting his nomination as a corresponding member of the Academy of Sciences of the Moldavian SSR. The next two documents, written by Rien Kaashoek and by Rien Kaashoek and Leonid Lerer, respectively, present a review of Gohberg's mathematical work. The final document of this part concerns the Nathan and Lily Silver Chair of Mathematics at Tel-Aviv University which was created for Israel Gohberg especially and has played such an important role in his life after his immigration to Israel.

Part III. This part reproduces the Gohberg Miscellanea as published earlier on the occasion of Gohberg's 60th birthday in Operator Theory: Advances and Applications, Vol. 40. This biographical text was composed from reminiscences, notes, letters and speeches prepared by Gohberg's former students, colleagues and friends.

Part IV. This part contains the dinner speeches of the 1998 IWOTA meeting in Groningen as they appeared in Operator Theory: Advances and Applications, Vol. 124.

Part V. This part reproduces articles that were written or co-authored by Israel Gohberg himself. Some of those have the character of an *In Memoriam*, paying tribute to a dear colleague who has passed away. Others are *Recollections, Reminiscenses* or *Reviews* that highlight the work and personality of a friend celebrating a special occasion. These documents taken together give a fascinating and sometimes moving insight in the human factors that have influenced the development of the field.

Part VI. This part contains material connected with the honorary doctorates of Israel Gohberg. Included are laudatios, acceptance speeches and other related documents.

Part VII. This final part consists of short articles written by friends and colleagues in which they reflect on their experiences with Israel Gohberg. The contributions are comparable to those of Parts III and IV but written especially for the present occasion. They again underline the impact Israel had on the lives of those that worked with him.

As indicated several documents appearing in this book have been published before. At appropriate places references are given. Here and there typos have been corrected. In translations of some documents into the English language the editors have taken the liberty of rewording certain phrases in order to clarify their meaning. Finally, editorial comments have been added at a few places.

Acknowledgement: We want to express our gratitude to Efim Sigal for allowing us to use his artistic drawing of Israel Gohberg as front cover. Also thanks to Cor Koole who made many photographs in 2003 at a party in Bleiswijk, The Netherlands, celebrating Israel Gohberg's 75th birthday. Several of these photographs are reproduced in this volume. His help in arranging the pictures is highly appreciated. We are also grateful to Andrei Iacob, who helped us with the translations from Russian into English of some of the documents appearing in Part II. Finally, we thank our colleagues for their contributions to Part VII of this volume.

The editors: Harm Bart, Thomas Hempfling, Rien Kaashoek   May 2008

Congratulations from the Publisher

Thomas Hempfling

It is a great pleasure to congratulate Israel Gohberg on the occasion of his 80th birthday, in particular by producing this special commemorative book full of stories, reminiscences, speeches and eulogies. But it is at least such a great pleasure to celebrate his 30 years of cooperation with Birkhäuser.

In March 1978 the first issue of the journal *Integral Equations and Operator Theory (IEOT)* appeared. It started with one volume of four issues per year and a total of 30 articles. In fact, it had been one year earlier, in 1977, that Israel Gohberg got in touch with Birkhäuser for the first time, in order to sign a contract for the German edition of the Gohberg/Krupnik book on singular integral equations – which appeared in May 1979.

Already in 1982 the number of issues IEOT published was increased to 6, and in 1993 it was decided to produce two volumes per year and a total of 8 issues, and only one year later it reached 12 issues per year within three volumes, i.e., one issue per month. The total number of pages was extended to 1800 in 2005 and is kept at this level since. To give a few more figures: up to now about 1600 articles have been published in IEOT, since 2003 all articles are published *Online First*; in addition, since 2003 all articles that appeared so far have been retro-digitized and are available online, and from July 2008 all articles will be managed through the online system *Editorial Manager*. Thus, content-wise as well as following the most recent technical improvements, the journal always developed well.

In 2001 Sjoerd Verduyn Lunel joined Israel Gohberg as editor of the journal. He successfully managed the transition to the requirements of the online version, in particular the complete switch to LaTeX files, and he always took great care of even technical details. From July 2008 Christiane Tretter will be co-editor of Israel Gohberg, moving IEOT to the next phase. In addition, the Editorial Board will be substantially extended.

All these developments were supported by the work of Greta Riesel who served as editorial assistant for IEOT from almost the beginning up to 2002 when the technical responsibility for the journal moved to Sjoerd Verduyn Lunel. Her dedication and excellent work was most beneficial for the journal. IEOT was also

Israel Gohberg and Sjoerd Verduyn Lunel at Israel Gohberg's 75th birthday party in Bleiswijk, the Netherlands, 2003.

well supported by the administration of Tel Aviv University and the School of Mathematical Sciences.

Only one year after the creation of the journal, the book series *Operator Theory: Advances and Applications (OTAA)* was founded. What had been started as a series of occasionally appearing books became the most productive specialized series at Birkhäuser. Meanwhile it publishes an average of eight titles each year, comprising more than 180 volumes up to now. Looking at the proposals waiting for publication it can easily be foreseen that the publication of volume 200 is not very far away. Every single volume has been reviewed by Israel Gohberg, and in about one third of the volumes he was either co-author or co-editor. His ability to convince authors to write or finish a book and his high expectations with regard to the quality of the manuscripts, where he often discussed improvements in detail with the involved authors, led to a continuous success of OTAA.

Two subseries have been introduced to OTAA: *Advances in Partial Differential Equations* in 2001, with seven volumes published, and *Linear Operators and Linear Systems* in 2005, with four volumes published, both edited by distinguished subseries editors.

Working on IEOT and OTAA with Israel Gohberg always was extremely productive and pleasant. All discussions and developments have been carried out in a very professional and friendly style, keeping in mind the scientific, economic and technical aspects.

What to expect? Publishing business is always under further development, and so is Science. Israel Gohberg's commitment seems unbreakable – he usually has a very clear picture of possible next steps. We at Birkhäuser's are proud of this long-standing fruitful cooperation and look forward to further developments and publications in IEOT, OTAA or in other products related to the general themes of interest in research of Israel Gohberg and his many collaborators.

Part I
Mathematical Tales and
Philosophical-Mathematical Tales

This part begins with *Mathematical Tales*, a presentation given by Israel Gohberg at the 1988 Calgary Conference organized to celebrate his 60th birthday. It contains stories from Gohberg's career in mathematics, mostly from the times when he lived in the Soviet Union before immigrating to Israel. The paper is preceded by an introduction by Ralph Phillips. The second contribution, *Philosophical-Mathematical Tales: A personal account*, is a talk given by Gohberg in January 2002 at the University of West Timişoara, where he was awarded the degree of honorary doctor. It contains reflections on the general nature of mathematics and on the way mathematical research is done. Finally, there is *Dad's Mathematics*, an account of the two Gohberg daughters on how their father's mathematics came to them in their younger years. It gives an impression of Israel Gohberg's talent to convey the beauty of the field even to those lacking elaborate mathematical training.

Mathematical Tales

I. Gohberg

Introduction (by R.S. Phillips)

I am very gratified to have been asked to give this introductory talk for our hon-oured guest, Israel Gohberg. I should like to begin by spending a few minutes talking shop.

One of the great tragedies of being a mathematician is that your papers are read so seldom. On the average ten people will read the introduction to a paper and perhaps two of these will actually study the paper. It's difficult to know how to deal with this problem. One strategy which will at least get you one more reader, is to collaborate with someone. I think Israel early on caught on to this, and I imagine that by this time most of the analysts in the world have collaborated with him. He continues relentlessly in this pursuit; he visits his neighbour Harry Dym at the Weizmann Institute regularly, he spends several months a year in Amsterdam working with Rien Kaashoek, several weeks in Maryland with Seymour Goldberg, a couple of weeks here in Calgary with Peter Lancaster, and on the rare occasions when he is in Tel Aviv, he takes care of his many students.

I should remark that all this activity has not gone unnoticed: At the University of Tel Aviv he occupies the Silver Chair of mathematics (one must distinguish between the Silver Chair and a silver chair), he is a member of the Royal Dutch Academy and he has received the Israel Rothschild Prize in Mathematics, and, as I understand it, he is favoured to win the most travelled mathematician award.

This all began about 40 years ago, at which time he was a protégé of Mark Krein. The two of them formed one of the most successful collaborations in math-ematics, and the resulting books and papers are out standing. One of the con-sequences of this collaboration was that he became a member of the Moldavian Academy. However, I really believe that Israel blossomed after he left the Soviet Union in 1974. Since then his output and influence has been truly prodigious.

The feeling of good-will which pervades this conference is in good part due to the fact that we are celebrating Israel's sixtieth birthday. I suppose I should

Originally published in *Operator Theory: Advances and Applications* **40** (1989), 15–57.

say his approximate birthday since Don Sarason has cast some doubt as to the exact date of his birth. In any case, it is a great pleasure to present Israel an this occasion.

Mathematical Tales

I want to thank Ralph very much for his kind introduction. Now I am going to relate some stories from my career in mathematics, mostly from the times when I lived in the Soviet Union, before immigrating to Israel.

1. University Studies And My First Papers

The high school I graduated from in 1946 was in a village called Vasilevsky Sovhoz, not far from Frunze the capital of the Kirgizian Soviet Socialist Republic, where my mother, sister and myself lived during the time of World War Two. This republic borders on continental China. During my final three years I had a wonderful mathematics teacher, whom I will never forget. He was M.S. Shumbarsky, a young man recently graduated from the Warsaw University under the instructorship of Professor K. Borsuk.

During my first two years after high school I studied in Frunze, at the Pedagogical Institute. Apart from a small stipend given to all students by the authorities, I was supported by my mother. She worked very hard in a small country hospital where she was the only midwife, and consequently always on call. My mother had to be very inventive to be able to keep my sister and herself on her small salary, as well as to help me. This postwar period was a very difficult time. Food and clothing were rationed. I could not afford to eat at the student menza, so every week I brought food from home. I lived in a student residence together with seven other students in one room. My high school teacher, M.S. Shumbarsky, often visited me when I came home for weekends, and during long hours we would discuss the new material which I had studied and read during the past week. He always brought food with him, hidden under his coat (I think he did not want his wife to know.) They had no children. I had very good teachers (unusually so for such a remote place). I remember with gratitude G.Ja. Sukhomlinov (his name appears at the beginning of functional analysis) and G.Ja. Bykov (an expert in differential equations). My favourite professor of this period was D.L. Picus, who obtained his Ph.D. from V.F. Kagan in Moscow University. I will always remember his very interesting lectures and conversations. From him I learned about the most popular unsolved problems in mathematics.

In 1947 I won a Stalin stipend. It was a high award and nearly every academic institution of higher education awarded a small number of them. Financially my new stipend was three to four times larger than the old one, and I could now support myself, and even offer some help to my mother and sister. During the following academic year (1947/1948) I had my first small mathematical success. I invented my first theorem and found an interesting proof of a well known one. During this year I realized that I did not want to continue my studies at a Pedagogical Institute, where a lot of time was spent an training to be a high school teacher. I was more interested in a university education and in research. My professors also advised me to move to a university. There was no university in Frunze at that time, so I discussed the matter with my mother and she offered me the necessary moral and financial support. We also decided to take advantage of this situation to return to the area from which we originally came. In the summer of 1948 I left Frunze and the Stalin stipend, and moved back to the West. By coincidence, D.L. Picus was moving to Kishinev, and he helped me to transfer to Kishinev to continue my studies there. The following academic year I was busy adjusting to the university curriculum, and I also started to support myself partially by tutoring.

I was lucky; when I was in my fourth year at this university a young doctor from Leningrad came to work there. As he himself said, he taught us everything he knew. He was very knowledgeable as well as a good lecturer, and I listened to everything he had to say. This was I.A. Itscovitch, a former Ph.D. Student of S.G. Mikhlin, and I am grateful to him for introducing me to singular integral equations, to Fredholm operators, to index, and to all those topics in which later on I started to work actively. In 1949 I already had my first results which were published a little later, in 1951.

D.L. Pikus introduced me to S.M. Nikolsky, and advised me to send him my first results for evaluation. S.M. Nikolsky was interested in these topics and had made nice contributions to this area. He recommended my manuscript to A.N. Kolmogorov, who presented my first two papers to "Doklady". In 1951, they were published in this journal whilst I was still in my fifth year. At the same time, in another part of the world, in Ibadan (Nigeria), F.V. Atkinson obtained similar results. He published them in 1951 in Russian in "Mathematichesky Sbornik". Also in 1951, in the United States, a paper of B. Yood with similar results appeared. As B. Yood discovered, this story started with an earlier paper of J.A. Dieudonné, published in France in 1943, during the war, (see the paper: J.A. Dieudonné, *Integral Equations and Operator Theory*, 580–589, 1985). I am happy to share these results with good friends.

In the autumn of 1950, I decided to visit M.G. Krein. Somehow I had heard that Krein was a nice man and would talk to me if I came. So I decided to go to Odessa to visit him. I could travel by train, but there was the question of an hotel. Hotels in Odessa were always fall, so how could I find a place? One had to pay under the table and I didn't know how to do that. I had once heard from my uncle that his wife had family living in Odessa, so I wrote to my uncle. He wrote to the relatives in Odessa and they agreed to accept me for a couple of days. So I came to Odessa to these relatives.

I didn't even know the telephone number of Krein, but I knew where he was working; I already knew some of his work. He was at the Marine Engineering Institute in Odessa. This was a semi-military academic institution, so the professors wore uniforms and held ranks, which were denoted by the braid on their uniforms. An ordinary person could not enter; one needed to have some special identity card, which I for Bure did not have, but I saw some students running in waving something in their hands. I also ran in waving something, and so I came to Krein's office. At that time he held the chair in theoretical mechanics. Some time earlier he had been dismissed from the Odessa University. I came up to the secretary sitting at a deck and Said that I would like to meet Professor Krein. She said, "He will be here in an hour. There will be a seminar." I waited. After an hour I saw two men with all this braid and insignia on their uniforms – colonels, captains, whatever – two men, one looked very ordinary and the second fat and distinguished, who I was sure must be Krein. I went up to him and asked, "Are you Mark Grigorievich Krein?" "No", he said, "This one is." Krein invited me to his seminar and I listened to his talk. It was a very interesting talk on the behaviour of a stretched string, on extreme mass distributions. After his talk, he spoke to me. I showed him my first results; they were about Fredholm operators and index. He invited me to his home in the evening, and there we had our first long conversation. He drew my attention to two topics. The first was commutative normed rings, today we would say commutative Banach algebras. I did not know anything about it. "You don't know about this?" he said. This was in 1950. He said this was essential for a mathematical education. I felt terrible, non-educated. Then he showed me a couple of papers which had just appeared on infinite Toeplitz matrices. He said that they looked interesting to him. I am not sure if these two papers are known in the West, they are by Rappaport, who was the first to see the connection between Toeplitz matrices and singular integral equations. These topics became very important in my career.

At the end of our meeting I told M.G. Krein that I would be happy to listen to his lectures and to be under his direction. He answered that, unfortunately, he could not take me as a graduate student. Later I learned that his institution would not accept Jewish Ph.D. students. After World War Two M.G. had only two or three Jewish Ph.D. students, all war veterans. One of them was I.S. Iokhvidov who returned from the war with the rank of major, and the second M.A. Krasnoselsky, who, if I am not mistaken, returned with the rank of captain.

When I returned to my hosts I related to them what a great impression M.G. Krein had made on me. While I was leaving Odessa and thanking them for their hospitality, they asked me how much the professor had charged me. I told them, nothing. They could not believe this because they were used to dealing with professors of medicine. This meeting with M.G. Krein was a very important beginning.

A cartoon from my student years done by a friend and fellow student, Zobov. It is entitled "Twenty Years Hence." The two men at the right and left are the colonels who taught us military subjects one a week at the university. I am in the centre.

In 1951 I graduated from Kishinev University with a degree equivalent to a masters degree in the West. This was just before the tragedy of the Jewish doctors in the Soviet Union. During the last years of Stalin antisemitism grew rapidly and the situation for the Jews was fast worsening. I could not find a university which would accept me as a Ph.D. student. I tried hard. I asked in my own university, but the Dean did not even want to discuss it with me. I asked S.M. Nikolsky about possibilities in Moscow (he was then Vice Director of the Steklov Institute). He answered that for reasons which I would probably understand, it would be impossible. In some places they would get away with it by saying that strong students do not need to be accepted for the Ph.D. programme because they can write their thesis without any help; only the weak ones, who need help, have to be accepted to the Ph.D. programme. This policy enabled the universities to accept communist party activists who were weak students. I got a compulsory assignment to a high school in a small town on the border with Rumania (Ungeny). In Moldavia there is a lot of wine, and the previous teacher in mathematics drank too much and could no longer teach, so they sent me in his place. M.G. Krein wrote a highly complimentary review on my Diploma (master) thesis. This, together with a recommendation of my former professors, V.A. Andrunakievich and I.F. Volkov, made an impression on the Minister of Education of the Moldavian Soviet Republic, A.N. Krachun, who took the responsibility of changing my assignment. Finally I was sent to work at a two-year teachers' college in Soroki, a more interesting place. I never was a Ph.D. student; I wrote my Ph.D. thesis externally in my spare time and defended it in Leningrad. This became possible in 1954 after Stalin's death, when the situation improved.

The advice Krein gave me was very useful. In my diploma thesis, there was already a chapter on normed rings in which I found maximal ideals. I generated an algebra with all linear singular integral operators with continuous coefficients on a closed contour, and factored it by the compact operators (it was known that this is a commutative ring) and found the maximal ideals. This result enabled something new to be found, namely the conditions necessary for a singular integral operator to satisfy the F. Noether theorems. The proof was based on the fact that for a linear singular integral operator the F. Noether theorems hold if and only if they have a bounded regularizer, or what is the same, that the singular integral operator has a bounded inverse, modulo the ideal of compact operators. This was in 1950. Later, I showed these results to M.G. Krein, who liked them. I showed these results also to I.M. Gelfand, and they both advised me to publish them in Uspekhi, where they appeared later as a short paper. There was a problem with this paper which I am going to relate.

I was very proud of this paper. As I remember, it appeared in the second issue of 1952. I was so happy that I even subscribed to Uspekhi! I didn't have enough money, but I did it anyway, and so each number of the Journal was delivered to my home. When I received number six I could hardly believe my eyes. The title of a two page paper included in this issue of the Journal was "On a note of I.C. Gohberg", the author was F.D. Gakhov. Later I was told by some former students of Gakhov that my paper had discouraged some of his graduate students from working with nonbounded regularizers. To raise the morale of his students Gakhov had presented the matter in his paper in the following way: he wrote that in my paper I had introduced the artificial condition of boundedness of the regularizer just to be able to use theorems from abstract functional analysis. He had left out the crucial fact that this is the condition which is equivalent to the F. Noether theorems. The end of F.D. Gakhov's paper is ridiculous. It contains the conclusion that "the theorem proved by I.C. Gohberg has a conditional character and holds only under restrictions an the choice of the regularizer which the author made". This remark of his applies to any correct theorem in the world and thus all of them are conditional. Later I was told by M.G. Krein that he thought this little paper of Gakhov was directed somehow against abstract mathematics in general. I heard that A.N. Kolmogorov also got this impression, but it is obvious that they were all afraid to do or say anything officially; 1952 was a bad time. Till now the paper of Gakhov remains unchallenged. In general, I do not understand how a respected journal Ilke Uspekhi could publish such a paper.

Some of F.D.Gakhov's students were very good mathematicians. I have a high regard for I.B. Simonenko from Rostov, who made very important contributions to the theory of singular integral operators. I met him for the first time in the late fifties and many times thereafter. It was always a pleasure to discuss mathematics with him.

2. My First Years of Work

I was assigned to work at a teachers institute in Soroki, a small town in the north of Moldavia. This was a two year college type institute which trained teachers for the 5–7 grades of elementary school. A large number of the students were quite weak. The administration judged a teacher by the marks he gave his students. My work was even more complicated because a large part of my lectures had to be in Moldavian (a Rumanian dialect), and I had to start lecturing with an extremely limited knowledge of this language. I did not imagine then that some twenty-four years hence I would have the same problems with English in the United States and with Hebrew in Israel.

I worked for two years in Soroki, from 1951 to 1953. These were the most difficult years of Stalin's rule, and years of the severest antisemitism. A Jewish teacher could be dismissed, or worse, sent to jail, as a result of an accusation by a student who failed to pass an examination. Later I learned that I was about to be dismissed on just such grounds, but I was saved by Stalin's death.

The director in Soroki was a mathematician by the name of V.G. Cheban, a very pleasant man. Before World War Two he was a graduate student of A.N. Kolmogorov. He got his Ph.D. degree much later in mechanics at Moscow University. He supported my work with M.G. Krein and allowed me to travel to Odessa whenever possible.

In 1953 my director was promoted to director of the Pedagogical Institute at Beltsy (the third largest town in Moldavia). He had me transferred also to Beltsy where my job was much more interesting. This was a four year university type institution. Whilst working there I wrote my Ph.D. thesis externally. During the autumn semester of 1953 I taught a 40–46 hours a week (which was my yearly work load). This enabled me to spend the spring semester of 1954 in Leningrad, where I completed my thesis and defended it at the Leningrad Pedagogical Institute. The opponents (readers of the thesis) were G.P. Akilov and S.G. Mikhlin.

In Leningrad I met S.G. Mikhlin for the second time. I visited him for the first time in 1951 and it was a great pleasure to meet him again. He showed interest in my work and was always very kind to me and supportive. I remember with gratitude our interesting mathematical conversations and discussions. He and his papers had an important influence on me, and I consider him as one of my teachers. I was very glad to meet V.I. Smirnov, a famous mathematician and a very interesting man of high personal integrity. For a long time we corresponded and he discussed with me various problems which appeared whilst he was writing a new edition of his volume containing the theory of integral equations. I attended on a regular basis the seminar of V.I. Smirnov and the seminar of L.V. Kantorovich. The latter was studying the book of L. Schwartz on distributions. We were using a pirate edition of this book printed in the USSR. I recall with pleasure my meetings with M.S. Birman, V.A. Jakubovich, L.D. Faddeev (who was then a student), A.I. Koshelev, and O.A. Ladyzhenskaja. A few years later I also met M.Z. Solomjak and N.K. Nikolsky.

After the defence my salary doubled and I was able to bring my mother from Kirgizia to Beltsy and to arrange for my sister to transfer from Frunze to Kishinev. She was a student of medicine and such a move was no trivial matter. Mrs. A.N. Krachun, who was now in an even higher position (vice premier) in the government of Moldavia, helped me for the second time, and I remember her help with gratitude. We were always a dose family and after the death of my father we felt even closer, and it was very important for us to be together. (My sister and her family are now living in Israel. She is a pediatric surgeon and has a Ph.D. In 1985 she was awarded the title "Woman of the Year".)

The nearest USSR university town to Beltsy from the north was Chernovitsy, which had a well established university. I often visited this town because my sister lived there with her family. On each occasion that I visited I also met with my colleagues and good friends K.M. Fischman and S.D. Eidelman. We had many interesting mathematical discussions since our interests were quite dose and we understood each other well. We also studied interesting papers together. This very pleasant and useful connection continued for many years with visits and lectures, however we never produced anything jointly.

At the beginning of 1956 I.M. Gelfand and S.V. Fomin organized a conference in Functional Analysis in Moscow University. This conference was the first conference in which I participated and lectured. It was the time when I.M. Gelfand and his colleagues were writing the now famous volumes on generalized functions and their manifold applications. It was expected that the new theory would have a very strong influence and would introduce order into many fields.

The conference opened with an introduction by M.V. Keldysh and a talk by the famous physicist L.D. Landau. According to the programme he was to present the lecture "Functional analysis methods and theoretical physics". I think it was expected that he would explain how distributions solve difficult problems in physics. I remember how L.D. Landau started by remarking that he did not know what functional analysis was doing in the title of his talk and he would speak about the state of affairs in theoretical physics. I understood that the influence of distributions was already behind him.

I found the conference very interesting and I very much enjoyed many of the talks and the very pleasant meetings. There for the first time I met B.Sz.-Nagy. I knew about his important contributions from M.G. Krein, and I was very happy to be introduced to him. We also had common interests, and I tried to have a talk with him. Unfortunately I did not speak English and my German was not very good, but we soon discovered that we could talk in Rumanian, and this is how we communicated until I learned English. I very much liked B.Sz.-Nagy's papers, and on a few occasions I found in his work answers to questions which were bothering me. Later I followed with great interest his joint papers with C. Foias. Their important series of papers were often discussed at M.G. Krein's seminars. I met C. Foias for the first time ten years later at the congress in Moscow. I feel that my interests in mathematics are very dose to those of C. Foias and B.Sz.-Nagy, and it is a pleasure to have such good mathematical neighbours and friends.

At this conference I met a number of very active young mathematicians. Ju.M. Berezansky, S.V. Fomin, M.A. Krasnoselsky, A.G. Kostjuchenko, S.G. Krein (M.G.'s younger brother), G.E. Shilov and M.I. Vishik. I was very impressed by their talks and research plans. On the last day of the conference, following a talk on J. von Neumann algebras, Gelfand pondered in his remarks on how it could be that such a beautiful theory had no applications (which was the situation at that time). At the end of the conference in his concluding remarks, the famous topologist, P.S. Alexandrov, said that I.M. Gelfand had stated the law of the conservation of beauty. If a theory is beautiful then it must have applications.

After this conference I went to Leningrad where Bella and I got married. This was the best thing that ever happened to me, and I now understand fully just how lucky I was. My wife was a medical student in her final year. We had met a few years earlier in Frunze when my sister introduced me to her friend and classmate in the Institute. After the wedding Bella and I spent a few days together; this was the beginning of our honeymoon, which I think continues even today. My wife works as a medical doctor. We have two lovely daughters, Zvia and Yanina, married respectively to Nissim and Arie. They also have children – Tali, Jonathan and Keren (who was born soon after the Calgary conference). From Leningrad I went to Beltsy and Odessa, where I worked together with M.G. Krein on writing the first draft of the expository paper on Fredholm operators and index.

Later in Beltsy I got the chair in mathematics, and so to my regular duties of lecturing (18 to 24 hours a week) and research, was added also the responsibilities entailed in training teachers of mathematics.

In 1959 I was invited to join the Moldavian branch of the main academy in Kishinev. I applied and was accepted. I left the chair and a better salary and moved to Kishinev. This institution was trying to organize a strong mathematical group because it was preparing to be transformed into the Moldavian Academy of Science. Almost every republic already had its own academy. At the beginning we were attached to the Department of Physics, headed by T.I. Malinovsky, a well known crystallographer. Very soon this academy was inaugurated and an institute of mathematics was established. My colleagues and myself (as chairman) formed the department of functional analysis of the new institute. In the institute there were also two other departments of theoretical mathematics. One of these was the department of algebra, with the group of V.A. Andrunakievich, which was interested in the theory of rings and modules, and the group of the Tate V.D. Belousov which specialized in quasigroup theory. V.D. Belousov was a pleasant man and a very good mathematician. We worked together in Beltsy (he was born there) and we became friends. By the way, Belousov was the only one of my colleagues who had spent an academic year in the United States (Madison, University of Wisconsin). From him I got my first picture of academic life in a university in the United States. He was very excited about this period, but he had to be very careful when speaking about it at home. His stories had to be balanced, and he had to describe not only the positive, but also an exaggerated negative side of American life, otherwise he would be in trouble, especially since he was a communist party member. The other department was the department of differential equations and dynamical systems, with I.U. Bronshtein, K.S. Sibirsky and B.A. Stcherbakov as its senior people. Their interests were mostly in the theory of invariants of ordinary differential equations and topological theory of dynamical systems. V.A. Andrunakievich was appointed director of this institute. We were happy with this choice since he was a very good mathematician and a very intelligent person. During my student years V.A. Andrunakievich was a professor at Kishinev University where I took his courses in linear algebra and number theory and wrote a joint paper with him. In the early fifties he left Kishinev for Moscow, fearing that his mathematical career, and maybe much more, was in danger. The person he feared was the Dean of the Faculty of Physics and Mathematics at Kishinev University, an extremely weak mathematician, as well as a man of low morals, and an active party member. When the academy was inaugurated V.A. Andrunakievich (a full professor of the Mendeleev Institute in Moscow, and an ethnic Moldavian) was the only mathematician to be appointed a full member. The initial members of the academy were appointed by the government of the republic. V.A. Andrunakievich soon returned to Kishinev and this time the Dean was unable to harm him. Shortly after the Dean disappeared from Kishinev.

3. Working with M.G. Krein

The work I did together with M.G. Krein was an important part of my life, especially the two books we wrote together. At this point we already had the experience of writing together two large expository papers (the paper about Fredholm operators and the paper about Wiener-Hopf operators). In 1958 we started to write these two books as one book on perturbation theory. At the beginning I found it very difficult. I was not familiar enough with the background, and I had a great deal to learn. The discussions I had with M.G. Krein whilst we were working together were very important for me. In a very simple and elegant way he soon introduced me to new areas. Whatever he told me was nicely connected and motivated, and had an interesting historical perspective. Heuristic remarks always shed light on difficult formal considerations. In this way I was able to learn the background very quickly and to grasp the global picture.

In writing Mark Grigorievich is extremely demanding of himself and of everyone working with him. If, by a complete change in the writing of a paper or a chapter, he could improve it by even a small fraction, he would not think twice. It was never a question of whether it was worthwhile or not. If you can improve it you have to do so. Papers that were almost ready would lie for years until they attained perfection. He maintained that the writing of a paper is no less scientifically important than proving the result, because whilst polishing up a paper one can find new proofs, new presentations, new connections and new mathematics.

Probably some readers noticed that the paper on systems of Wiener-Hopf integral equations of I. Gohberg and M.G. Krein appeared before Krein's paper on a single Wiener-Hopf equation. At the same time the first published paper was a continuation of the second. The explanation of this is as follows. The 200 page typescript of Krein's paper was actually ready much before the paper of Gohberg and Krein was written. But M.G. decided that his paper needed a few small additions and some polishing. This work was not attractive, especially taking into consideration the fact that in such a huge paper every small change easily turns into a snowball of changes throughout the entire manuscript, so M.G. kept on postponing it. Work on systems was fresh and much more interesting, so he was more interested in working with me on systems. I remember this paper on systems was written in a very short time. In three or four months we produced a major part of the results and wrote the paper. We sent the manuscript to "Uspekhi" and it was accepted. M.G.'s paper was still waiting, and I proposed to M.G. that I help him to finish his paper. Some time later I came to Odessa for a month for this purpose, and at the end of this period the fat manuscript was sent to "Uspekhi". The work we had to do was mostly technical. The whole family was worried about this manuscript because the situation had a bad effect on M.G. and I remember how happy Raisa Lvovna (M.G.'s wife) was when we showed her the receipt from the Post Office. I also remember a celebration. After M.G.'s paper on Wiener-Hopf equations was published he presented me with a reprint with the inscription "To the midwife of this paper..." Some of M.G.'s manuscripts were never published. I would especially like to mention his large manuscript on the theory and applications of cones in Banach space written before World War Two. This manuscript was very popular in Voronezh and was used by M.A. Krasnoselsky and his students when they developed the theory of positive solutions of operator equations. Shortly after World War Two M.G. Krein wrote a large manuscript together with M.S. Brodsky on representation of topological groups. This manuscript was never submitted for publication because M.G. thought it needed some additions and some polishing.

Even today in the Soviet Union the formulae in typescripts are written in by hand. At the beginning of our collaboration M.G. wrote the formulae in the first copy of the paper; he did not trust me enough to do this. When I said "But I think

I can do it", he answered, "Yes, but I like to write them in myself because while doing so I again think over the proofs and the whole material".

The plans of the books changed many times; the whole concept was changed, we made unexpected additions and extensions, including results which were still fresh. At a later stage we developed the theory of singular values, the very same singular values which are now so popular even for matrices. Very few people knew of them at that time and only a few papers had appeared in the literature. It was very time consuming, but later it was extremely enjoyable and satisfying.

We decided to include in the book a number of very interesting recent results of M.V. Keldysh on completeness and distribution of eigenvalues for certain classes of nonselfadjoint operators. These results were published without proofs in a four page short note in Doklady. We tried to get the proofs from the author, but were informed that they do not exist. The detailed manuscript with the complete proofs disappeared. At that time M.V. Keldysh was President of the Academy of Sciences of USSR. This post can be compared with the post of a member of the government responsible for science. He was also responsible for the theoretical part of the USSR cosmos project. In official reports and the reminiscences of the first cosmonauts (as the USSR astronauts are called), there were two persons who were not called by their real names. These were the "chief designer" and the "chief theoretician". We all knew that M.V. Keldysh was the chief theoretician, and S.P. Koroleov the chief designer, but it was forbidden to publish their names in the press. This information was disclosed officially much later. During these years M.V. Keldysh was very busy and there was no chance that he would have time to recover the proofs of his results. With the help of some colleagues we recovered the proofs of an important part of these results. So the first time these results of M.V. Keldysh with complete proofs appeared was in our first book. I met M.V. Keldysh personally in 1966 at the Congress in Moscow. He told me that he liked the exposition of his results in our book and that it differs from his original presentation. The original manuscript of M.V. Keldysh was found much later (I do not know the circumstances) and was published in Uspekhi in 1971.

In 1961 I took part in the Fourth Soviet Mathematical Congress in Leningrad. V.B. Lidsky, who is well known for his excellent contributions in operator theory, was scheduled to present a joint talk with M.V. Keldysh about the theory of nonselfadjoint operators. A day before he had to give this important invited talk he felt that he was loosing his voice. He feared that he would loose it completely during bis talk, so he asked me to prepare myself to take over should he not be able to continue. The talk was not written down, so V.B. spent a couple of hours instructing me. The following day, after a long evening of work, I was really ready to take over at any point. Fortunately it was not necessary for me to do so, and V.B. Lidsky gave a very nice talk. But the incident was very useful and gave me an opportunity to go over the theory and to discuss the details with V.B. Lidsky under the pressure which usually precedes lecturing. He was a very good friend and I enjoyed discussing mathematics with him. Unfortunately our interests later went different ways.

This photograph was taken around 1958 when work was begun on the book with M.G. Krein. On the right G.N. Chebotarev, Professor of Mathematics at Kazan University son of N.G. Chebotarev, who was M.G. Krein's teacher. M.G. Krein is in the centre, I am on the left.

There were areas which did not exist before we started to write the book, for instance the theory of factorization of operators (this theory generalizes the multiplicative lower upper triangular decomposition of matrices to the infinite dimensional case). The same is true for the theory of transformation of triangular truncation with applications to differential operators. All of these results were obtained during the writing of the book. It was very satisfactory for us to see later in papers of T. Kailath and his colleagues how these theories work beautifully in applications. In general our research paths and those of T. Kailath and his colleagues oft en Cross, and it is always a pleasure when this happens.

When selecting material or discussing expositions or proofs, we sometimes had differing opinions, and this usually led to an argument. In general M.G. is a fighter and he can make himself very convincing. But he would also listen attentively to my arguments. During such discussions we sometimes exchanged positions – I would adopt his position and he mine. But we always ended up on a cheerful note with a common point, jokes often being used to illustrate important arguments.

I think it took us around five to six years to write the books, initially as one volume. I spent two or three months a year in Odessa for joint work, part of this time during my vacation, on M.G. Krein's dacha on the shores of the Black Sea.

On occasion we would even allow ourselves an hour off for a swim. Sometimes my family would join me in Odessa and I had some difficulty in dividing my time betweeen them and M.G., but most of the time we were working in Krein's apartment. A small room there got the official title of Izia's office. Whilst I was in Odessa M.G. spent most of his time writing the book and discussing it with me. He would never allow me to eat out, this wonderful family always insisted that I eat with them. Members of the family would answer the telephone, and the caller would be told that he could speak with M.G. only if it were urgent since M.G. is busy, Izia is here. Some of my Odessa friends threatened to beat me up since I took up too much of Mark Grigorievich's attention.

When I started to work together with M.G. Krein and our first joint papers appeared I received congratulations and a warning; a friend warned me that readers would only remember that M.G. Krein had written the paper, and the name of the coauthor would not be remembered, not even its existence. He also thought that in reviews and references I would get little credit for joint work with M.G. Krein. Now I see he was wrong (maybe with very rare exceptions). ¿From the very outset I understood that working together with M.G. was the best thing that could have happened to me. I had heard many tales about what a difficult person M.G. Krein was, but during the twenty four years during which we worked together I found him to be otherwise. True he was demanding, and it may be for this reason that he got the name of being difficult. He set very high standards for himself, and I found his demands of others to be fare and reasonable.

After we wrote our first paper on Fredholm operators and Index I suggested that his name appear before mine on the manuscript. I explained to M.G. that this would be to the benefit of the paper, which would be more appreciated and attract more attention. He refused, maintaining that he would not write a joint work with a person who could not be an equal partner, and that he always used the alphabetical order of names in joint papers and any change will cause speculation in the mind of the readers. After a while I drew his attention to a paper by "M.G. Krein and M.A. Krasnoselsky". He laughed and remarked that the difference is only of the third degree. In our joint work we never gave specific indications of who did what. M.G. considered joint work as the work of a team, and after deciding that we would work together on certain topics, we then would not disinguish who did what in these topics. In research it is not always obvious what is the more important. Sometimes it is an idea or the choice of topics or the timing, sometimes the conjecture, or the proof, sometimes the concept, sometimes a definition, a remark or finding an error, and a counterexample. M.G. would compare joint research with a team of sportsmen, especially in a long term collaboration. Each player passes the ball on to another player, and it is impossible to delineate which pass was the most important.

These principles worked very well and during the twenty four years of our collaboration we never had any misunderstanding.

While writing the books we always worked with other people, and there were some topics where we acted as contractors. For instance, if we needed something in an area where our colleagues were better than us, then we would ask them to prove a certain theorem for us. They also kept us informed about new results and new improvements, so the material just grew before us. Among those colleagues who helped us were the following: M.S Birman, M.S. Brodsky, Ju.P. Ginzburg, G.E. Kisilevsky, S.G. Krein, B.Ja. Levin, V.B. Lidsky, Ju.I. Lyubich, V.I. Macaev, A.S. Markus, L.A. Sakhnovich, E.M. Semenov, Ju.L. Shmuljan, and M.Z. Solomyak.

The writing of the book took up more and more time. I would return home to Kishinev and my wife would ask if the book was finished yet. "There is still a little more work to be done", I would tell her. Each time, "Did you finish?" and each time I would reply that a little more remains to be done. As I remember, my wife finally stopped asking. She probably gave up all hope. It is still told in Odessa, now as a joke, though at the time it was true, how Sahnovich met me and asked, "How is the book going?" "Well, it's eighty five per cent ready", I replied. "Then why do you look so sad? That's wonderful." "Yes", I answered, "but if you had asked me yesterday I would have said it was ninety five per cent ready". Our students and friends made fun of us; I.S. Iokhvidov even wrote a little poem on the eve of the new year in 1963. I am grateful to Chandler Davis for translating this poem into English:

> Around the festive table, all our friends
> Have come to mark our new book's publication.
> The fresh and shiny volume in their hands,
> They offer Izia and me congratulations.
> The long awaited hour is here at last.
> The sourest sceptic sees he was mistaken,
> And, smiling, comes to cheer us like the rest,
> And I'm so delighted, ... I awaken.

From M.G. Krein's dream, New Year's Eve, 1963.

In the spring of 1963 Krein agreed to allow me to fill in the formulae in the first copy. I filled in almost 1500 pages of formulae; more than 1000 pages of manuscript for the book, then came 250 pages which were my second thesis which I was preparing at the same time. There were also a few papers which we were submitting. All in all I filled in 1500 pages. At that time my family lived in two small rooms. Our daughters were small then, and at ten o'clock everyone went to sleep. From ten in the evening till two in the morning was my time for filling in formulae. My friends helped me to fill in the second and third copies. Soon I went to Moscow to submit the book. The publisher said it was too long, so we had another problem. We divided it into two and made two volumes, the first on non-selfadjoint operators, and the second on Volterra operators.

On the same trip to Moscow I submitted my second thesis to Moscow University. The Dean was N.V. Efimov, a famous geometer and a very nice man, but he was not interested in having me or coping with my defence. One extra Jew to defend in Moscow University brings only difficulties. He tried to convince me that I really did not want to defend my thesis in Moscow University. By chance Professor A.G. Kurosch, the famous algebraist, a very straight forward man, dropped into Efimov's office. Whenever Efimov came up with yet another argument to convince me, he would intervene. "You don't sound convincing", he would tell Efimov, "Why should you not take this thesis?" Finally Efimov took the thesis, but said that I would have to wait my turn as there was a long line of candidates. In February of 1964, I defended my thesis in Moscow University. This was no formality but a real defence, with presentations and questions put and a secret ballot held at the end at a meeting of a highly qualified scientific council. Four abstained, still more than two thirds voted for me, and so I passed approval. In my case the opponents (the official readers of my thesis) were S.G. Mikhlin, M.A. Naimark, and G.E. Shilov. A special review came from the Institute of Applied Mathematics and was signed by M.V. Keldysh and V.B. Lidsky.

I continued working an the book together with M.G. Krein. After dividing it into two, it went to the copy editor. With the copy editor we were both lucky and unlucky. He was F.V. Shirokov, a university professor in mathematics, with the Russian language as well as mathematics at his finger tips. Shirokov was, I think, the best copy editor I have ever seen. He was extremely demanding. First of all he insisted that I sit with him while he read the manuscript. I had no alternative. Secondly he would try to understand all the proofs in detail. He would suddenly hurst out, "I don't like this proof". "So give another one!" "No, no, I don't like this proof". Sometimes he forced us to work very hard. It would take us weeks to change a proof till it met with bis satisfaction. Sometimes bis demands lead us to a bettet understanding of the material or new results. He had good taste and the exposition was thereby improved essentially. In general he acted as an attorney for the reader. "The reader will not like that. The reader will not understand." We did not understand where this power of attorney came from, but we took his remarks very seriously. Eventually the books appeared, the first in 1965 and the second in 1967.

A. Feinstein translated the books into English for the AMS Tranlations programme. He did an excellent job. Professor Ando, who made a second English translation, sent us a list of misprints. Feinstein also had a list of misprints, and we ourselves also found some. Each of the three parties found around 80 misprints, and surprisingly enough there was almost no intersection between these lists. I was told that the first book in English is an AMS bestseller, and a third printing came out this year.

At the Mathematical Congress held in Moscow in 1966 a mathematician by the name of Gokr appeared there for the first time. M.G. Krein gave a plenary talk, and he had to refer frequently to our joint papers, and somehow it was awkward for him to have to repeat our two names each time, so he introduced Gokr as a short form of Gohberg-Krein. "This is done by Gokr, this is a paper of Gokr", he would say. I remember M.V. Keldysh, turning to ask M.I. Vishik, who was sitting near him in the audience, "Who is this Gokr?" I had helped M.G. Krein to prepare this talk, and it took some weeks of hard but interesting work. Part of the talk was based on our books, but it contained many other things as well.

At this congress I met C. Foias for the first time, during the preparation of M.G. Krein's talk (he helped in its translatiort into French). At this congress I also met R.G. Douglas, P.D. Lax, N. Levinson, R.S. Philips and I.J. Schoenberg. All of these meetings were very pleasant and very useful.

For many years M.G. Krein ran a mathematical seminar at the Scientists House in Odessa. I lectured there and listend to many excellent talks. It was always interesting with a very pleasant atmosphere. During my visits to Odessa I also attended other seminars of M.G. Krein at the various institutions where he worked. At these seminars I met very nice people and excellent mathematicians, among them the famous old generation of Krein students: M.S. Livsic, author of the theory of characteristic operator functions; D.P. Milman who coauthored the extremal points theorems; V.P. Potapov, known for his contributions to the theory of matrix functions and their multiplicative decompositions, and M.A. Rutman, the coauthor of the theory of cones in Banach spaces. I would like to include M.S. Brodsky also in this group. In the late fifties he started to work on the theory of nonselfadjoint operators and made very important contributions. Among the younger generation I met V.M. Adamjan, D.Z. Arov, M.L. Brodsky, Ju.P. Ginzburg, I.S. Iokhvidov, I.S. Kac, K.R. Kovalenko, B.R. Mukminov, A.A. and L.A. Nudelman, G.Ja. Popov, L.A. Sakhnovich, Ju.L. Schmuljan, and V.G. Sizov. All were formal or informal Ph.D. students of M.G. Krein, except Mukminov and Sakhovich who were students of Livsic, and Ginzburg of Potapov; Sizov I do not remember. H. Langer from Dresden and V.I. Macaev from Kharkov were oft en guests of M.G. and lectured at his seminars. Langer once spent a full academic year in Odessa and was already considered a native "Odesid"; we called him Heinz Kurtovich.

I also met M.G.'s Armenian students, V.A. Javrjan, F.E. Melik-Adamjan and Sh.N. Saakjan. They were sent for extended periods of time from Erevan to Odessa to write their Ph.D. thesis with M.G.

So I was introduced to an interesting group whose achievements I was able to follow. The interests of this group were wide and extended my horizon. In this way I learned about operator theory in spaces with indefinite scalar products, interpolation theory, theory of cones in Banach space, theory of operator functions, theory of characteristic functions, scattering theory and other areas of analysis. Later this knowledge proved to be very important for me. As a matter of fact, no one from Odessa University attended M.G. Krein's seminars. They probably did not even realize what they were missing. I think it was one of the best seminars in analysis and its applications in which I have participated. As I already mentioned, M.G. Krein was dismissed from Odessa University earlier on, and all of those mathematicians mentioned above worked in other academic institutions.

For more than twelve years I worked intensively with M.G. Krein. The remaining years he was mostly busy with other projects and we worked less intensively, but we always met and thought about joint plans and joint projects. We also planned a third book. Working with M.G. was a joy and an important school for me. I never was a formal student of his, but he was more than an ordinary instructor for me. He was a great teacher and is a very good friend.

4. Gelfand's Seminar

I used to like to visit Moscow University and to stay as a guest at the student residence which occupies the wings of the skyscraper. The middle part is used for academic purposes and consists mostly of halls, classrooms, libraries and offices. (Professors do not have their own offices in the universities of the USSR.) Concentrated on four or five floors of this building are to be found one of the best mathematics libraries in the world, very good book stores, and halls in which many important seminars and meetings take place. There are also caffeterias, menzas, restaurants, a concert hall and cinemas. The library was especially good; almost every mathematical Journal and book from all over the world was to be found there. Newly published books in mathematics made their first appearance in the book stores in this building. One could meet Soviet and foreign mathematics stars, and attend many important seminars, lectures and conferences. The Moscow Mathematical Society made this their meeting place. Even the bureaucratic procedure to register as a guest at the student residence was pleasant. First one had to get the recommendation of the head of the Chair of Functional Analysis, which for a long time was held by D.E. Menshov. One then needed the recommendation of the Chairman of the Department of Mathematics, P.S. Alexandrov. After that permission of the Dean (for a long time A.N. Kolmogorov), was needed. Finally one had to get the approval of the Rector, I.G. Petrovsky. For a young mathematician, as I was then, all of this was like touching the history of mathematics. In the libraries, halls and caffeterias, etc., one could meet friends and colleagues from the whole of the Soviet Union, hear the latest mathematical news, rumours and gossip, and one could also work quietly. And all of this within one building. Unfortunately it was not always possible to obtain a place in this residence. The entrance to the building was strictly guarded and a special identity card was needed to enter, but without giving away any secrets, I can say that it was possible to stay at the residence without having an officially approved place.

In 1962 I visited Moscow University and I was fortunate to be able to stay at the student residence. At the invitation of G.E. Shilov I gave a course in the theory of nonselfadjoint operators. This course was based an the rough version of the manuscript of the Gohberg and Krein book. I had a good audience; I remember B.C. Mitjagin (at that time a Ph.D. student of G.E. Shilov), and D.A. Raikov sitting in at this course. I enjoyed this work very much, and this course also served as a useful experiment for us in process of writing the book.

Gelfand's seminar is a seminar which is held at Moscow University already more than 45 years, and I am certain there has never been another seminar quite like it. First of all at least 50 people holding a second doctors' degree attended, and many others, including very strong high school students. Gelfand's behaviour was very unusual. Sometimes he would say who should sit in the first row and who in the second, and he would move people around. He would criticize freely, but not only criticize. Here is an example. The Moscow University professor, G.E. Shilov, one of his former students as well as a close friend and collaborator, is lecturing and writing on the blackboard a linear differential equation with coefficient depending on t. "Yura", Gelfand exclaims, "cross out the t." "But Israel Moiseevich", Shilov explains, "It's more general this way, the proofs are much more complicated. I went through them and I want to keep it." "Yura, cross out the t!" "Israel Moiseevich, it's more difficult, and very non trivial. I don't want to do it." "Yura, don't do cheap generalizations, cross out the t." I also recall other examples of this type. In some cases Gelfand gave the lecturer three minutes in which to state some result. He calculated thus: "Look, there are 50 professors here. You get three minutes. That's 3×50, or 150 minutes of highly qualified professors' time. That's a lot of time!" The meetings of the seminar were very interesting – many different modern topics were discussed, sharp questions asked, with illuminating discussions and comments. I often had the feeling that I was being enriched. On the other hand the seminar somehow had the flavour of a circus; it was a show and somewhat intimidating because someone could be hurt. There were those who could not take it. Not everyone was criticized by Gelfand. When a foreign speaker was talking, the seminar would take on the form of a regular seminar. He would not criticize foreigners, and he refrained from criticizing S.V. Fomin, as well as some others.

Whilst visiting Moscow University I attended Gelfand's seminar regularly, and he invited me to speak there. In general the seminar met formally on Mondays at 6.30. Everyone would turn up at 6.30, but Gelfand would come a little later. People would stand around, Gelfand speaking to one, to another, to a third, and so on. Meantime everyone is standing around and waiting. Around eight o'clock the seminar would start, and talks would continue up to ten. The same thing was repeated each time. On the day he invited me to give a talk, there were many other talks, and ten minutes before ten he said, "All right, who else? Ah, yes, Gohberg." I opened my talk with two general sentences: "There are different classes of linear non-selfadjoint operators, some of them, by their properties, close to selfadjoint, and some of them very different. For instance the Volterra operators which have only one point of spectrum." I don't think I had said anything else before Gelfand got up and started to develop his philosophy. In his opinion there has to be a better definition of a linear operator and a better definition of a space, and I thought he expected the whole picture would change and be nicer. I had heard such remarks from him on previous occasions, but I never heard constructive suggestions. I had spoken for two minutes, he the remaining eight, and this was the end of my talk. I was shattered. Somehow I felt I was presenting not only my results, but also the results of others, and I had done such a terrible job of it. These others were not there, and could do nothing to defend themselves. I thought about this a lot and decided that whether he allowed me to speak or not, I would get up at the next meeting and say that what he had said was wrong, that there are a lot of deep results of many people, and so on. During the following week, while planning my revolt, I met Shilov and told him of my intention. He started to laugh, "Don't do it. You'll be the laughing stock! Nobody pays any attention, that's his style. Don't do it, it's wrong", and so on. Finally he convinced me and so I kept quiet.

I did lecture in Gelfand's seminar two or three times. He even invited me once to lecture for a full session. In general Gelfand would ask you to lecture, and after ten minutes he would take over. That was usual practice. But when I lectured – I gave a talk on factorization – I spoke for almost the whole two hours. He put questions, wrote things down in a notebook, and discussed with other people. I thought that I had made a very good impression, but later I found out that I was wrong.

Much later, closer to 1970, I had a very interesting experience. I was visiting Moscow and Gelfand, seeing me in the audience, promptly invited me to lecture. "I'm not ready", I remonstrated. "I don't want to lecture." I did not want to experience those pains again. He insisted. "But I'm not ready." And still he insisted. He then put forward a theory that you should lecture when you are not ready, because then you will say only what you remember and will not give all the details and nonsense which no one wants to hear. So finally I lectured. I decided that this time I would say what I was not given a chance to say on the first occasion. But without philosophizing. I started by simply stating theorems, and it came much better. He put questions, showed interest and discussed results, and I enjoyed it very much.

I had some interesting conversations with I.M. Gelfand during a conference on Operator Theory held in Tihany (Hungary) in the summer of 1970, and on the way from Budapest to Moscow. We even discussed working together, but unfortunately it never worked out.

In general the conference in Tihany (which was held directly after the congress in Nice) was very important for me. There I met for the first time a large number of Western colleagues; L. Cobum, Ch. Davis, P. Halmos, H. Helson, J.W. Helton, M.A. Kaashoek, P. Massani, J.D. Pincus. Many of these acquaintances helped me very much later on.

It was a pleasure to meet Gelfand recently when he visited Tel Aviv to receive his Wolf Prize. At the same time he received the Wolf Prize on behalf of both Kolmogorov and Krein.

5. Department of Functional Analysis in Kishinev

During my last ten years in Kishinev (1964–1974) I headed the Department of Functional Analysis of the Institute of Mathematics at the Moldavian Academy. Around 100 to 200 researchers worked in this institute, as well as some Ph.D. students, whilst in the department there were no more than ten persons. The senior researchers in this department, apart from myself, were I.A. Feldman and

A.S. Markus, both of whom were more or less former students of mine. The remaining colleagues held positions of junior researchers, post doctoral fellows, or Ph.D. students. A seminar, also attended by mathematicians from other institutions, was held regularly. N.Ja. Krupnik, from Kishinev University, at one time an informal student of mine, was another senior participant in this seminar. A pleasant, friendly and enthusiastic atmosphere pervaded the department and the seminar.

We worked mainly on problems of nonselfadjoint operators and operator functions, Wiener-Hopf operators, Toeplitz operators, singular integral operators, and equations of transport. One of our specialities was factorization of operator functions and operators. It was during these years that the books of I.A. Feldman with I.Gohberg, and I. Gohberg with N.Ja. Krupnik were written. We had a large number of graduate students of Moldavian origin (this was required), as well as graduate students from other republics and countries. I am proud of R.V. Duduchava from Tbilisi, now one of the leading young professors there, and of G. Reinig and J. Leiterer from GDR, both of them outstanding experts in their areas. All three were my Ph.D. students. I am happy that students of mine now play an important role in Kishinev in higher education, in research and in other areas.

The research group was very critical, though with good will and good taste. All results obtained were thoroughly discussed and criticized with good humour. All members of the group worked hard and results were appreciated. I think that in certain areas of operator theory we had a very good reputation, both inside and outside the Soviet Union. I am happy that our results of these years are often used and quoted, and that some of them have interesting applications to different areas, including electrical engineering. In obtaining these results important contributions were made also by our younger colleagues. I would like to mention M.A. Barkar, V.M. Brodsky, M.S. Budjanu, I.S. Chebotaru, R.V. Duduchava, V.M. Eni, V.D. Frolov, L.S. Goldenstein, G. Heinig, J. Leiterer, L.E. Lerer, V.I. Levchenko, I.V. Mereutsa, V.I. Paraska, B.A. Prigorsky, G.I. Russu, A.A. Semencul (Sementsul), E.M. Shpigel, E.I. Sigal, O.I. Soibelman, V.P. Soltau, N. Vizitei, M.K. Zambitsky, and V.A. Zolotarevsky.

The institute published a journal "Matematicheskie Issledovanija". Formally I was an associate editor of this journal, but in practice I organized the journal and edited the analysis section. On looking back I see that we published many important papers in operator theory. The journal was exchanged with many Western universities, and it is a pity it no longer exists.

Members of the department were also interested in matrix theory and numerical methods, and we obtained good results in these areas. For a while we worked actively in combinatorial geometry, and our results attracted the interest of other mathematicians. Based on this research, jointly with V.G. Boltjansky, I wrote two small books on combinatorial geometry, now translated into many languages. I enjoyed this work very much. Boltjansky is a very strong mathematician with a very broad knowledge, as well as an excellent lecturer and a very pleasant person with whom to work. Along with all of this, he has a special talent for writing mathematics, and his first draft is, as a rule, the final work. (By the way, I worked together with Boltjansky on the index problem. We stopped when the first paper of M. Atiyah and I. Singer appeared.) V.G. often visited us and we also met many times in Moscow.

Colleagues from various cities throughout the USSR, as well as a number of foreign mathematicians, visited us. We interacted with many of them, and here are some examples. We had good connections with groups of mathematicians from Chernovitsy, Kharkov, Leningrad, Moscow, Odessa, Rostov, Tbilisi, and Voronezh. V.I. Macaev (Matsaev) was very oft en our guest. Our paths first crossed in Odessa when he was a Ph.D. student (Kharkov) and lectured in M.G.'s Seminar. His contributions to operator theory were very impressive, some of which are included in the first book of Gohberg and Krein. For a long time V.I. worked together with A.S. Markus, and they produced very interesting results in the theory of operator polynomials and operator functions. Macaev and Markus were customers of J. Leiterer and myself. We proved for them theorems of factorization of operator functions, and they used these theorems. E.M. Semenov (Voronezh) visited us a number of times. He was a former student of S.G. Krein, and we had common. interests. It was while he was helping us with the book that we first met. In Voronezh there was a strong school of functional analysis and we felt dose to this group. Members of the department and our students oft en participated in the famous mathematical winter schools in the neighbourhood of Voronezh. I remember the visit to us of M.A. Shubin (Moscow) whom I met for the first time in Moscow University when he was a Ph.D. student of M.I. Vishik and many times later. A very useful visit was the visit of M.V. Maslennikov Moscow), a well known expert in transport equations. Some of these equations can be transformed into Wiener-Hopf equations. We were very interested in these connections, L.A. Feldman especially so. He had very nice results in this area and it was planned that he would use them for his second doctoral thesis. B.S. Mitjagin (Moscow) and N.K. Nikolsky (Leningrad) were also among our important visitors. Our first visitors from the West were Chandler Davis, Seymour Goldberg and Israel Halperin. Our first visitors from Eastern European countries were H. Baumgaertel, S. Proessdorf, H. Langer and J. Bognar.

During the last ten years I also lectured at the Kishinev University, in a part time position. Each year I gave a year long course on special topics for senior undergraduate and Ph.D. students. lt contained chapters on operator theory with different applications to differential and integral equations and numerical methods. I was a member of the chair of cybernetics and computer science headed by P.S. Soltan. He is a very enthusiastic and inventive mathematician, and an interesting person. He was a Ph.D. student of P.S. Alexandrov and V.G. Boltjansky and studied in Moscow University. Our common interests were in combinatorial geometry.

As a result of these activities, within a short period of time and with very modest means available, a centre of analysis was created in Kishinev. Our connections with the university made it easier for us to prepare and to choose our future students. The speciality of this centre was operator theory, integral equations and their applications. M.G.'s help and influence was of great importance.

We experienced difficulties in having Jews accepted to this department; the usual semi-official response was that the percentage of Jews in our department was already too high. There was also a problem when it came to accepting a Jewish Ph.D. student. The usual procedure was that a Jew applying for the Ph.D. programme would be rejected automatically at the entrance examination an Marxism, unless the ground was well prepared. This preparation took the form of under the table bargaining with the administration, and there were instances where it was successful. There was also a serious problem when it came to finding an appropriate job for a Jew after he received his degree. The scenario was usually as follows: I would call people whom I knew and suggest that they hire a Ph.D. student of mine who was graduating. Having described the student, the first reaction would usually be: "Sure, well take him, we have had good experience with your students, let him come tomorrow." Usually they would take him, if the student was not Jewish. In the Gase of a Jewish student I would get a call: "Sorry, the administration will not give permission, I already took a Jewish student two years ago." In one year four students of mine of various ethnic origins graduated, all of them very good. In one day I found jobs for the three non Jews, but for the Jew it took more than a year.

In 1968 things worsened. In the Soviet Union the defence of the first or second doctoral degree has to obtain additional approval by a special committee sitting in Moscow (Higher Certification Commission). This was usually only a formality especially for a good Ph.D. thesis, but from 1968 it was almost impossible for Jews to pass; the second doctoral degree was no longer available at all to Jews, except in some very special cases. When the policy of this committee changed, A.G. Kurosh and G.E. Shilov resigned from it. During these bad years this committee was chaired first by V.A. Ilyin (professor at Moscow University), and then later by V.S. Vladimirov (now director of the Steklov Institute, academician, and hero of socialist labour). Accounts I have heard about discussions which took place in this committee are unbelievable. As an example see the book of G. Freiman, "It Seems I am a Jew", Fefer and Simons Inc., London, Amsterdam, 1980.

A.S. Markus defended an excellent second doctorate in Voronezh University, and it was highly acclaimed by main experts in the area. It was passed by a unanimous secret vote of a very qualified scientific council. However this special certification committee in Moscow did not approve Markus' thesis, rejection based an one negative review, instigated by the committee itself. It took the committee a very long time to find a person willing to write such a review, but finally they found one. A very good thesis of Ju.L. Shmuljan from Odessa was also rejected, and there were many similar cases. I was very disappointed. I considered this a personal injury and I did my utmost to fight it. It was a drawback for all the department, generally affecting the whole atmosphere within the department and the Institute. A.S. Markus never made another attempt, I.A. Feldman, who had excellent results an Wiener-Hopf and transport equations with which to form the second thesis never completed it, and only recently was N.Ja. Krupnik able to defend his second thesis in Tbilisi.

In the late sixties and early seventies it became much more difficult to publish the work of Jewish mathematicians, and sometimes even of Ph.D. students of Jewish instructors. An editorial board, headed by L.S. Pontrjagin, was set up to control the editorial policy in mathematics throughout the entire Soviet Union. I am sure that many excellent books did not appear because of the discriminatory policy of this board. These changes were very painful for us.

In 1968 to 1970 I was allowed to visit some Eastern European countries. Later I was not so lucky; in the early seventies I was not allowed to go to Poland or to Bulgaria, and my Jewish friends are not allowed to travel abroad even today. By the way, in each of the lucky cases I had to set out from Moscow, because there I had to pick up my passport and to receive instructions. I also had to return home via Moscow in order to return the passport and to hand in a report, entailing a lot extra travelling. For instance to spend four or five days in Rumania, which was only a short distance from my home, I had to travel thousands of kilometres.

6. Elections to the Academy

The Academy of Science of the USSR is the main academy in the Soviet Union. However each republic has its own Academy of Science. Election to an academy is very prestigious, bringing with it many benefits, among them essential financial benefits. From 1968, the year of the death of S.N. Bernstein, until 1984 no Jews were full members in the Department of Mathematics of the main academy. I.M. Gelfand was a corresponding member. Only in 1984 was he elected to full membership; this was possible only after the death of I.M. Vinogradov. M.G. Krein is a corresponding member of the Ukrainian Academy (since 1939). There is a joke in this connection: "The Ukrainian Academy must be the strongest in the world because M.G. Krein is only a corresponding member there." The Academy of the Moldavian Republic is one of the youngest and smallest among the academies. I was elected to it as a corresponding member at the beginning of 1970.

Full and corresponding members are elected to the academies by a secret vote held at a meeting of all members. Prior to the elections there is a very complicated preparatory procedure. The first part of the procedure is carried out in secret, the second part is more open. The most important part is probably the first: one has to be proposed and then the candidate has to obtain the approval of a number of institutions, including party offices, KGB, academy offices on the republic level and on the Moscow level. A list of those candidates who passed the first part successfully is published in the press. The second part of the procedure for the republic academies consists of two secret votes, the first in Moscow in the Department of Mathematics of the main academy. These results are supposed to play an advisory role only, but in reality they have much more weight. The last vote is held at meetings of the academy.

The first time I was proposed in 1965 my name was dropped at the end of the first part of the procedure. A friend who knew the details told me later that it was not for reasons of myself personally, but was connected with my father who was arrested by the Soviets on 23rd August, 1940, my birthday, two months after they took over Bessarabia. He was sentenced by a troika to eight years imprisonment, and I never saw him again. My father died in the Gulag a couple of years later, exactly where or when we do not know. We had to keep the story of my father a secret otherwise I would not have been accepted to the university or be able to work at the academy, and my sister would have had the same sort of difficulties. This information was passed on to the academy by the KGB when my candidacy was discussed, and as a result I was rejected. In 1966 a court which reexamined my father's case found that he had not committed any crime. His case was closed and, as with many others, he was posthumously rehabilitated. In 1969 I was proposed to the academy for the second time. I had very good recommendations and my candidacy somehow passed the initial procedures successfully. I have no details on this. I had thought that the rest would go through smoothly, but I was wrong. It soon became apparent that there was a problem; that someone was making trouble. I do not know precisely how things developed, but I was told that M.V. Keldysh became involved. He knew some of my work and supported me. M.V. Keldysh was informed about my case by P.S. Alexandrov. I was also strongly supported by the following members of the main academy: I.M. Gelfand, L.V. Kantorovich, N.I. Muskhelishvili, V.I. Smirnov, S.L. Sobolev, and I.N. Vekua. They were all familiar with at least some of my work. Later a vote took place at a meeting of the Department of Mathematics in the main academy in Moscow, and I passed with a two third majority. At this point it seemed as if a happy end was in sight. The final elections in Kishinev were scheduled for the 2nd January 1970, and I was the only candidate. I learned later that three or four days prior to the elections a letter reached the Moldavian Academy (the same letter was also sent to the Department of Science of the Central Committee of the Communist Party of the USSR). This letter was signed by the Director of the Steklov Institute, academician I.M. Vinogradov, hero of socialist labour. He wrote that in his opinion I am not suited to the position of a corresponding member of the Moldavian Academy because there is some doubt as to whether my papers with M.G. Krein are correct (he referred to a letter of G.E. Shilov in "Uspekhi" 1966, which has nothing to do with our work), and also that I do not have any outstanding results of my own. It appears to me that Vinogradov did not expect such a development and his letter was a last minute blow below the belt. I am sure he did not have any idea about my work or joint work with M.G. Krein. Probably this letter was based on a report. By the way, I was told recently that, after his death, a suitcase full of reports was found in a safe place in his home. Fortunately for me the letter of Vinogradov arrived too late. Everything was already officially approved, and it was tauch easier for the bureaucrats to go ahead. The elections went through and I was elected unanimously.

It is well known that antisemitism was an important issue in the activities of Director Vinogradov. He was very successful and under his directorship, the Steklov Institute in Moscow – the central institute of mathematics of the USSR – became Judenfrei. B.N. Delone told me during a visit to Kishinev, that any topic discussed with I.M. Vinogradov always comes down to antisemitism. I know that Vinogradov did not forget me, and in 1976 (two years after my emigration) he continued to make unpleasant remarks about me.

I did not know I.M. Vinogradov personally. I had met him face to face only once briefly at the Steklov Institute in Moscow at the beginning of 1960, when I was approved for the degree of senior researcher. I knew he was famous for his outstanding contributions to number theory, and at the university I studied his text book on this subject. I was surprised later when to learn that he had practically no Ph.D. students, and is a poor lecturer. I had attended a rather strange lecture which he had given to the USSR Mathematical Congress in 1956. This was a unique plenary lecture delivered in the largest hall of the Moscow University, before an audience of around 1500 to 2000. I think it was intended to be the lecture of the world's greatest mathematician, but it turned out differently. Whenever he used the blackboard he would forget about the microphone, so it was rather like watching a silent movie. Then he suddenly took the microphone and spoke into it very loudly, reading out formulae from a manuscript. He was confused and it was impossible to follow him so people started to fidget and get up and leave. He also overstepped the allotted time. Later I heard that he blamed P.S. Alexandrov, who chaired that particular session, for the fiasco. During the final part of the lecture P.S. Alexandrov signalled to him; he interpreted these signs to mean "Beautiful, go on", but they were in fact meant to convey to him, "Finish, your time is over".

In 1971 I visited Tbilisi and participated in the Ph.D. defence of my student R.V. Duduchava. There I met for the first time N.I. Muskhelishvili, a famous mathematician. I have a high regard for him and his school, and for their contributions to the theory of integral equations with singular kernels. N.I. Muskhelishvili was a very interesting man, one of the very few who was a member of the Supreme Soviet (the parliament of the USSR) from the beginning (1936) till his death. He was very influential and held many important posts in the main and Georgian academies. I visited him and thanked him for his support in my election to the Moldavian Academy. He answered modestly that he had done nothing exceptional, only fulfilled his duty. I told him that I had heard that his support was important, to which he replied that it may be so because some "bad characters had to be balanced". He presented me with a copy of the latest edition of his book "Singular Integral Equations", which I like very much. During this visit I also met other well known Georgian mathematicians: G.S. Chogoshvili, B.V. Khvedelidze and G.F. Mandzhavidze. Khvedilidze and I had mutual interests and we were friends for a long time.

In 1972 I was nominated as a candidate for full membership of the academy. This time Vinogradov and his friends were more careful. At a very early stage a special committee which was sent to Kishinev was also instructed to oversee my candidacy. This committee came to the conclusion that the Moldavian Academy did not need a full member, and that they would be better off with two corresponding members instead (my name was not mentioned at all). I followed developments regarding my nomination, and friends told me that progress went well before this special committee arrived. It was made up of N.N. Bogolubov, Chairman, A.N. Tichonov, A.A. Dorodonitsyn, A.N. Shirshov, and S.M. Nikolsky – a very powerful committee. They came to their decision and I was disappointed. But one can neuer know what is for the best. Had I been elected, who knows how many years I would have had to wait for permission to emigrate.

I would like to relate how my career as a member of the Moldavian Academy ended. When I left for Israel I could not take with me any documents showing that I was a corresponding member of the academy. I had a special identification card (which helped me in hotels, or to go to a movie without standing in line, or to a restaurant). I wanted to keep this identification, but I was not allowed to do so. When I arrived in Israel I asked the Ministry of Foreign affairs to get some document for me showing that I was indeed a corresponding member of the Moldavian Academy. At that time I thought it important for me to have such a document. By the way, during the almost 15 years since then I have never been asked for this identification. The Ministry sent a letter to the Dutch embassy in Moscow, which represents Israel in USSR, and the Dutch embassy asked for this document. After six months I received a reply from the Minister of Foreign Affairs saying that they had received a letter from the Soviet Union in which they state that they are unable to give me a document showing that I was indeed a member of the Academy of Science in Moldavia since I no longer appear on the list of members of this academy.

7. Towards Emigration

For a long time I had thought of emigrating to Israel, but the idea was more a dream than a real possibility. In 1969 things started to change, and more and more people were able to emigrate for reasons given as reunification of families. We had many relatives in Israel, uncles, aunts, and entire families. With the agreement of all my family I started to plan our emigration. At the same time I also took the necessary steps at my place of work. I refrained from taking on any new students, and I saw to it that almost all of my students would finish their theses before I applied for emigration. By the way, two of my Ph.D. students from Kishinev University defended their theses after I applied for an exit visa, my name as instructor was replaced by another, and the students were not allowed to mention me at all. I stopped working (part time) at Kishinev University. I tried to complete joint work which was in process, and many other things had to be taken care of. I followed closely the progress of the emigration process of people in a position similar to mine, and I remember what an impression the case of Boris Korenblum (Professor of Kiev Polytechnic Institute) made on me. The following is his story.

In order to apply for a visa at that time one had to prepare various papers, one of them a reference from one's employer. An initial meeting of one's colleagues would be arranged where some of those present (often communist party members) would discuss the applicant's behaviour and as a rule come to the conclusion that since he is a person who intends to leave the Socialistic Homeland for a capitalistic country, then he is a traitor. After this meeting the administration would give a letter of reference based on what had been said at this meeting. Where Boris Korenblum was concerned the authorities decided to make a show case. The meeting was planned as if a session of the Scientific Council (Senate). The Rector was the first to speak. He said that Professor Korenblum had received a free education in the Soviet Union, and just when he could be useful and serve bis country, he wishes to leave it and to emigrate to another country, and not even a friendly one. The Rector closed by asking "Is this fair?" B. Korenblum was allowed to answer the Rector. He said that he wanted to begin by stating some facts from his biography. He graduated from high school before the Second World War and was accepted to the university. When war broke out he volunteered and joined the Red Army, and spent the whole of the war fighting at the front. After he was demobilized at the end of the war he was allowed, as a veteran, to graduate from the university externally (by sitting examinations without actually attending classes). He never was a graduate student. He wrote his Ph.D. thesis in his spare time and defended it externally. In the same way he obtained also his second doctoral degree. B. Korenblum closed by asking if the Rector could perhaps explain how it is that he, Korenblum, who had volunteered in the World War Two to fight and to endanger his life for the Soviet Union, no longer accepted this country as his own and planned to emigrate to another. I do not remember any more details of this meeting which was a complete fiasco for the Rector. The result was that B. Korenblum applied for a visa and was given an extremely short time (muck less than others) in which to leave the USSR. By the way my family and myself, apart from my mother, had to go through such meetings. Even my twelve year old daughter. Worst of all was the meeting at the university to discuss my elder daughter. She was given an extra haranguing an my behalf also since I had already stopped working there and no longer in their hands.

During these years various stories were circulated. Once a university colleague asked me if the following rumours which were being spread were true. According to these rumours the Rector of Kishinev University had suggested to me that I change my patronymic names, Israel Tsudikovich. He maintained that students had complained that my name was too Zionistic and they felt uncomfortable pronouncing it. My reputed answer to the rector was that I refused to make any changes, explaining that the first name was given me by my parents, and the second is the name of my father, who is no Tonger alive. I also said that it was not the business of my students what my name is. Such an answer was considered extremely bold for that time, and my colleague told me that it gave him great satisfaction to hear of it. I had to disappoint him – this story was all fiction. Others also asked me similar questions about my name. I can only guess that my name was the subject of discussion at a certain administrative level, but probably no decision was taken as to any action.

Soon after I applied for a visa to Israel my sister was told by someone who neither knew that she was my sister, nor was close to academic circles, that the following was reputed to have happened in my case: I applied for a visa to a conference. When it was refused I threw the paper with the refusal into the face of the authorities, threatening that I would not let it go at that. I decided to leave the Soviet Union, and when I applied for an emigration visa, people standing in a long line waiting for visas allowed me and my family to make the application without standing in line, and I am probably already in Israel.

In reality the process of obtaining a visa for emigration was much more difficult and painful for my family and myself, with many refusals and a long waiting time. I was demoted and subjected to all kinds of discriminations, and as a result I resigned from my job.

On another occasion my wife was told that she should look out for my safety. Someone had overheard in a conversation at the academy that a group of people was planning to prevent my emigration by the use of brute force. There were many cases where people who had applied for an exit visa were brutally attacked, and after hearing this my family never let me out of the house without someone accompanying me. At the time I wondered about this, but it was only when we reached Israel that my wife told me the whole story.

During this difficult period of being a refusenik my family and I received a lot of support from the West. A businessman from England, Gerald Wise, often called us on the telephone. This gave us great moral support. We felt as if the entire Jewish people, the entire free world, was behind him. Gerald informed my colleagues in the West of my difficulties, and shortly after I received an important call from Chandler Davis, who became very active in my case. We received letters and telephone calls from other colleagues and friends in the United States and Israel. I was told that I was also supported at a meeting of the Bourbaki seminar.

During the darkest period, when the future of my family and myself was in real danger, two couples from the United States visited Kishinev for a few days in order to help us and other refuseniks. One couple was the New York lawyer, Alvin Hellerstein and his wife, and a special prosecutor of the State of New York, Maurice Nadjari and his wife. It is impossible to describe how much this visit meant to us. My family and myself will always remember with gratitude all those who helped us in our struggle to emigrate from the Soviet Union to Israel.

I had thought that my emigration would not greatly harm the Department of Functional Analysis in Kishinev or my friends, and I was sure that my friends would be able to replace me. Unfortunately they were not allowed to do so, and within a short time the Department of Functional Analysis was closed and its members dispersed among other departments. This does not mean that my colleagues stopped working. Recently I was very happy to see two excellent new books, one by N.Ja. Krupnik, "Banach Algebras with Symbol and Singular Integral Operators", Kishinev, Stiintsa, 1984; the English translation appeared in the OT series of Birkhäuser Verlag, 1987 (OT 26). The other book is by A.S. Markus, "Introduction to the Spectral Theory of Polynomial Operator Bundles", Kishinev, Stiintsa, 1986, English translation appeared in the AMS translation series in 1988. The journal was closed down and a series of brochures, virtually unavailable in the West, is being published in its place. In publications of the Moldavian Academy (as well as of the Ukrainian Academy) it is forbidden to quote my papers or even to mention my name, so my colleagues cannot quote their own joint papers with me. This applies also to some publications of the main academy, even when the editor is G.I. Marchuk, President of the Academy of Science of the USSR. Soon after I emigrated, V.M. Bychkov and K.S. Sibirsky wrote a brochure "Development of Mathematics in Moldavian SSR". A complete section is devoted to functional analysis and integral equations, and my name does not appear there at all (see review by Ch. Davis in Historia Mathematica, 1976, pages 235–236). More than that, rumours are occasionally spread in Kishinev that I had died. My relatives discovered that the source of these rumours was the Institute of Mathematics, where I had been working. At the end of one of their weekly meetings, it was announced that they had been informed that I could not find a job in Israel and that I had died. It was suggested they stand in silence in my memory. Why these rumours were spread I do not know, but obviously they were exaggerated.

8. Epilogue

I arrived in Israel with my family at the end of July 1974, and it was not at all clear to me how my career would progress. I had left behind 26 Ph.D.s (candidates) whom I had educated, my friends, and my teacher, Mark Gregorievich Krein. I knew I would miss all of them and I realized that it would probably take a very long time before I would have a group to work together with. In the West things developed much better than I had expected, and as a result I have a home country. This is Israel, and Tel Aviv University is my home university. I hold the Silver Chair donated by those wonderful people, Nathan and Lily Silver. And I have even more; I have second homes and wonderful friends, groups of colleagues and students with whom I work. My friends and colleagues now include not only mathematicians, but also engineers. I feel at home in the Netherlands at the Free University of Amsterdam, in the United States at the University of Maryland in College Park. I regularly visit the Weizmann Institute in Israel, and the University of Calgary in Canada, as well as many other places. Last but not least I would like to mention my visits to Basel, Switzerland, to my publisher, Birkhäuser Verlag.

9. Acknowledgements

The material presented above is based on a talk given at the Conference on Operator Theory held in Calgary, August 22 to August 28, 1988, to mark my 60th birthday. This talk preceded the birthday banquet. The other talks given at this conference were of a much more serious nature.

I would like to thank my colleagues and friends, former and present students, and all participants from many countries for attending this conference and making it such a success. I would like to thank the organizing committees and my very good friends H. Dym, S. Goldberg, M.A. Kaashoek and especially P. Lancaster, for their efforts and friendship. Thanks to all those who conveyed to me their congratulations.

My sincere gratitude is addressed to the University of Calgary, the Department of Mathematics and the Conference Department, for the excellent organization of this conference. Support of the conference by United States organizations, institutions of Canada, and Nathan and Lily Silver from Israel, is highly appreciated. My thanks to the Lancaster family for their kindness, friendship and outstanding hospitality.

I am happy that my family was present at this conference, sharing with me the warmth and wonderful atmosphere. I am grateful to them for their patience, and for sharing with me my problems and difficulties, and for bearing with my frequent travels. I am only very sorry that my mother could not be with us. She died in Israel five years ago. She brought my sister and myself up an her own, under extremely difficult conditions. She was my first and most important teacher.

Tel Aviv, December 1988

Philosophical-Mathematical Tales
A Personal Account

I. Gohberg

I am not an expert in philosophy. More than that I disliked philosophy in my student years, when a third of my study time was spent studying Marxist-Leninist philosophy. During my studies for the third degree, I was forced to read a considerable amount of works of Marx, Lenin, Stalin and Mao.

In general I was disappointed by their philosophies. I did not like the way they substantiated their main statements. I could not accept proofs such as "it cannot be because it can never be". I did not like the main practical conclusions. I thought they were wrong. At that time in the Soviet Union the main statements of this philosophy were considered to be the absolute truth and for any disagreement expressed publicly in a seminar a student would be punished and could end up in jail. Because of this philosophy, I lost my father when he was 42 years old. All of this had a strong influence on me and I decided to distance myself from philosophy in general, even when it was no longer dangerous. Unfortunately, I could not do this because every mathematician must be at least a little philosopher. Other scientists in their research have a direct connection to the real world via experiments. In this way they are able to check all aspects of their research, starting with the choice of topics, and ending with the analysis of the results and applications and to make changes if necessary. This connection is in general not available for a mathematician. His only tool is logic which is far from ideal. Restrictions are in order. Even the greatest mathematicians did not know how to behave in this regard. For instance, P.L. Tschebyshev never used complex numbers and complex analysis. This was already in the 19th century. The distance between mathematics and sciences is widening; mathematics is becoming more abstract and more formal. Society is less and less able to understand mathematics and this also creates problems in the social area as well as in the area of education. Mathematicians need at least some philosophical principles for protection and to serve them as a compass and a help in dealing with all of these difficult problems.

The first difficulty appears already in the description or the definition of the subject. We learned the definition of mathematics given by F. Engels, that

A general audience lecture given in Timişoara, June 2002.

the subject of mathematics is "quantitive relations and spacial forms of the real world".

Already in my student years I understood that this definition is very narrow. Where should formal logic or noneuclidian geometry, or complex numbers be included? But this was the accepted official definition and any critical remark at that time would be heavily punished because F. Engels could not be wrong. The famous mathematician A.N. Kolmogorov wrote a very nice paper for the Soviet encyclopedia for the word "mathematics". He had to write this paper in such a way that it would justify the definition of Engels. It was a very difficult and dangerous assignment but he did it nicely.

In the Soviet Union at that time, every department of mathematics had a methodogical seminar. At the meetings of these seminars mostly Marxist-Leninist philosophy of mathematics and sciences were studied. Once, when I visited Moscow University I learned that at the next meeting of the methodological seminar the subject for discussion was to be the definition of mathematics and the main talk would be given by G.E. Shilov. This came soon after the appearance of the Russian translation of the paper of Bourbaki. "The architecture of mathematics". This was some time in the early sixties when the situation in the Soviet Union was a little more liberal. I was interested to know how such discussions went in MGU. We also has such a methodological seminar in Kishinev and it was always boring. I remember one year studying the book "Mathematical manuscripts of K. Marx". It was a complete disappointment. I thought that this book and the discussions harmed the name of the famous philosopher. The majority of the manuscripts in fact were notes made by K. Marx during his study of calculus. He studied using an old and mediocre book from that time. I remember pages and pages of discussions and philosophical explanations on how to understand the relations $0/0, \infty/\infty$ and others. In general the study was not interesting and it was a waste of time.

I am returning to the talk of G.E. Shilov. He was an outstanding mathematician; I respected him and highly appreciated his work. I was interested to hear how he would put together peacefully the definition of F. Engels (it was obvious to me that he could not avoid it) and the article of Bourbaki. He really did it in a beautiful way, maintaining a calm atmosphere in the seminar. He presented the article of Bourbaki as the description of a town from within, with the layout of its streets, avenues and squares, with its architecture (structures and hierarchy of structures), with the explanation of the unity of mathematics from within. One town, one mathematics. The definition of mathematics of F. Engels was given by G.E Shilov as a description of the same town from a distance, when you do not see the streets, avenues and squares, when you do not see the details of the architecture. One describes only what is seen in general with its unity. The talk was illustrated by many examples and it was warmly received. The seminar was certainly very interesting and famous mathematicians took part in the discussion. At the same time it did not enrich me with any new ideas. I did not feel that these definitions helped me to understand mathematics any better.

Mathematics is often very abstract and mathematicians have to find ways to check their research. A mathematician has to be convinced that his choice is interesting, useful and logically correct. Let me start with the choice of topics for research. The problems have to be well chosen. It is is too easy it is not interesting. If it is too difficult very often one cannot solve it. In general especially appreciated are problems which have been well known for a long time and which many researchers were unsuccessful in solving. A nice example is Fermat's last theorem. For more than 350 years many mathematicians tried to prove the statement made by Fermat – that there do not exist three integers such that the sum of a certain integer greater than two power of two of them are equal to the third in the same power. An entire army of Fermatists dedicated their lives to this problem. Usually their knowledge was insufficient to solve this problem and their efforts ended up with nothing. The problem was solved recently by A. Wiles who worked for seven years using very sophisticated tools and theories.

In joint work with M.C. Krein at the beginning he took responsibility for the choice of topics, motivations and applications. I remember one period when we were working on trace class operators and on von Neumann-Schatten classes I had doubts that this material was the right choice, and M.G. had to convince me. It was the same when we were studying the asymptotics of eigenvalues and other issues: M.G.'s power to convince was outstanding and his choice was always very good and motivated. In this collaboration I really learned how to choose topics for research. Later this experience helped me tremendously. Very often one well chosen topic leads in a natural way to the next, but sometimes this way leads to the wrong decisions also. One becomes used to a certain area and finds it difficult to change even when new plans are not so impressive. Once M.I. Vishik told me that he is following such a principle: if in a certain topic or area one can prove every day a new theorem, then this is an indication that one has to change the topic. I used this principle a few times. In general, we worked very seriously on making new plans and choosing new topics for research. M.G. would say that in his research all topics become connected even if at the beginning they look far apart. The problem in the theory of nonselfadjoint operators and factorization problems were at the beginning quite a distance from one another. The first topic was more of a spectral character and the second was more connected with Wiener-Hopf equation and Rieman-Hilbert problems. We were very happy when it turned out that the factorization problem appeared in the abstract operator theory.

M.G. paid very serious attention to connections between infinite dimensional problems and their finite dimensional analogues, to connections between operator theory and finite matrix theory. For instance, the operator factorization problem has as its finite dimensional analogue the Gauss elimination for systems. There are problems in operator theory which do not have finite dimensional analogues. This was also an important indication in the choice of problems. One source for producing interesting problems are generalizations from finite dimensions to infinite dimensions. Much later I learned that conversely is also interesting. Namely, some infinite dimensional problems lead sometimes to new interesting finite dimensional problems. For instance, the problem of positive extensions of functions led to the problem of extensions of band matrices to positive definite ones. Another interesting example is the finite dimensional analogue of the formulas for inversion of Wiener-Hopf equations and their connections with fast algorithms for inversion of structured matrices.

I.M. Gelfand once told about a very capable student colleague of his who, most of his time, was busy solving elementary geometry problems where the solution had to be constructed by line and compass. Gelfand criticized him and told him that with his capabilities and ingenuity he could solve much more important and interesting problems. He answered that their still remained many problems of construction in elementary geometry which he is planning to work on.

I would like to tell a short story on a personal note. This happened during the time when M.G. Krein and myself were working on the two books on nonselfadjoint operators. During one of the weekends I worked at home very hard and in the end I understood that all I had done was wrong. I gave the pile of scribbled paper to my mother telling her to throw it out. My mother took the pile of paper to my sister telling her that something was wrong with me that I had worked all the weekend long and now I am throwing all my work out. They decided to ask me what was going on. The following day, they asked me what Krein and I are doing. M.G. Krein was highly regarded in our family. I told them that we are inventing problems and trying to solve them. My mother immediately answered, "So why don't you invent simple problems?

Unfortunately nobody would share with his colleagues the negative experience in research so that others would not repeat mistakes and bad choices. Once I even planned to start a mathematical journal which would publish this negative experience but this did not work out.

There re other criteria for the choice of topic for research. I will not discuss in detail those criteria. In general it is very difficult to tell a priori if the problem will lead to something difficult or beautiful, or with connections to other areas or applications or nothing will come out of it.

The criteria for beauty is very important in mathematics. The following story happened during the 1956 conference on functional analysis in Moscow organized by I.M. Gelfand. After a talk on von Neuman algebras Gelfand made some remarks. In one of the remarks he wondered why such a beautiful theory as that of von Neuman algebras has no applications, that such beauty cannot be lost. This talk was one of the last talks, and at the end of this session the famous topologist, P.S. Alexandrov, closed the session with a nice speech (he was a very good speaker). In his speech he said in particular that today we were all witnessing an important discovery, namely the discovery of the law of conservation of beauty (it cannot be lost). Now these algebras have plenty of applications.

Let me now discuss another philosophical-mathematical topic: when is a problem solved? Let me start with an example. Let us talk about the problem of computing the surface of plane figures. The problem can be reduced to one or a few problems of integration. The first solution of this problem was roughly speaking by numerical integration by one of the methods like the rectangle or others. This is in general a solution which is very difficult to do by hand. But historically this method was first used only for very special cases. Later was discovered the Newton-Leibnitz formula which gives the solution of the problem of integration. Namely first find the primitive of the function of integration and then the integral is equal to the difference of the values of the primitive function in the end points of the interval. So the problem is really beautifully solved and research is going ahead in this direction. In fact this is possible only if the primitive can be found, but if not, this solution does not solve anything. Such a situation happens in many cases. Some well defined new notion or reduction helps to claim that a problem is solved.

In general there are different ways to present the solution of a problem:

via proofs of existence, via constructive proofs, via algorithms for solutions, via computational stable algorithms without error accumulation, via compute aided proof, and others.

As an example let me recall that a few years ago the problem of four colors was solved with the help of computer aided proofs, but some mathematicians do not accept such proofs (as law courts do not accept videos as evidence). Computational algorithms often do not allow one to reach a solution. The reasons are error accumulation. The errors accumulate for different reasons, namely, for the round up of numbers in the operation for the wrong design of the algorithms and many others. In the case of error accumulation the solution has nothing in common with the real solution. Each problem has its natural form of solution. In many cases we may end up with an approximate or numerical solution and it may be quite sufficient. There are problems for which approximate or numerical solutions do not make sense (for instance Borsuk's problem).

I did not touch at all the philosophy of mathematics of the creativity process and many other aspects. There is plenty of literature on these topics. I do not intend or want to give recipes on how to choose the theme or topic for research; I am sure that recipes for a general choice for everybody do not exist. Each researcher must decide for himself. It is a difficult and important part of the work. This choice is strongly based on our intuition, experience, mathematical taste and our demands. My aim was to share with you my thoughts and my experience.

Finally I would like to describe in short three papers. They are three isolated papers, not very difficult but very important in the development of operator theory. I have in mind the following three wonderful papers.

I. Fredholm, Sur une classe d'equations functionelles, Acta Mathematics, 27; 1903, 365–390. F. Riesz, Über lineare Functionalgleichungen, Acta Mathematica, 41, 1916, 81–93. F. Noether, Über eine Klasse singulärer Integralgleichunden, Math. Ann. 82, 1921, 42–63.

The first is a paper in which the theory of integral equations was obtained. It contains the generalization of Cramer's rules for integral operators, solutions for homogeneous equations, spectral descriptions, and is based on a new technique of determinants for infinite dimensional operators. It solved a problem in which many distinguished mathematicians were interested.

The second contains the general abstract theory of linear compact operators, including the abstract Fredholm theorems. The paper is written beautifully. To me this paper looks ideal.

The third is the paper where for singular integral equations was introduced for the first time the index and associated properties. Many distinguished mathematicians worked in this area. The index was discovered only here.

These papers had a tremendous influence which is felt even today. I think that from such papers as these three we should learn how to choose problems and how to solve them.

Our Dad's Mathematics

Zvia Faro-Gohberg and Yanina Israeli-Gohberg

Our Dad is Professor Israel Gohberg. It is not so easy to be a daughter of a famous mathematician. As little girls we slept in Dad's library, where along the walls stood bookshelves. They were filled with mathematical books with strange words on the covers. We did not understand the titles of the books and always were very curious to know what they meant. When we learned to read, we tried to find out the answers, and even read through the first couple of pages, but there were no pictures inside and it did not clarify anything. On the contrary, our curiosity increased and even more than before we desired a simple explanation to the question: what are all these books about?

While the content of the books did not make any sense to us, the names of the authors were familiar. We met many of them when they were Father's guests and visited our house.

We always understood that those books where Dad's treasure. When we left for Israel it was very important for him to bring them with us. This was not a trivial task: on the Soviet border, custom officers opened with suspicion every single book, checking every page and the binding; it took almost an entire day. All those books and many more are in Dad's study in Ra'anana, Israel, and to this day they are among his favorite possessions and biggest treasures.

The titles of the books often contained the following phrases: Functional Analysis, Theory of Operators, Combinatory Geometry, Banach Spaces, Hilbert Spaces, Differential and Integral Equations as well as many others.

Zvia remembers how her second grade teacher asked for an example of a word with double consonant. She raised her hand and said: "differential". The teacher reacted in a strange way and came up with a more familiar word like "support".

The book by Gohberg and Krein: "Introduction to the Theory of the Linear Nonselfadjoint Operators in Hilbert Space" was published in 1965. It was translated into many languages. Dad showed us with pride the English translation of this book, published by the American Mathematical Society . But unfortunately we understood the English version even less than the Russian one.

A general audience lecture given in Timişoara, June 2002.

At school a lot was expected of us: being the daughters of a mathematician we were supposed to display ingenious mathematical abilities. We loved math and were good at it, but by no means we were genii. The older one of us studied Mathematics and Computer Science at Tel Aviv University. She even listened to Professor Gohberg's lectures, which brought her closer to Father's abstract world. The younger studied art in Bazalel, (Israeli Academy of Art in Jerusalem), she had to listen attentively to Fathers explanations and ask a lot of questions in order to understand this complex science.

Our education was very important to Dad. He always found time for us, helping with school work, as well as explaining many extra curricular subjects. In High School we often asked for help with complex mathematical problems. No matter how busy he was, he would drop anything he was doing and listen attentively. It is needless to say that in a couple of seconds he always came up with an elegant and clear solution..., but at that time we did not know that. He asked us questions first to make sure we understand the subject and he expected us to come up with a solution ourselves, pointing towards the right direction. Only later he would share his solution with us. We must say that at that time we were not very happy; we expected him to solve the problem immediately. Why waste time, when he can tell us the answer? This was a very valuable lesson we learned, that before asking questions one needed to read the textbook and understand the subject.

Our family spent summer vacations at the sea shore or in the mountains, where we often went with Dad for long walks. We loved going with him to the fishermen, watching them cook and eating with them, climbing the mountains, crossing the stormy rivers, snorkeling and swimming through the caves. On these walks we learned a lot in many different areas of life, science and agriculture. During the war as a young boy our Father lived for a long time on the farm in a kolhoz, and as a result he knows a lot about agriculture. We often walked along the water on the beach, through the woods, vineyards, walnut groves, corn fields. During those trips we learned about different plants, irrigation systems, new agricultural discoveries and wine making. Dad taught us how to pick and eat walnuts, how corn flour is made, how to pick ripe, sweet melons and survive in the forest by watching the signs of nature, eating berries and edible herbs to quench the thirst. Sometimes we went fishing, first catching the worms, then the fish and later cooking the fish on the open fire. We often gathered mushrooms. Dad taught us the names of many mushrooms and how to distinguish the good mushrooms from the poison ones. He had several large aquariums at home and passionately told us about tropical fish and alga.

At times we discussed politics, news he heard on the international radio, samizdat books or his dream to immigrate to Israel. But these topics were confidential and strictly between us; we knew we could not tell anyone about them, the birds were our only witnesses.

Of course, often the stories were about math and physics. When we were very little, he taught us arithmetic and logic, later binary and hexadecimal number systems, often writing formulas and calculations with the stick in the sand. He answered all of our curious questions including the one about all those mysterious books in his library and when we were old enough to understand he explained to us "his math".

Below is our Father's explanation of "his mathematics", or the way we understood this explanation and remember it:

Imagine a straight line: it represents a one-dimensional space. Every point on the line is characterized by a single real number. As everyone knows the plane is a two-dimensional space. Every point on a plane is characterized by two real numbers called coordinates. The space in which we live is three-dimensional and every point is characterized by three numbers, also called coordinates. In mechanics often four-dimensional spaces are used, where the coordinates represent movement in space, one coordinate measures time, and three other describe location in space at this specific moment of time. Solutions to some mathematical problems require the use of multi-dimensional spaces.

At this point we always tried to imagine a multi-dimensional space, as we have imagined the three dimensional one. We tried very hard, even shut our eyes, but unfortunately never succeeded.

Dad continued: *It is not possible to imagine a multi-dimensional space with dimension higher than the way we visualize the three dimensional space.*

(We were relieved and felt much better at this point.)

We can not imagine, but we can use our intuition to build geometry in these spaces. Geometry helps us to better imagine and understand these spaces, even though they are abstract and we can not see them. For example, we can prove the Pythagorean Theorem in every dimension that is bigger or equal to two. Sometimes the results discovered in multi-dimensional spaces are strange, not even resembling the familiar situations in two or three dimensional space: e.g. in a four-dimensional space we can separate two intertwined ordinary rings without tearing them apart.

This is still hard to understand, but things do get interesting! In a four dimensional space our Dad, like a magician, can separate linked rings without tearing them apart. Let's hear some more:

Geometrical imagination helps us solving many complex problems; some of them in infinite-dimensional spaces. Intuitively we understand that in the above spaces every point is characterized by an infinite sequence of coordinates.

At this point we would always ask: "Why do you need infinite dimensional spaces?"

Infinite-dimensional spaces are used in Functional Analysis, Quantum Mechanics and other areas of mathematics and physics, and they are extremely useful. Infinite-dimensional spaces with a certain notion of distance between two points and certain natural properties of this distance are called functional spaces.

So, what is Functional Analysis?

Functional Analysis studies functional spaces and their transformations. Numerous complicated differential and integral equations are interpreted via transformations in functional spaces and, in this way they can be solved or analyzed with the use of geometrical intuition in functional spaces.

What is an Operator?

Transformations in functional spaces are called operators.

OK, we have some idea of our Dad's scientific environment; now sophisticated words, like Functional Analysis, Operator Theory, multi-dimensional and infinite-dimensional spaces are starting to make sense. It is not very intuitive and not immediately clear; one has to get used to new ideas, concepts and develop abstract thinking, imagination and even the appropriate mentality.

Our Dad is most certainly a specialist in geometry, but this is not the geometry we learned in high school. "Dad's geometry" is related to multi-dimensional and even infinite-dimensional spaces, where the problems are solved with the help of geometrical intuition. Now we understand that when it looks like Dad is dreaming, he actually is concentrating and transitioning into those abstract generalized spaces, so distant to us, trying to understand their geometry. In his dissertation Dad studied and researched Fredholm Operators or transformations. Ivar Fredholm was a famous Swedish mathematician and those operators are named after him.

So, "Dad's Math" is a little more familiar now, but not completely clear yet. We continue asking the question: "How can we imagine what you do?" Dad patiently continues his explanation, he tells us about one of the episodes from his scientific experience. This example characterizes a small area of his research. Here is his story:

There is a clear correlation between the geometric characteristics of a space and its dimension. In two-dimensional space two perpendicular lines (90 degrees angle) can be constructed through every point. These two lines can be constructed in many different ways, but there is no way to construct three or more perpendicular lines through one point.

In three-dimensional space through every point one can construct three pair wise perpendicular lines, but not four. In N-dimensional space one can construct N (pairwise) perpendicular lines, but not $N+1$. So by the number of perpendicular lines that can be constructed through every point we can determine the dimension of the space. In the infinite-dimensional spaces, so called Hilbert spaces, an infinite number of perpendicular lines can be constructed through every point.

While researching functional spaces, Dad realized:

In all the analyzed cases, the dimension of a space was connected with another number: the covering number, which is determined by covering convex figures or bodies with smaller similar ones. In order to explain in more details this observation, we will have to formulate several definitions. We will try to avoid precise mathematical formulations, which are too complicated at times and are beyond the scope of this article. For the readers with mathematical background we shall provide sources where these precise definitions can be found. Although most of our explanations will be based on the readers' intuition, below is given the definition for the convex figures, which may be familiar from high school geometry lessons.

Here is the definition is:

A convex figure is one that contains all points of a segment joining points belonging to it. This means that if we connect any two points inside the convex figure with a straight line (segment), this segment will be located inside this convex figure. In two-dimensional space examples of convex figures are: a square, a parallelogram, ellipses, regular polygons. You can see examples of convex figures in Picture 1 below.

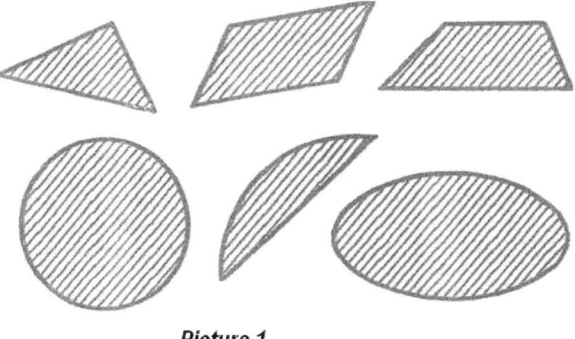

Picture 1

There are figures which are not convex: they are called non-convex figures. Examples of non-convex figures are shown below in Picture 2.

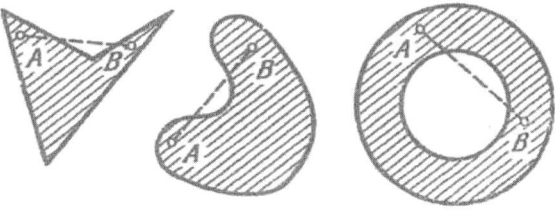

Picture 2

The examples show that the segment connecting 2 points inside the figure is not entirely located inside the figure.

Many of us remember the definition of similar figures from the school lessons in geometry. There is a precise mathematical definition for similarity of figures, but we will again rely on our intuition. In the present context two figures will be called similar if one of them can be obtained from the other by parallel translation and/or by shrinking with a constant factor. Similar figures have the same shape but they may be of different sizes.

Let's look at the circle: a figure similar to a given circle is another circle, which can be smaller, larger or equal in size to the given circle. A figure similar to a given square is another square with the sides parallel to the original one. A *figure similar to a parallelogram is another parallelogram, with sides proportional and parallel to the original one. Picture 2a shows two similar ovals. Similarity for the oval figures means, that $bh/qh = ah/ph = k$, where h is the center of similarity and the constant k is the coefficient of similarity.*

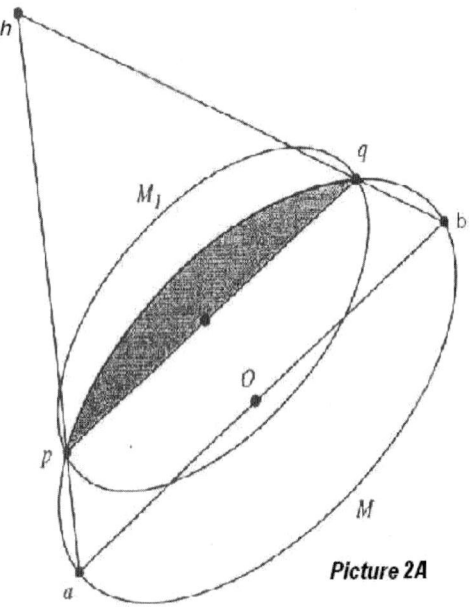

Picture 2A

Now, after having discussed convexity and similarity of figures, let's imagine a given circle and a couple of smaller circles. We would like to cover the large circle with the smaller ones, as if we were sowing a blanket from the smaller circles that would entirely cover the large given circle. How many smaller circles are required to cover all the points inside the given circle? The answer to the question is: at least three circles of a smaller diameter are required to cover any given circle. Two circles are not enough! See Picture 3.

Any given parallelogram can be covered with four similar parallelograms, but three is not enough. See Picture 4.

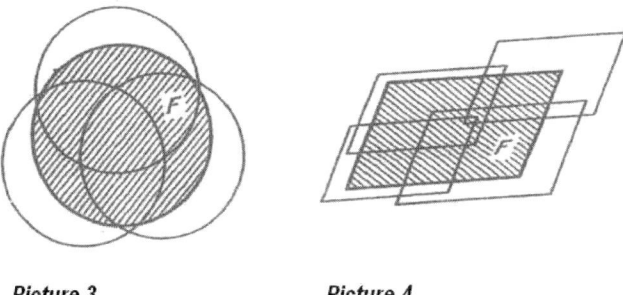

Picture 3   **Picture 4**

What is this number for the more complex convex polygons? We do not know the answer. And who knows, what is this number for a polygon with one million sides?

Dad made an amazing suggestion: For any given convex figure in a two-dimensional space, the number of similar smaller figures that will entirely cover a given figure equals three, with the exception of the parallelogram, for which this number equals to four.

Cool! This is unexpected, and not what we thought the answer would be. Dad has discovered two things:

One, that in a two dimensional space at most four smaller similar convex figures are needed to cover any given convex figure The second discovery was, that when a micron (or less) is sliced from one of the parallelogram angles, while the resulting figure still remains convex, the number of convex similar figures to cover it will be three.

Dad was excited about this discovery and shared it with his colleague and friend A.S. Marcus. In a short period of time they obtained two proofs of Dad's conjecture: one was analytical and another was geometric. This discovery became known as the *Gohberg-Marcus Theorem.*

Picture 5 illustrates the idea of a covering for an arbitrary convex figure.

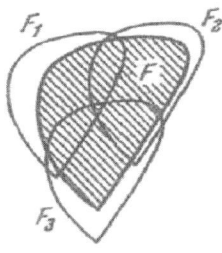

Picture 5

The proof of this theorem is quite complicated, it has to be generalized to and made valid for any convex figure. We can imagine the complexity, because there are so many different convex figures out there. An analogous theorem formulated differently was proven earlier by P. Levi. The Gohberg-Marcus theorem can be derived from the Levi Theorem and vice versa. Simple examples involving convex bodies in three dimensional space suggest that there these covering numbers can be 4, 5, 6, 7 or 8. The number for the sphere is equal to 4.

Picture 6 displays two spheres, each one is divided into four smaller parts. Intuitively we understand that each one of those parts can be inserted into a sphere with diameter smaller than the diameter of the initial sphere. This means that we can cover a sphere with 4 smaller spheres.

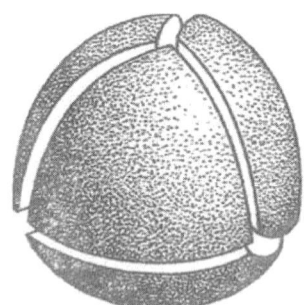

Picture 6

Nobody proved so far that the five numbers 4, 5, 6, 7 and 8 mentioned above are indeed the (only) covering numbers for convex bodies in three-dimensional space, and there are no others. These unproven statements are called conjectures.

The above results show that even in the three-dimensional space familiar to us there are still features and secrets that as yet we can not prove or disprove.

We keep asking questions: "What is a conjecture?" *A conjecture is a scientific statement based on certain observations or on intuition which is not proven (yet), but has a certain likelihood. Once the conjecture is proven, it becomes a theorem or a lemma (in mathematics) and ceases to exist as a conjecture. Conjectures can be proven wrong; in that case the disproved conjecture becomes a false assumption and ends its existence as well.*

The Gohberg-Marcus Conjecture states: For every convex body in three-dimensional space, the minimal number of smaller similar figures to completely cover the given figure is 4, 5, 6, 7 or 8 (and nothing else).

This number is called the covering number (a term already used in passing before).

What is happening in N-dimensional space? For N-dimensional space it is thought that the possible covering numbers are $N + 1, N + 2, \ldots, 2N$. For an N-dimensional sphere the covering number is $N + 1$, and for a cube this number equals to $2N$. This issue is not settled to this day and it is called the N-dimensional Gohberg-Marcus Conjecture.

While working on this problem, Dad became friends with an outstanding mathematician V.G. Boltyansky. Boltyansky formulated another problem: illuminating the surface of convex figures.

They tried to find how many light sources are needed to illuminate (in an appropriate sense tangent rays are not counted) the surface of a convex figure or body.

This makes more sense to us now! Let's say we need to illuminate a sphere, meaning that every point on a sphere has to be illuminated by at least one source of light.

What is the minimal number of the required sources of illumination? Boltyansky proved that for a convex figure or body this minimal number equals the Gohberg-Marcus covering number. This number is called the Boltyansky number. So Boltyansky has proved the following result, which is called the Boltyansky Theorem:

For a given convex figure or body the Boltyansky number and the Gohberg-Marcus number are identical (regardless whether or not the values of these numbers are known). See Pictures 7 and 8.

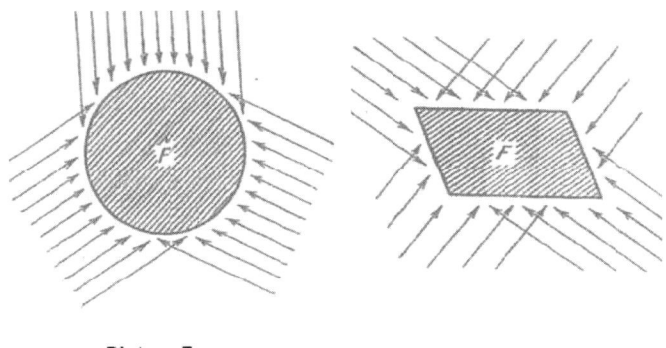

Picture 7   **Picture 8**

Boltyansky also proved that for a convex figure or body with smooth boundary *this number, is equal to $N + 1$, where N is the dimension of the space. See Picture* 6.

Here are some examples of convex bodies with smooth boundaries: a sphere, a balloon and an egg. Cubes, cones, cylinders and pyramids do not have smooth boundaries.

V.G. Boltyansky and our Dad published two booklets dedicated to the above subject. This area of math is called "combinatorial geometry".

The readers with some mathematical experience, who are interested in additional information, are advised to consult the books quoted in bibliography at the end of this article.

We keep asking questions: What happened with the conjecture of Gohberg Marcus?

At this point nobody has proven it yet in general for any dimension larger than 2.

Then, what is its practical value today?

Today it may not have a practical value, but numerous mathematical theories started without a practical value. Let's take for example tomography, today it is widely used in biology, archeology, oceanography, geophysics and other areas of science. It is hard to imagine modern medicine without tomography. Tomography is used in X-rays, angiography, CT and MRI. It helps uncover and diagnose many diseases and medical problems. Probably, not everyone knows today that tomography is based on integral transformation published by an Austrian mathematician Johann Radon in 1917.

So, what will it take to prove this conjecture?

This is not a simple task and some proofs take hundreds of years. As an example, take the Fermat theorem. The Fermat theorem states:

If an integer n is larger than 2, *the equation* $a^n + b^n = c^n$ *has no solutions in non-zero integers a, b, and c.*

In 1637 *Pierre de Fermat wrote in his copy of the Arithmetica of Diophantus that he has a marvelous proof for this theorem, but that the margin of the book was too narrow to write it down. Famous mathematicians all over the world tried to prove this theorem for* 357 *years and although some came close, only Andrew Wiles proved it in* 1995. *Wiles has dedicated* 7 *years of his life to the proof of the last Fermat theorem.*[1]

We are telling Dad that we have no intention to attempt to prove Gohberg-Marcus Conjecture, but nevertheless we are asking him one more question: "What if we did solve it?"

Dad does not have to think long, and with his characteristic sense of humor answers: "Well, then it will be called the Theorem of Gohberg's Daughters."

References

[1] V. Boltyansky and I. Gohberg, *Results and problems in combinatorial geometry.* Translated from Russian. Cambridge University Press, Cambridge, 1985.

[2] V. Boltyansky and I. Gohberg, *Decomposition of figures in into smaller parts.* Translated from Russian. The Chicago University Press, Chicago and London, 1980.

[3] V. Boltyansky, H. Martini and P. Soltan, *Excursions into Combinatorial Geometry.* University Text, Springer-Verlag, Berlin, 1997.

---

[1]http://www-groups.dcs.st-and.ac.uk/~history/HistTopics/Fermat's_last_theorem.html; *Mathematical Topics*, Article by J.J. O'Connor and E.F. Robertson.

Part II
Work and Personalia

This part consists of eight documents. It contains Gohberg's curriculum vitae, the list of his publications and a list of his Ph.D students. Also included are a translation of a letter of reference written by M.G. Krein, when Gohberg was a master student, and translations of letters and telegrams supporting his nomination as a corresponding member of the Academy of Sciences of the Moldavian SSR. The next two documents, written by Rien Kaashoek and by Rien Kaashoek and Leonid Lerer, respectively, present a review of Gohberg's mathematical work. The final document concerns the Nathan and Lily Silver chair of Mathematics of which Israel Gohberg has been the incumbent from 1981 to 1998.

Curriculum Vitae of Israel Gohberg

Work Address:	School of Mathematical Sciences
	Raymond and Beverly Sackler Faculty
	of Exact Sciences
	Tel Aviv University
	Ramat Aviv 69978
	Israel
E-mail:	gohberg@post.tau.ac.il
Date and place of birth:	August 23, 1928, Tarutino, USSR
Marital status:	Married
No. of children:	2

Education

Two years studies in the State Pedagogical Institute,
Frunze, Soviet Republic of Kirgizia, USSR,
Date of award: July 1948

M.Sc. Mathematics
Kishinev University, Kishinev,
Moldavian Soviet Republic, USSR
Date of award: July 1951

Kandidat of Sciences, Ph.D. Mathematics
Leningrad Pedagogical Institute, USSR
Date of award: April 1954

Doctor of Sciences, Mathematics
Moscow State University, USSR
Date of award: February 1964

Academic and Professional Experience

1951–1953	Assistant Professor, Soroki Teacher's Institute, Moldavian Soviet Republic, USSR
1953–1959	Assistant Professor, Associate Professor, Chairman of the Department of Mathematics, Beltsky Pedagogical Institute,

	Moldavian Soviet Republic, USSR
1959–1974	Senior Researcher and Head of the Department of Functional Analysis, Institute of Mathematics of the Academy of Science, Moldavian Soviet Republic, USSR (part-time)
1996–1973	Professor at Kishinev University, USSR (part-time)
1974–1996	Professor, Tel Aviv University, Israel
1996–	Professor Emeritus, Tel Aviv University, Israel
1975–1983	Professor, Weizmann Institute of Science, Rehovot, Israel (part-time)
1981–1998	Incumbent of the Nathan and Lily Silver Chair in Mathematical Analysis and Operator Theory, Tel Aviv University
1983–1998	Professor, Vrije Universiteit, Amsterdam, The Netherlands (part-time)
1980- 2002	Professor, University of Maryland, College Park, MD. U.S.A. (part-time)
1974–2000	Visiting and adjunct professor for various extended periods at State University of New York at Stony Brook, New York, U.S.A. University of Calgary, Alberta, Canada University of Georgia, Athens, GA, U.S.A. Vrije Universiteit, Amsterdam, The Netherlands. Virginia Polytech Institute and State University, Blacksburg, Virginia, U.S.A. University of Connecticut, Storrs, CT, USA University of Regensburg, Regensburg, Germany University of Mainz, Mainz, Germany A. Humboldt University, Berlin, Germany Institut fur Angewandte Analysis und Stochastik, Berlin, Germany

Grants and Awards

1970	Elected corresponding member of the Academy of Sciences of MSSR, USSR
1974	Removed from the list of members of the Academy of Sciences of MSSR, USSR

1996	Reinstated as a corresponding member of the Academy of Sciences of Moldova
1985	Elected Foreign Member of the Royal Netherlands Academy of Arts and Sciences
1976	Awarded Landau Prize in Mathematics (Israel)
1986	Awarded Rothschild Prize in Mathematics (Rothschild Foundation, Israel)
1992	Awarded A. Humboldt Research Prize (Humboldt Foundation, Germany)
1994	Awarded Hans Schneider Prize in Linear Algebra (International Linear Algebra Society)
1997	Awarded Honorary Doctoral Degree, (Technische Hochsule, Darmstadt, Germany)
2001	Awarded Honorary Doctoral Degree Vienna University of Technology (Technische Universitat Wien, Austria)
2002	Awarded Honorary Doctoral Degree, (Universitatea de Vest din Timisoara, Romania)
2002	Awarded Honorary Doctoral Degree, (Universitatea de Stat din Moldova, Cisinau, Moldova)
2002	Awarded Honorary Doctoral Degree, (Universitatea de Stat "Aleco Russo" din Balti, Balti, Moldova)
2007	M.G. Krein Prize in Mathematics for 2007, awarded by the Ukrainian National Academy of Sciences
2008	Awarded Honorary Doctoral Degree of Science, (Technion, Israel Institute of Technology, Haifa, Israel)

Editorial Work

1978-	Founder and editor of the international journal "Integral Equations and Operator Theory" (Birkhäuser)
	Founder and editor of the book series "Operator Theory: Advances and Applications" (Birkhäuser)
	Member of editorial board of the journals: "Asymptotic Analysis" (North-Holland); "Applied Mathematics Letters" (Pergamon Press); "Calcolo" (Springer Verlag); "Complex analysis and Operator Theory" (Birkhäuser);

"Georgian Mathematical Journal" (Heldermann Verlag)
"Indagationes Mathematicae" (North-Holland);
"Mediterranean Journal of Mathematics" (Birkhäuser)

1995-   Advisory editor of
                          "Methods of Functional Analysis and Topology" published
                          by the Institute of Mathematics, Ukrainian National Acad-
                          emy of Sciences

2004-   Honorary editor of the
                          "Journal of Spectral Mathematics and Applications" pub-
                          lished by Northern Illinois University, De Kalb

Doctoral Students Supervised

Supervised 40 doctoral students (see the separate List of
Ph.D. students in this volume).

Date: March 2008

List of Publications of Israel Gohberg

Books

1. I. Gohberg, M. Krein
Vvedenie v Teoriju Linejnyh Nesamosopriazhennyh Operatorov v Gilbertovom Prostranstve. Nauka, Moscow, 448 pages (Russian), 1965.

1a. I. Gohberg, M. Krein
Introduction to the theory of linear nonselfadjoint operators. American Mathematical Society, Providence, 378 pages (translated from Russian), 1969; second printing 1978, third printing 1983, fourth printing 1988.

1b. I. Gohberg, M. Krein
Introduction a la Theorie des Operateurs Lineaires Non Autoadjoints dans un Espace Hilbertien. Dunod, Paris, 372 pages (French, translated from Russian), 1971.

2. I. Gohberg, M. Krein
Tories Volterovhy Operatorov v Gilbertovom Prostranstve i Ejo Prilozhenia. Nauka, Moscow, 508 pages (Russian), 1967.

2a. I. Gohberg, M. Krein
Theory and Applications of Volterra Operators in Hilbert Space. American Mathematical Society, Providence, 430 pages (Translated from Russian), 1970. Second printing, 2006.

3. I. Feldman, I. Gohberg
Proektionnye Metody Reshenia Uravnenij Wiener-Hopfa. Akademija Nauk MSSR, Kishinev, 164 pages (Russian), 1967.

4. I. Feldman, I. Gohberg
Uravnenija v Sveortkah i Proektionnye Metody ih Reshenia. Nauka, Moscow, 352 pages (Russian), 1971.

4a. I. Feldman, I. Gohberg
Faltungsgleichungen und Projektionsverfahren zu ihrer Lösung. Akademie-Verlag, Berlin, 276 pages (German, translated from Russian), 1974.

4b. I. Feldman, I. Gohberg
Faltungsgleichungen und Projektionsverfahren zu ihrer Lösung. Mathematische Reihe 49. Birkhäuser Verlag, 275 pages (Translated from Russian), 1974.

4c. I. Feldman, I. Gohberg
Convolution Equations and Projection Methods for their Solution. American Mathematical Society, Providence, 262 pages (Translated from Russian), 1974. Second printing 2006.

5. I. Gohberg, N. Krupnik
Vvedenije v Teoriju Odnomernyh Singuliarnyh Integralnyh Operatorov. Shtiinta, Kishinev, 428 pages (Russian), 1973.

5a. I. Gohberg, N. Krupnik
Einführung in die Theorie der eindimensionalen singulären Integraloperatoren. Birkhäuser Verlag, 379 pages (German, translated from Russian), 1979.

6. V.G. Boltyanskii, I. Gohberg
Teoremy i Zadachi Kombinatornoj Geometrii. Nauka, Moscow, 108 pages (Russian), 1965.

6a. V.G. Boltyanskii, I. Gohberg
Sätze und Probleme der Kombinatorischen Geometrie. Deutscher Verlag der Wissenschaften, Berlin, 128 pages (German, translated from Russian), 1972.

6b. V.G. Boltyanskii, I. Gohberg
Tetelek es Faladtok A Kombinatorikus Geometriabol. Tankoyvkiado, Budapest, 112 pages (Hungarian, translated from Russian), 1970.

6c. V.G. Boltyanskii, I. Gohberg
Results and Problems in Combinatorial Geometry. Cambridge University Press, 108 pages (Translated from Russian), 1985.

7. V.G. Boltyanskii, I. Gohberg
Razbienie Figur na Menshie Chasti. Nauka, Moscow, 88 pages (Russian), 1971.

7a. V.G. Boltyanskii, I. Gohberg
Division de Figuras en Partes Menores. Mir, Moscow, 106 pages, (Spanish, translated from Russian), 1973.

7b. V.G. Boltyanskii, I. Gohberg
The Decomposition of Figures into Smaller Parts, 75 pages, University of Chicago Press, 75 pages (Translated from Russian), 1980.

7c. V.G. Boltyanskii, I. Gohberg
Alakzatok Felbontasa Kisebb Reszekre. Tankonyvkiad, Budapest, 93 pages, (Hungarian, translated from Russian), 1976.

8. H. Bart, I. Gohberg, M.A. Kaashoek
Minimal Factorization of Matrix and Operator Functions. Operator Theory: Advances and Applications, Vol. 1. Birkhäuser Verlag, 236 pages, 1979.

9. I. Gohberg, S. Goldberg
Basic Operator Theory. Birkhäuser Verlag, 285 pages, 1981.

10. K. Clancey, I. Gohberg
Factorization of Matrix Functions and Singular Integral Operators. Operator Theory: Advances and Applications, Vol. 3. Birkhäuser Verlag, 234 pages, 1981.

11. I. Gohberg, P. Lancaster, L. Rodman
    Matrix Polynomials. Academic Press, 409 pages, 1982.
12. I. Gohberg, P. Lancaster, L. Rodman
    Matrices and Indefinite Scalar Products. Operator Theory: Advances and Applications, Vol. 8. Birkhäuser Verlag, 374 pages, 1983.
13. I. Gohberg, P. Lancaster, L. Rodman
    Invariant Subspaces of Matrices with Applications. Canadian Math. Soc. Series of Monographs and Advanced Texts, John Wiley & Sons, 629 pages, 1986. Second printing, Classics in Applied Mathematics, SIAM, 2006
14. J.A. Ball, I. Gohberg, L. Rodman
    Interpolation of Rational Matrix Functions. Operator Theory: Advances and Applications, Vol. 45. Birkhäuser Verlag, 605 pages, 1990.
15. I. Gohberg, S. Goldberg, M.A. Kaashoek
    Classes of Linear Operators, Vol. 1. Operator Theory: Advances and Applications, Vol. 49. Birkhäuser Verlag, 468 pages, 1990.
16. I. Gohberg, N. Krupnik
    One-Dimensional Linear Singular Integral Equations, Vol. 1. Introduction. Operator Theory: Advances and Applications, Vol. 53. Birkhäuser Verlag, 263 pages, 1992.
17. I. Gohberg, N. Krupnik
    One-Dimensional Linear Singular Integral Equations, Vol. 2. General Theory. Operator Theory: Advances and Applications, Vol. 54. Birkhäuser Verlag, 232 pages, 1992.
18. I. Gohberg, S. Goldberg, M.A. Kaashoek
    Classes of Linear Operators, Vol. 2. Operator Theory: Advances and Applications, Vol. 63. Birkhäuser Verlag, 562 pages, 1993.
19. I. Gohberg, M.A. Kaashoek, F. van Schagen
    Partially Specified Matrices and Operators: Classification, Completion, Applications. Operator Theory: Advances and Applications, Vol. 79. Birkhäuser Verlag, 333 pages, 1995.
20. C. Foias, A.E. Frazho, I. Gohberg
    M.A. Kaashoek *Metric Constrained Interpolation, Commutant Lifting and Systems.* Operator Theory: Advances and Applications, Vol. 100. Birkhäuser Verlag, 587 pages, 1998.
21. I. Gohberg, S. Goldberg, N. Krupnik
    Traces and Determinants of Linear Operators. Operator Theory: Advances and Applications, Vol. 116. Birkhäuser Verlag, 270 pages, 2000.
22. R.L. Ellis, I. Gohberg
    Orthogonal Systems and Convolution Operators. Operator Theory: Advances and Applications, Vol. 140. Birkhäuser Verlag, 236 pages, 2003.
23. I. Gohberg, S. Goldberg, M.A. Kaashoek
    Basic Classes of Linear Operators. Birkhäuser Verlag, 423 pages, 2003.
24. I. Gohberg, Peter Lancaster, Leiba Rodman
    Indefinite Linear Algebra and Applications. Birkhäuser Verlag, 2005.

25. H. Bart, I. Gohberg, M.A. Kaashoek, A.C.M. Ran
    Factorization of Matrix and Operator Functions. The State Space Method.
    Operator Theory: Advances and Applications, Vol. 178. Birkhäuser Verlag,
    407 pages, 2007.

Articles

1. I. Gohberg
   On linear equations in Hilbert space. Dokl. Acad. Nauk SSSR, 76, no. 4, 9–12
   (Russian) (1951); MR 13, 46 (1952).
2. I. Gohberg
   On linear equations in normed spaces. Dokl. Akad. Nauk SSSR, 76, no. 4,
   447–480 (Russian) (1951); MR 13, 46 (1952).
3. I. Gohberg
   On linear operators depending analytically upon a parameter. Dokl. Akad.
   Nauk SSSR, 78, no. 4, 629–632 (Russian) (1951); MR 13, 46 (1952).
4. V.A. Andrunakievich, I. Gohberg
   On linear equations in infinite-dimensional spaces. Uch. zap. Kishinev. Univ.,
   Vol. V, 63–67 (Russian) (1952).
5. I. Gohberg
   On an application of the theory of normed rings to singular integral equations.
   Uspehi Mat. Nauk 7, 149–156 (Russian) (1952); MR 14, 54 (1953).
6. I. Gohberg
   On the index of an unbounded operator. Mat. Sb 33(75), I, 193–198 (Russian)
   (1953). MR 15, 233 (1954).
7. I. Gohberg
   On systems of singular integral equations. Uc. Zap Kisinevsk. Univ. 11, 55–60
   (Russian) (1954); MR 17, 163 (1956); MR 17, 75 (1956).
8. I. Gohberg
   On zeros and zero elements of unbounded operators. Dokl. Akad. Nauk SSSR
   101, 9–12 (Russian) (1955); MR 17, 284 (1956).
9. I. Gohberg
   Some properties of normally solvable operators. Dokl. Akad. Nauk SSSR 104,
   9–11 (Russian) (1955); MR 17, 647 (1956).
10. I. Gohberg, A.S. Markus
    On a characteristic property of the kernel of a linear operator. Dokl. Akad.
    Nauk. SSSR 101, 893–896 (Russian) (1955); MR 17, 769 (1956).
11. I. Gohberg
    The boundaries of applications of the theorems of F. Noether. Uch. zap
    Kishinev Univ, Vol. 17, 35–43 (Russian) (1955).
12. I. Gohberg, A.S. Marcus
    On stability of certain properties of normally solvable operators. Mat. Sb. 40
    (82), 453–466 (Russian) (1956); MR 19, 45 (1958).

13. I. Gohberg, M.G. Krein
    On the basic propositions of the theory of systems of integral equations on a half-line with kernels depending on the difference of arguments. Proc. III Math. Congr. SSSR, Vol. 2 (Russian), 1956.

14. I. Gohberg, M.G. Krein
    The application of the normed rings theory to the proof of the theorems of solvability of systems of integral equations. Proc. III Math. Cong. SSSR, Vol. 2, l956.

15. I. Gohberg
    On the index, null elements and elements of the kernel of an unbounded operator. Uspehi Mat. Nauk (N.S) 12, No.1 (73), 177–179 (Russian) (1957); MR 19, 45 (1958); English Transl. Amer. Math. Soc. Transl. 2(16) 391–392 (1960); MR 22 #8374.

16. I. Gohberg, L.S. Goldenstein, A.S. Markush
    Investigations of some properties of linear bounded operators with connection to their q-norm. Uch. zap. Kishinev Univ. Vol. 29 (Russian), 1957.

17. I. Gohberg, M.G. Krein
    The basic propositions on defect numbers root numbers and indices of linear operators. Uspehi Mat. Nauk 12, No. 2 (74), 43–118 (1957); English transl. Amer. Math. Soc. Transl. (2) 13, 185–264 (1960); MR 20 #3459; MR 22 # 3984.

18. I. Gohberg, M.G. Krein
    Systems of integral equations on a half-line with kernels depending on the difference of arguments. Uspehi Mat. Nauk 13, No.2 (80), 3–72 (1958); English transl. Amer. Math. Soc. Transl. 2 (14), 217–287 (1960); MR 21 #1506; MR 22 #3954.

19. I. Gohberg, M.G. Krein
    On the stability of a system of partial indices of the Hilbert problem for several unknown functions. Dokl. Akad. Nauk SSSR 119, 854–857 (1958); MR 21 #3547.

20. I. Gohberg
    On the number of solutions of a homogeneous singular integral equation with continuous coefficients. Dokl. Akad. Nauk SSSR 122, 327–330 (Russian) (1958); MR 20 #4748 (1959).

21. I. Gohberg
    Two remarks on index of a linear bounded operator. Uch. zap. Belz Pedagog. Inst. No. 1, 13–18 (Russian) (1959).

22. I. Gohberg, M.G. Krein
    On a dual integral equation and its transpose I. Teoret. Prikl. Mat. No. 1, 58–81, Lvov (Russian) (1958); MR 35 #5877.

23. I. Gohberg
On bounds of indexes of matrix-functions. Uspehi Nat. Nauk 14, No. 4 (88), 159–163 (Russian) (1959); English Transl. Operator Theory: Advances and Applications, Volume 7, pp. 243–274, Birkhäuser Verlag (1983); MR 22 #3993.

24. I. Gohberg, A.S. Markus
Two theorems on the gap between subspaces of a Banach space. Uspehi Mat. Nauk 14, No. 5 (89) 135–140 (Russian) (1959); MR 22 #5880.

25. I. Gohberg, M.G. Krein
On completely continuous operators with spectrum concentrated at zero. Dokl. Acad. Nauk SSSR, 128, No. 2, 227–230 (Russian) (1959); MR 24 #A1022.

26. I. Gohberg, A.S. Markus
Characteristic properties of the pole of a linear closed operator. Uch. zap Belz. Pedagog. Inst. No. 5, 71–75 (Russian) (1960).

27. I. Gohberg
A remark of standard factorization of matrix-functions. Uch. zap. Belz. Pedegog Inst. No. 5, 65–69 (Russian) (1960).

28. I. Gohberg, A.S. Markus
Characteristic properties of certain points of spectrum of bounded linear operators. Izv. Vyso. Ucel Zaved. Matematik No. 2 (15) 74–87 (Russian) (1960); MR 24 #A1626.

29. I. Gohberg, L.S. Goldenstein
On a multidimensional integral equation on a half-space whose kernel is a function of the difference of the arguments, and on a discrete analogue of this equation. Dokl. Acad. Nauk SSSR 131, No. 1, 9–12 (Russian) (1960); Soviet Math. Dokl. 1, 173–176 (1960); MR 22 #8298.

30. I. Gohberg
On the theory of multidimensional singular integral equations. Dokl. Acad Nauk SSSR 133, No. 6, 1279–1282 (Russian) (1960); Soviet Math. Dokl. 1, 960–963 (1961); MR 23 #A2015.

31. I. Gohberg
Some topics of the theory of multidimensional singular integral equations. Izv. Mold. Akad. Nauk No. 10 (76), 39–50 (Russian) (1960).

32. I.A. Feldman, I. Gohberg, A.S. Markus
On normally solvable operators and ideals associated with them. Bul. Akad. Stiince RSS Moldoven. No. 10 (76), 51–70 (1960); English transl., Amer. Math. Soc. Transl. (2) 61, 63–84 (1967); MR 36 #2004.

33. I. Gohberg, A.S. Markus
One problem on covering of convex figures by similar figures. Izv. Mold. Acad. Nauk 10 (76), 87–90 (Russian) (1960).

34. I. Gohberg, A.S. Markus
Some remarks about topologically equivalent norms. Izv. Mold. fil. Acad. Nauk SSSR 10 (76), 91–95 (Russian) (1960).

35. I. Gohberg, M.G. Krein
   On the theory of triangular representations of non-selfadjoint operators. Dokl. Acad Nauk SSSR 137, No. 5, 1034–1037 (1961); Soviet Math. Dokl. 2, 392–395 (1961); MR 25 #3370.

36. I. Gohberg, M.G. Krein
   On Volterra operators with imaginary component in one class or another. Dokl. Acad. Nauk SSSR 139, No. 4, 779–782 (1961); Soviet Math. Dokl. 2, 983–986 (1961); MR 25 #3372.

37. I. Gohberg, M.G. Krein
   The effect of some transformations of kernels of integral equations upon the equations' spectra. Ukrain. Mat. Z 13, No. 3, 12–28 (1961); English transl., Amer. Math. Soc. Transl. 2(35) 263–295 (1964); MR 27 #1788.

38. I. Gohberg, A.S. Markus
   On the stability of bases in Banach and Hilbert spaces. Izv. Moldavsk. Fil. Akad. Nauk SSSR, No. 5, 17–35 (Russian) (1962); MR 37 # 1955.

39. I. Gohberg, A.S. Markus
   On some inequalities between eigenvalues and matrix elements of linear operators. Izv. Moldavsk. Fil. Akad. Nauk SSSR, No. 5, 103–108 (Russian) (1962).

40. I. Gohberg
   Tests for one-sided invertibility of elements in normed rings and their applications. Dokl. Akad. Nauk SSSR 145, No. 5, 971–974 (1962); Soviet Math. Dokl. 3, 1119–1123 (1962); MR 25 #6147.

41. I. Gohberg
   A general theorem concerning the factorization of matrices-functions in normed rings, and its applications. Dokl. Akad. Nauk SSSR 146, No. 2, 284–287 (1962); Soviet Math. Dokl. 3, 1281–1284 (1962); MR 25 #4376.

42. I. Gohberg, M.G. Krein
   On the problem of factorization of operators in Hilbert space. Dokl. Akad. Nauk SSSR 147, No. 2, 279–282 (1962); Soviet Math. Dokl. 3, 1578–1582 (1962); MR 26 #6777.

43. I. Gohberg
   On factorization of operator-functions. Uspehi Mat. Nauk 18, No. 2, 180–182 (Russian) (1963).

44. I. Gohberg
   On relations between the spectra of the Hermitian components of nilpotent matrices and on the integral of triangular truncation. Bul. Akad. Stiiuce RSS Moldoven, No. 1, 27–37 (Russian) (1963); MR 35 #2168.

45. I. Gohberg
   On normal resolvability and the index of functions of an operator. Izv. Acad Nauk Mold. SSR, No. 11, 11–24 (Russian) (1963); MR 36 #6965.

46. M.S. Brodskii, I. Gohberg, M.G. Krein, V. Matsaev
On some new investigations on the theory of non-self-adjoint operators Proc.
IV All-union Math. Congress, Vol. 2, 261–271 (Russian) (1964); MR 36
#3153.

47. I. Gohberg, A.S. Markus
Some relations between eigenvalues and matrix elements of linear operators.
Mat. Sb. 64 (106), No. 4, 481–496 (1964); English transl. Amer. Math. Soc.
transl. (2), 52, 201–216 (1966); MR 30 #457,

48. I. Gohberg, M.G. Krein
Criteria for completeness of the system of root vectors of a contraction.
Ukrain. Mat. Z. 16, No. 1, 78–82 (1964); English transl. Amer. Math Soc.
Transl. (2), 54, 119–124 (1966); MR 29 #2651.

49. I. Gohberg, M.G. Krein
On factorization of operators in Hilbert space. Acta Sci. Math., Szeged, 25,
No. 1–2, 90–123 (1964); English transl. Amer. Math. Soc. Transl. (2), 51,
155–188 (1966); MR 29 #6313.

50. I. Gohberg
A factorization problem in normed rings, functions of isometric and symmetric operators and singular integral equations. Uspehi Mat. Nauk 19, No. 1
(115), 71–124; Russian Math. Survey 19, No. 1, 63–144 (1964); MR 29 #487.

51. I. Gohberg
The factorization problem for operator functions. Izv. Akad. Nauk SSSR, Ser.
Mat. 28, No. 5, 1055–1082 (Russian) (1964); MR 30 #5182.

52. V.G. Ceban, I. Gohberg
On a reduction method for discrete analogues of equations of Wiener–Hopf
type. Ukrain Mat. Z 26, No. 6, 822–829 (1964); English transl. Amer. Math.
Soc. transl. (2), 65, 41–49 (1967); MR 30 #2244.

53. M.S. Budjanu, I. Gohberg
A general theorem about factorization of matrix–functions. Studies in Algebra and Math. Anal Izd. "Kartja Mold".Kishinev, 116–121, 1965; MR 36
#726.

54. I. Gohberg, M.K. Zambitskii
On normally solvable operators in spaces with two norms. Bull. Akad. Nauk
Mold. SSR, No. 6, 80–84 (Russian) (1964). MR 36 #3143.

55. I.A. Feldman, I. Gohberg
On approximative solutions of some classes of linear equations. Dokl. Akad.
Nauk SSSR 160, No. 4, 750–753 (1965); Soviet Math. Dokl. 6, 174–177 (1965);
MR 34 #6572.

56. I. Gohberg, M.G. Krein
On the multiplicative representation of the characteristic functions of operators closed to unitary ones. Dokl. Acad. Nauk SSSR 164, No. 4, 732–735
(1965); Soviet Math. Dokl. 6, 1279–1283 (1965); MR 33 #571.

57. I.A. Feldman, I. Gohberg
    On reduction method for systems of Wiener–Hopf type. Dokl. Akad. Nauk
    SSSR 165, No. 2, 268–271 (1965); Soviet Math. Dokl. 6, 1433–1436 (1965);
    MR 32 #8085.

58. I. Gohberg, M.K. Zambitskii
    On the theory of linear operators in spaces with two norms. Ukrain. Mat. Z.
    18, No. 1, 11–23 (1966); MR 33 #4676.

59. I. Gohberg
    A generalization of theorems of M.G. Krein of the type of the Wiener–Levi
    theorems. Mat. Issled. 1, No. 1, 110–130 (Russian) (1966); MR 34 #3366.

60. M.A. Barkar, I. Gohberg
    On factorization of operators relative to a discrete chain of projections in
    Banach space. Mat. Issled. I, No. 1, 32–54 (1966); English transl. Amer.
    Math. Soc. Transl. (2), 90, 81–103, (1970); MR 34 #6539.

61. M.A. Barkar, I. Gohberg
    On factorization of operators in a Banach space. Mat. Issl. 1, No. 2, 98–129
    (Russian) (1966); MR 35 #780.

62. I.A. Feldman, I. Gohberg
    On truncated Wiener–Hopf equations. Abstracts of Short Scientific Reports,
    Intern. Congress of Math. (Moscow), Section 5, 44–45 (Russian) (1966).

63. I. Gohberg, M.G. Krein
    On triangular representations of linear operators and on multiplicative rep-
    resentations of their characteristic functions. Dokl. Acad. Nauk SSSR, 175,
    No. 2, 272–275 (1967); MR 35 #7157.

64. I.A. Feldman, I. Gohberg
    On indices of multiple extensions of matrix functions. Bull. Akad. Nauk Mold.
    SSR, No. 6, 76–80 (Russian) (1967); MR 37 #4658

65. I. Gohberg, M.G. Krein
    On a description of contraction operators similar to unitary ones. Funkcional
    Anal. i Priloz, 1, No. 1, 38–68 (1967); Functional Anal. and Appl., Vol. 1, 1,
    38–60 (1967); MR 35 #4763

66. I. Gohberg
    On Toeplitz matrices composed of the Fourier coefficients of piece–wise con-
    tinuous functions. Funkcional Anal. i Priloz., 1, No. 2, 91–92 (1967); Function
    Anal. Appl. 1, 166–167 (1967); MR 35 #4763.

67. M.S. Budjanu, I. Gohberg
    On factorization problem in abstract Banach algebras I. Splitting algebras.
    Mat. Issled 2, No. 2, 25–61 (1967); English transl. Amer. Math Soc. Transl.
    (2); MR #5697.

68. M.S. Budjanu, I. Gohberg
    On factorization problem in abstract Banach algebras II. Irreducible algebras.
    Mat. Issled 2, No. 3, 3–19 (1967); English transl. Amer. Math. Soc. Transl.
    (2); MR 37 #5698.

69. I. Gohberg, O.I. Soibelman
    Some remarks on similarity of operators. Mat. Issled. 2, No. 3, 166–170 (Russian) (1967); MR 37 #3387.

70. M.S. Budjanu, I. Gohberg
    On multiplicative operators in Banach algebras. I. General propositions. Mat. Issled 2, No. 4, 14–30 (Russian) (1967); MR 379 #1972; English transl Amer. Math. Soc. Transl. (2), 90, 211–223 (1970).

71. I. Gohberg, N. Ia. Krupnik
    On the norm of the Hilbert transform in L^p spaces. Funkcional Anal. i Priloz, 2, No. 2, 91–92 (1968); Functional Anal. Appl. 2, 180–181 (1968).

72. I. Gohberg, N.Ia. Krupnik
    On the spectrum of one-dimensional singular integral operators with piecewise continuous coefficients. Mat. Issled. 3, No. 1 (7), 16–30 (1968); English transl. Amer. Math. Soc. Transl. (2), 103, 181–193 (1973); MR 41 #2469.

73. M.S. Budjanu, I. Gohberg
    General theorems on the factorization of matrices-functions I. The fundamental theorem. Mat. Issled 3, No. 2 (8), 87–103 (1968); English transl. Amer. Math. Soc. Transl. (2), 102, 1–14 (1973); MR 41 #4246a.

74. M.S. Budjanu, I. Gohberg
    General theorems on the factorization of matrices-functions II. Some tests and their consequences. Mat. Issled 3, No. 3 (9), 3–18 (1968); English transl. Amer. Math. Soc. Transl. (2), 102, 15–26 (1973); MR 41 #4246b.

75. I.A. Feldman, I. Gohberg
    On Wiener–Hopf integral difference equations. Dokl. Akad, Nauk SSSR 183, No. 1, 25–28 (1968); Soviet Math. Dokl. 9, 1312–1316 (1968); MR 44 #3096.

76. I. Gohberg, N. Ia. Krupnik
    On the spectrum of singular integral operators in L^p spaces. Studia Math. 31, 347–362 (Russian) (1968); MR 38 #5068.

77. N.N. Bogolyubov, I. Gohberg, G.E. Shilov
    Mark Grigorevich Krein (on his sixtieth birthday). Uspehi Mat. Nauk, 23, No.3, 197–214 (1968); Russian Math. Surveys, 23, No. 3, 177–192 (May–June 1968); MR 37 #5077.

78. I. Gohberg, N. Ia. Krupnik
    On the spectrum of singular integral operators in L^p spaces with weight. Dokl. Acad. Nauk. SSSR 185, No. 4 745–748 (1969); Soviet Math. Dokl. 10, 406–410 (1969); MR 40 #1817.

79. I. Gohberg, N.Ia. Krupnik
    On an algebra generated by Toeplitz matrices. Funkcional Anal. i Priloz, 3, No. 2, 46–59 (1969); Functional Anal. Appl. 3, 119–127 (1969); MR 40 #3323.

80. I.A. Feldman, I. Gohberg
    Integro–difference Wiener–Hopf equations. Acta Sci. Math., Szeged, 30, No. 3–4, 199–224 (Russian) (1969); MR 40 #7880.

81. I. Gohberg, N.Ia. Krupnik
Systems of singular integral equations in L^p spaces with a weight. Dokl. Akad. Nauk SSSR 186, No. 5, 998–1001 (1969); Soviet Math. Dokl. 10, 688–651 (1969); MR 40 #1818.

82. I. Gohberg, N.Ia. Krupnik
On quotient norm of singular integral operators. Mat. Issled 4, No. 3, 136–139 (Russian) (1968); MR 41 4306. English transl. Amer. Math. Soc. Transl. (2), Vol. III, 117–119 (1978).

83. I. Gohberg, N.Ia. Krupnik
On the algebra generated by Toeplitz matrices in hp spaces. Mat. Issled 4, No. 3, 54–62 (Russian) (1969).

84. V.M. Brodskii, I. Gohberg, M.G. Krein
General theorems on triangular representations of linear operators and multiplicative representations of their characteristic functions. Funk. Anal. i Priloz, 3, No. 4, 1–27 (1969); MR 40 #4794.

85. I. Gohberg, N.Ia. Krupnik
On composite linear singular integral equations. Mat. Issled. 4, No. 4, 20–32 (1969); English transl. Amer. Math. Soc. Transl. (2), Vol. III, 121–131 (1978); MR 43 #996.

86. I. Gohberg, N.Ia. Krupnik
The symbols of one-dimensional singular integral operators on an open contour. Dokl. Akad Nauk SSSR 191, 12–15 (1970); Soviet Math. Dokl. 11, 299–303 (1970); MR 41 #9060.

87. V.M. Brodskii, I. Gohberg, M.G. Krein
The definition and basic properties of the characteristic function of a knot. Funkt. Anal. i Proloz, No. 4, 1, 88–90 (1970).

88. I. Gohberg, N.Ia. Krupnik
On the algebra generated by the one-dimensional singular integral operators with piecewise continuous coefficients. Funkcional Anal. i Priloz 4, No. 3, 26–36 (1970); Functional Anal. App. 4, 193–201 (1970); MR 42 #5057.

89. I. Gohberg, M.G. Krein
New inequalities for the eigenvalues of integral equations with smooth kernels. Mat. Issled 5, No. 1 (15), 22–39 (Russian) (1970); English transl. Operator Theory: Advances and Applications, Vol. 7, 275–293, Birkhäuser Verlag, 1983; MR 44 #818.

90. I. Gohberg, N.Ia. Krupnik
Singular integral equations with continuous coefficients on a composite contour. Mat. Issled No. 5, 1, 22–39 (Russian) (1970).

91. I. Gohberg, N.Ia. Krupnik
On singular integral equations with unbounded coefficients. Mat. Issled 5, No. 3 (17), 46–57 (Russian) (1970); MR 45 #980.

92. I. Gohberg, A.A. Sementsul. Toeplitz matrices composed of the Fourier coefficients of functions with discontinuities of almost periodic type. Mat. Issled. 5, No. 4, 63–83 (Russian) (1970); MR 44 #7379.

93. I. Gohberg, N.Ia. Krupnik
Banach algebras generated by singular integral operators. Colloquia Math. Soc. Janos Bolyai 5. Hilbert space operators, Tihany (Hungary), 239–267 (Russian) (1970).

94. I. Gohberg, E.M. Spigel
A projection method for the solution of singular integral equations. Dokl. Akad Nauk SSSR 196, No. 5, 1002–1005 (1971); Soviet Math. Dokl. 12, 289–293 (1971); MR 43 # 3755.

95. V.M. Brodskii, I. Gohberg, M.G. Krein
On characteristic functions of an invertible operator. Acta Sci. Math. Szeged, No. 32, 1–2, 141–164 (Russian) (1971).

96. I. Gohberg, N.Ia. Krupnik
Singular integral operators on a composite contour. Proc. Georgian Akad Nauk 64, 21–24 (Russian, Georgian and English summaries) (1971); MR 45 #4223.

97. I. Gohberg
On some questions of spectral theory of finite-meromorphic operator-functions. Izv. Arm. Akad. Nauk, No. 6, 2–3, 160–181 (Russian) (1971).

98. I. Gohberg
The correction to the paper "On some questions of spectral theory of finite-meromorphic operator-functions". Izv. Arm. Acad. Nauk, No. 7, 2, 152 (Russian) (1972).

99. I. Gohberg, N.Ia. Krupnik
Singular integral operators with piecewise continuous coefficients and their symbols. Izv. Akad. Nauk SSSR. Ser. Mat. 35, 940–964 (Russian) (1971); MR 45 #581.

100. I. Gohberg, V.I. Levchenko
Projection method for the solution of degenerate Wiener-Hopf equations. Funkcional Anal. in Priloz., 5, No.4, 69–70 (Russian) (1971); MR 44 #7317.

101. I. Gohberg, E.I. Sigal
An operator generalization of the logarithmic residue theorem and Rouche's theorem. Mat. Sb. (N.S.) 84 (126), 607–629 (1971); English transl. Math. USSR, sb. 13, 603–625 (1971). MR 47 #2409.

102. I. Gohberg, E.I. Sigal
Global factorization of a meromorphic operator-function and some of its applications. Mat. Issled. 6, No. 1 (19), 63–82 (Russian) (1971); MR 47 #2410.

103. I. Gohberg, E.I. Sigal
The root multiplicity of the product of meromorphic operator functions. Mat. Issled. 6, No. 2 (20), 30–50, 158 (Russian) (1971); MR 46 #2461.

104. I. Gohberg, E.M. Spigel
On the projection method of solution of singular integral equations with polynomial coefficients. Mat. Issled. 6, No. 3, 45–61 (Russian) (1971); MR 44 #7380.

105. I. Gohberg, V.I. Levchenko
     On the convergence of a projection method of solution of a degenerated
     Wiener–Hopf equation. Mat. Issled. 6, No. 4, 20–36 (Russian) (1971); MR
     45 # 918.
106. I. Gohberg, J. Leiterer
     The canonical factorization of continuous operator functions with respect to
     the circle. Funk Anal. i Priloz 6, No. 1 73–74 (1972); Functional Anal. Appl.
     6, 65–66 (1972); MR 45 #2519.
107. I. Gohberg, J. Leiterer
     On factorization of continuous operator functions with respect to a contour
     in Banach algebras. Dokl. Akad. Nauk SSSR, 206, 273–276 (1972); English
     transl. Soviet Math. Dokl. 13, 1195–1199 (1972).
108. I. Gohberg, J. Leiterer
     Factorization of operator functions with respect to a contour I. Finitely mero-
     morphic operator functions. Math. Nachrichten 52, 259–282 (Russian) (1972).
109. I. Gohberg, N. Ia. Krupnik
     A formula for the inversion of finite Toeplitz matrices. Mat. Issled 7, No. 2,
     272–283 (Russian) (1972).
110. I. Gohberg, A.A. Sementsul
     On the inversion of finite Toeplitz matrices and their continuous analogues.
     Mat. Issled 7, No. 2, 201–223, (Russian) (1972).
111. I. Gohberg, V.I. Levchenco
     On a projection method for a degenerate Wiener–Hopf equation. Mat. Issled.
     7, No. 3, 238–253 (Russian) (1972).
112. I. Gohberg. J. Leiterer
     General theorems on a canonic factorization of operator-functions with re-
     spect to a contour. Mat. Issled. No. 7, 3, 87–134 (Russian) (1972).
113. I. Gohberg, J. Leiterer
     On holomorphic vector-functions of one variable. Mat. Issled. No. 7, 4, 60–84
     (Russian) (1972).
114. I. Gohberg, J. Leiterer
     The factorization of operator-functions with respect to a contour II. Canonic
     factorization of operator-functions closed to unit ones. Math. Nachrichten,
     No. 54, 1–6, 41–74 (Russian) (1972).
115. I. Gohberg, J. Leiterer
     The factorization of operator-functions with respect to a contour III. Factor-
     ization in algebras. Math. Nachrichten, No. 55, 1–6, 33–61 (Russian) (1973).
116. I. Gohberg, J. Leiterer
     On co-cycles, operator-functions and families of subspaces. Mat. Issled. No.
     8, 2 (1973).
117. I. Gohberg, J. Leiterer
     On holomorphic functions of one variable II. Functions in a domain. Mat.
     Issled. No. 8, 1, 37–58 (Russian) (1973).

118. I. Gohberg, N. Ia. Krupnik
On algebras of singular integral operators with a shift. Mat. Issled No. 8, 2 (Russian) (1973).

119. I. Gohberg, N. Ia. Krupnik
On one-dimensional singular integral operators with a shift. Izv. Arm. Acad. Nauk, No. 1, 3–12 (Russian) (1973).

120. I. Gohberg, J. Leiterer
Criterion of the possibility of the fatorization of operator-function with respect to a contour. Dokl. Acad. Nauk SSSR, 209, 3, 529–532 (Russian) (1973).

121. I. Gohberg, J. Leiterer
General theorems on the factorization of operator-functions with respect to a closed contour. I. Holomorphic functions. Acta. Sci. Math., Szeged, 30, 103–120 (Russian) (1973).

122. I. Gohberg, J. Leiterer
General theorems on the factorization of operator-functions II. Generalizations. Acta. Sci. Math. Szeged (Russian) (1973).

123. I. Gohberg, J. Leiterer
On a local principle in the problem of the factorization of operator-functions. Funk. Anal. i Priloz. No. 7, 3 (Russian) (1973).

124. I. Gohberg, J. Leiterer
The local principle in the problem of the factorization of continuous operator-functions. Revue Rounainie de Math. Pures et Appl., XIX, 10 (Russian) (1973).

125. I. Gohberg, N. Ia. Krupnik
On a symbol of singular integral operators on a composite contour. Proc. Tbilisi Simp. Mech. Sploshnich Sred, Tbilisi, (Russian), 1973.

126. I. Gohberg, N. Ia. Krupnik
On the local principle and on algebras generated by Toeplitz's matrices. Annalele stiintifice ale Univ. "Al. I. Cuza", Iasi, section I a) Matematica, XIX, F. I, 43–72 (Russian) (1973).

127. I. Gohberg, J. Leiterer
Families of holomorphic subspaces with removable singularities. Math. Nachrichten 61, 157–173 (Russian) (1974).

128. I. Gohberg, G. Heinig
On the inversion of finite Toeplitz matrices. Math. Issled. No. 8, 3, 151–155 (Russian) (1973).

129. I. Gohberg, G. Heinig
Inversion of finite Toeplitz matrices composed from elements of a non-commutative algebra. Rev. Roum. Math. Pures et Appl. 20, 5, 55–73 (Russian) (1974).

130. I. Gohberg, G. Heinig
On matrix-valued integral operators on a finite interval with matrix kernels which depend on the difference of arguments. Re. Roumaine Math. Pures Appl., 20, 1. 55–73 (1975).

131. I. Gohberg, G. Heinig
     Matrix resultant and its generalizations, I. The resultant operator for matrix valued polynomials. Acta Sci. Math. (Szeged), T.37, 1–2, 1975, 41–61.

132. I. Gohberg, S. Prössdorf
     Ein Projektionsverfahren zur Lösung entarteter Systeme von diskreten Wiener–Hopf-Gleichungen. Math. Nachr., Band 65, 19–45 (1975).

133. I. Gohberg, G. Heinig
     Matrix resultant and its generalizations, II. Continual analog of the resultant operator. Acta Math. Acad. Sci. Hungar. T. 28 (3–4), 189–209 (1976).

134. I. Gohberg, J. Leiterer
     Uber algebren steitiger operator functionen. Studia Math., T. 57, 1–26 (1976).

135. I. Gohberg, M.A. Kaashoek, D.C. Lay
     Spectral classification of operators and operator functions. Bull. Amer. Math. Soc. 82, 587–589 (1976).

136. I. Gohberg, L.E. Lerer
     Resultant of matrix polynomials. Bull. Amer. Math. Soc. 82, 4 (1976).

137. K. Clancey, I. Gohberg
     Local and global factorizations of matrix-valued functions. Trans. Amer. Math. Soc., Vol. 232, 155–167 (1977).

138. I. Gohberg, M.A. Kaashoek, D.C. Lay
     Equivalence, linearization and decomposition of holomorphic operator functions. J. Funct. Anal., Vol. 28, No. 1, 102–144 (1978).

139. I. Gohberg, P. Lancaster, L. Rodman
     Spectral analysis of matrix polynomials, I. Canonical forms and divisors. Linear Algebra and its Appl., 20, 1–44 (1978).

140. I. Gohberg, P. Lancaster , L. Rodman
     Spectral analysis of matrix polynomials, II. The resolvent form and spectral divisors. Linear Algebra and its Appl. 21, 65–88 (1978).

141. I. Gohberg, L.E. Lerer
     Resultant operators of a pair of analytic functions. Proceedings Amer. Math. Soc., Vol. 72, No. 1, 65–73 (1978).

142. I. Gohberg, L.E. Lerer
     Singular integral operators as a generalization of the resultant matrix. Applicable Anal., Vol. 7, 191–205 (1978).

143. I. Gohberg, M.A. Kaashoek, F. van Schagen. Common multiples of operator polynomials with analytic coefficients. Manuscripta Math., 25, 279–314 (1978).

144. I. Gohberg, M.A. Kaashoek, L. Rodman
     Spectral analysis of families of operator polynomials and a generalized Vandermonde matrix, I. The finite-dimensional case. Topics in Functional Analysis. Adv. in Math., Supplementary Studies, Vol. 3, 91–128 (1978).

145. I. Gohberg, M.A. Kaashoek, L. Rodman
Spectral analysis of families of operator polynomials and a generalized Vandermonde matrix II. The infinite dimensional case. J. Funct. Anal., Vol. 30, No. 3, 358–389 (1978).

146. I. Gohberg, P. Lancaster , L. Rodman
Representations and divisibility of operator polynomials. Canad. J. Math., Vol. XXX, No. 5, 1045–1069 (1978).

147. I. Gohberg, L. Rodman
On spectral analysis of non-monic matrix and operator polynomials, I. Reduction to monic polynomials. Israel J. Math., Vol. 30, Nos. 1–2, 133–151 (1978).

148. I. Gohberg, L. Rodman
On spectral analysis of non-monic matrix and operator polynomials, II. Dependence on the finite spectral data. Israel J. Math., Vol. 30, No. 4, 321–334 (1978).

149. I. Gohberg, L. Lerer, L. Rodman
Factorization indices for matrix polynomials. Bull. Amer. Math. Soc., Vol. 84, No. 2, 275–277 (1978).

150. I. Gohberg, L. Lerer, L. Rodman
On canonical factorization of operator polynomials, spectral divisors and Toeplitz matrices. Integral Equations Operator Theory, 1, 176–214 (1978).

151. H. Bart, I. Gohberg, M.A. Kaashoek
Operator polynomials as inverses of characteristic functions. Integral Equations Operator Theory, 1, 1–18 (1978).

152. H. Bart, I. Gohberg, M.A. Kaashoek
Stable factorizations on monic matrix polynomials and stable invariant subspaces. Integral Equations Operator Theory, 1, 496–517 (1978).

153. I. Gohberg, S. Levin
Asymptotic properties of Toeplitz matrix factorization. Integral Equations Operator Theory, 1, 518–538 (1978).

154. I. Gohberg, M.A. Kaashoek
Unsolved problems in matrix and operator theory, I. Partial multiplicities and additive perturbations. Integral Equations Operator Theory, 1, 278–283 (1978).

155. K. Clancey, I. Gohberg
Localization of singular integral operators. Math. Z. 169, 105–117 (1979).

156. H. Dym, I. Gohberg
Extensions of matrix valued functions with rational polynomial inverses. Integral Equations Operator Theory, 2, 503–528 (1979).

157. I. Gohberg, P. Lancaster, L. Rodman
Perturbation theory for divisors of operator polynomials. SIAM J. Math. Anal., Vol. 10, No. 6, 1161–1183 (1979).

158. I. Gohberg, L. Rodman
     On the spectral structure of monic matrix polynomials and the extension
     problem. Linear Algebra Appl., 24, 157–172 (1979).
159. I. Gohberg, P. Lancaster, L. Rodman
     On selfadjoint matrix polynomials. Integral Equations Operator Theory, 2,
     434–439 (1979).
160. I. Gohberg., L. Lerer
     Factorization indices and Kronecker indices of matrix polynomials. Integral
     Equations Operator Theory, 2, 199–243 (1979).
161. I. Gohberg, M.A. Kaashoek
     Unsolved problems in matrix and operator theory, II. Partial multiplicities
     for products. Integral Equations Operator Theory, 2, 116–120 (1979).
162. I. Gohberg, S. Levin
     On an open problem for block Toeplitz matrices. Integral Equations Operator
     Theory, 2, 121–129 (1979).
163. E. Azoff, K. Clancey, I. Gohberg
     On the spectra of finite-dimensional perturbations of matrix multiplication
     operators. Manuscripta Math. 30, 351–360 (1980).
164. I. Gohberg, L. Lerer, L. Rodman
     Stable factorizations of operator polynomials. I.Spectral divisors simply be-
     haved at infinity. J. Math. Anal. Appl. 74, 401–431 (1980).
165. I. Gohberg, L. Lerer, L. Rodman
     Stable factorizations of operator polynomials. II. Main results and applica-
     tions to Toeplitz operators. J. Math. Anal. Appl. 75, 1–40 (1980).
166. I. Gohberg, M.A. Kaashoek, F. van Schagen
     Similarity of operator blocks and canonical forms, I. General results, feedback
     equivalence and Kronecker indices. Integral Equations Operator Theory, 3,
     350–396 (1980).
167. I. Gohberg, P. Lancaster, L. Rodman
     Spectral analysis of selfadjoint matrix polynomials. Annals of Mathematics,
     112, 33–71 (1980).
168. H. Dym, I. Gohberg
     On an extension problem, generalized Fourier analysis, and an entropy for-
     mula. Integral Equations Operator Theory, 3, 143–215 (1980).
169. H. Bart, I. Gohberg, M.A. Kaashoek, P. van Dooren
     Factorizations of transfer functions. SIAM J. Control and Optimization, 18,
     No. 6, 675–696 (1980).
170. I. Gohberg, M.A. Kaashoek, F. van Schagan
     Similarity of operator blocks and canonical forms. II. Infinite dimensional
     case and Wiener–Hopf factorization. Operator Theory: Advances and Appli-
     cations, 2, 121–170 (1981).
171. I. Gohberg, M.A. Kaashoek, L. Lerer, L. Rodman
     Common multiples and common divisors of matrix polynomials, I. Spectral
     method. Indiana University Mathematics Journal, 30 No. 3, 321–356 (1981).

172. H. Dym, I. Gohberg
Extensions of band matrices with band inverses. Linear Algebra and its Applications, 36, 1–14 (1981).
173. I. Gohberg, L. Rodman
Analytic matrix functions with prescribed local data. Journal d'Analyse Mathematique, 40, 90–128 (1981).
174. H. Bart, I. Gohberg, M.A. Kaashoek
Wiener–Hopf integral equations, Toeplitz matrices and linear systems. Toeplitz Centennial, Operator Theory: Advances and Applications, Vol. 4, 85–135 (1982).
175. I. Gohberg, P. Lancaster, L. Rodman
Factorization of selfadjoint matrix polynomials with constant signature. Linear and Multilinear Algebra, 11, 209–224 (1982).
176. E. Azoff, K. Clancey, I. Gohberg
Singular points of families of Fredholm integral operators. Toeplitz Centennial, Operator Theory: Advances and Applications, Vol. 4, 57–65 (1982).
177. E. Azoff, K. Clancey, I. Gohberg
On line integrals of rational functions of two complex variables. Proceedings American Mathematical Society, 88, 229–235 (1982).
178. I. Gohberg. L. Lerer
On non-square sections of Wiener–Hopf operators. Integral Equations Operator Theory, Vol. 5, No. 4, 518–532 (1982).
179. H. Bart, I. Gohberg, M.A. Kaashoek
Convolution equations and linear systems. Integral Equations and Operator Theory, Vol. 5, No. 3, 283–340 (1982).
180. H. Dym, I. Gohberg
Extensions of triangular operators and matrix functions. Indiana University Mathematics Journal, Vol. 31, No. 4, 579–606 (1982).
181. H. Dym, I. Gohberg
Extensions of matrix valued functions and block matrices. Indiana University Mathematics Journal, Vol. 31, No. 5, 733–765 (1982).
182. I. Gohberg, M.A. Kaashoek, F. van Schagen
Rational matrix and operator functions with prescribed singularities. Integral Equations Operator Theory, Vol. 5, No. 5, 673–717 (1982).
183. I. Gohberg, P. Lancaster, L. Rodman
Perturbations of H-selfadjoint matrices, with applications to differential equations. Integral Equations and Operator Theory, Vol. 5, No. 5, 718–757 (1982).
184. I. Gohberg, M.A. Kaashoek, L. Lerer, L. Rodman
Common multiples and common divisors of matrix polynomials, II. Vandermonde and resultant matrices. Linear and Multilinear Algebra, 159–203 (1982).
185. I. Gohberg, S. Goldberg
Finite Dimensional Wiener–Hopf equations and factorizations of matrices. Linear Algebra and its Applications, 48, 219–236 (1982).

186. I. Gohberg, L. Rodman
Analytic operator valued functions with prescribed local data. Acta Sci. Math., Szeged, 45, 189–199 (1983).

187. I. Gohberg, L. Lerer, L. Rodman
Wiener–Hopf factorization of piecewise matrix polynomials. Linear Algebra and its Applications, Vol. 52–53, 315–350 (1983).

188. H. Dym, I. Gohberg
On unitary interpolants and Fredholm infinite block Toeplitz matrices. Integral Equations Operator Theory, Vol. 6, 863–878 (1983).

189. H. Dym, I. Gohberg
Extensions of kernels of Fredholm operators. Journal d'Analyse Mathematique, 42, 51–97 (1982/83).

190. H. Dym, I. Gohberg
Unitary interpolants, factorization indices and infinite Hankel block matrices. Journal of Functional Analysis, Vol. 54, No. 3, 229–289 (1983).

191. H. Dym, I. Gohberg
Hankel integral operators and isometric interpolants on the line. Journal of Functional Analysis, Vol. 54, No. 3, 290–307 (1983).

192. I. Gohberg, P. Lancaster, L. Rodman
A sign characteristic for selfadjoint meromorphic matrix functions. Applicable Analysis, 16, 165–185 (1983).

193. H. Bart, I. Gohberg, M.A. Kaashoek
The coupling method for solving integral equations. Operator Theory: Advances and Applications, 12, 39–73 (1984).

194. I. Gohberg, M.A. Kaashoek, L. Lerer, L. Rodman
Minimal divisors of rational matrix functions with prescribed zero and pole structure. Operator Theory: Advances and Applications, 12, 241–275 (1984).

195. I. Gohberg, M.A. Kaashoek
Time varying linear systems with boundary conditions and integral operators, I. The transfer operator and its properties. Integral Equations and Operator Theory, 7, 325–391 (1984).

196. I. Gohberg, S. Goldberg
Extensions of triangular Hilbert–Schmidt operators. Integral Equations and Operator Theory, 7, 743–790 (1984).

197. I. Gohberg, M.A. Kaashoek, F van Schagen. Non-compact integral operators with semiseparable kernels and their discrete analogues: Inversion and Fredholm properties. Integral Equations and Operator Theory, 7, 642–703 (1984).

198. H. Bart, I. Gohberg, M.A. Kaashoek
Wiener–Hopf factorization and realization. Lecture Notes in Control and Information Sciences, Mathematical Theory of Networks and Systems, Springer-Verlag, 58, 42–62 (1984).

199. I. Gohberg, P. Lancaster, L. Rodman
     A sign characteristic for selfadjoint rational matrix functions. Lecture Notes
     in Control and Information Sciences, Mathematical Theory of Networks and
     Systems, Springer-Verlag, 58, 263–269 (1984).
200. J.A. Ball, I. Gohberg
     A commutant lifting theorem for triangular matrices with diverse applica-
     tions. Integral Equations and Operator Theory, 8, 205–267 (1985).
201. H. Bart, I. Gohberg, M.A. Kaashoek
     Fredholm theory of Wiener–Hopf equations in terms of realization of their
     symbols. Integral Equations and Operator Theory, 8, 590–613 (1985).
202. I. Gohberg, T. Kailath, I. Koltracht
     Linear complexity algorithms for semiseparable matrices. Integral Equations
     and Operator Theory, 8, 780–804 (1985).
203. I. Gohberg, P. Lancaster, L. Rodman
     Perturbation of analytic hermitian matrix functions. Applicable Analysis, 20,
     23–48 (1985).
204. R.L. Ellis, I. Gohberg, D. Lay
     Factorization of Block Matrices. Linear Algebra and its Applications 69, 71–
     93 (1985).
205. I. Gohberg, I, Koltracht
     Numerical solution of integral equations, fast algorithms and Krein–Sobolev
     equation. Numer. Math. 47, 237–288 (1985).
206. J.A. Ball, I. Gohberg
     Shift invariant subspaces, factorization, and interpolation for matrices. I. The
     canonical case. Linear Algebra and its Applications, 1–64 (1985).
207. H. Dym, I. Gohberg
     A maximum entropy principle for contractive interpolants. Journal of Func-
     tional Analysis 65, 83–125 (1986).
208. I. Gohberg, L. Rodman
     Interpolation and local data for meromorphic matrix and operator functions.
     Integral Equations and Operator Theory, 9, 60–94 (1986).
209. I. Gohberg, M.A. Kaashoek
     On minimality and stable minimality of time–varying linear systems with
     well-posed boundary conditions. Int. J. Control, 43, 5, 1401–1411 (1986).
210. I. Gohberg, L. Rodman
     On distance between lattices of invariant subspaces of matrices. Linear Alge-
     bra Appl. 76, 85–120 (1986).
211. I. Gohberg, T. Kailath, I. Koltracht
     Efficient solution of linear systems of equations with recursive structure. Lin-
     ear Algebra and its Applications 80, 81–113 (1986).
212. H. Bart, I. Gohberg, M.A. Kaashoek
     Wiener–Hopf factorization, inverse Fourier transforms and exponentially di-
     chotomous operators. Journal of Functional Analysis 68, No. 1, 1–42 (1986).

213. I. Gohberg, M.A. Kaashoek
     Similarity and reduction of time varying linear systems with well-posed boundary conditions. SIAM J. Control and Optimization 24, No. 5, 961–978 (1986).

214. I. Gohberg, S. Rubinsteinv Stability of minimal fractional decompositions of rational matrix functions. Operator Theory: Advances and Applications, 18, 249–270 (1986).

215. J.A. Ball, I. Gohberg
     Classification of shift invariant subspaces of matrices with Hermitian form and completion of matrices. Operator Theory: Advances and Applications, 19 23–85 (1986).

216. R.L. Ellis, I. Gohberg, D.C. Lay
     The maximum distance problem in Hilbert space. Operator Theory: Advances and Applications, 19, 195–206 (1986).

217. H. Bart, I. Gohberg, M.A. Kaashoek
     Wiener–Hopf equations with symbols analytic in a strip. Operator Theory: Advances and Applications, 21, 39–74 (1986).

218. I. Gohberg, M.A. Kaashoek, L. Lerer, L. Rodman
     On Toeplitz and Wiener–Hopf operators with contourwise rational matrix and operator symbols. Operator Theory: Advances and Applications, 21, 75–127 (1986).

219. I. Gohberg, M.A. Kaashoek
     Minimal factorization of integral operators and cascade decompositions of systems. Operator Theory: Advances and Applications, 21, 157–230 (1986).

220. H. Bart, I. Gohberg, M.A. Kaashoek
     Explicit Wiener–Hopf factorization and realization. Operator Theory: Advances and Applications, 21, 235–316 (1986).

221. H. Bart, I. Gohberg, M.A. Kaashoek
     Invariants for Wiener–Hopf equivalence of analytic operator functions. Operator Theory: Advances and Applications, 21, 317–355 (1986).

222. H. Bart, I. Gohberg, M.A. Kaashoek
     Multiplication by diagonals and reduction to canonical factorization. Operator Theory: Advances and Applications, 21, 357–372 (1986).

223. I. Gohberg, P. Lancaster, L. Rodman
     Quadratic matrix polynomials with a parameter. Advances in Applied Mathematics 7, 3, 253–281 (1986).

224. R.L. Ellis, I. Gohberg, D.C. Lay
     Band extensions, maximum entropy and permanence principle. In Maximum Entropy and Bayesian Methods in Applied Statistics (J. Justice, Ed.), Cambridge University Press (1986).

225. I. Gohberg, P. Lancaster, L. Rodman
     On Hermitian solutions of the symmetric algebraic Riccati equation. SIAM Journal of Control and Optimization 24, 6, 1323–1334 (1986).

226. I. Gohberg, M.A. Kaashoek, L. Lerer
     Minimality and irreducibility of time-invariant boundary-value systems. Int.
     J. Control 44, 2, 363–379 (1986).
227. J.A. Ball, I. Gohberg
     Pairs of shift invariant subspaces of matrices and nonconical factorization.
     Linear and Multilinear Algebra 20, 27–61 (1986).
228. A. Ben-Artzi, R.L. Ellis, I. Gohberg, D.C. Lay
     The maximum distance problem and band sequences. Linear Algebra and its
     Applications 87, 93–112 (1987).
229. J.A. Ball, I. Gohberg, L. Rodman
     Minimal factorization of meromorphic matrix functions in terms of local data.
     Integral Equations and Operator Theory, 10, 3, 309–348 (1987).
230. I. Gohberg, M.A. Kaashoek
     An inverse spectral problem for rational matrix functions and minimal divis-
     ibility. Integral Equations and Operator Theory, 10, 437–465 (1987).
231. H. Bart, I. Gohberg, M.A. Kaashoek
     The state space method in problems of analysis. Proceedings of the First
     International Conference on Industrial and Applied Mathematics (ICIAM
     87), 1–16, (1987).
232. I. Gohberg, S. Rubinstein
     Cascade decompositions of rational matrix functions and their stability. Int.
     J. Control 46, 2, 603–629 (1987).
233. I. Gohberg, M.A. Kaashoek, L. Lerer
     On minimality in the partial realization problem. System & Control Letters
     9, 97–104 (1987).
234. I. Gohberg, I. Koltracht, P. Lancaster
     Second order parallel algorithms for Fredholm integral equations with con-
     tinuous displacement kernels. Integral Equations and Operator Theory, 10,
     577–594 (1987).
235. I. Gohberg, T. Kailath, I. Koltracht, P. Lancaster
     Linear complexity parallel algorithms for linear systems of equations with re-
     cursive structure. Linear Algebra and its Applications 88/89, 271–315 (1987).
236. R.L. Ellis, I. Gohberg, D.C. Lay
     Invertible selfadjoint extensions of band matrices and their entropy. SIAM J.
     Alg. Disc. Meth. 8, 3, 483–500 (1987).
237. I. Gohberg, S. Goldberg
     Semi-separable operators along chains of projections and systems. Journal of
     Mathematical Analysis and Applications 125, 1, 124–140 (1987).
238. I. Gohberg, M.A. Kaashoek
     Minimal representations of semiseparable kernels and systems with separable
     boundary conditions. J. Math. Anal. Appl. 124, 2, 436–458 (1987).
239. I. Gohberg, M.A. Kaashoek, F. van Schagen
     Szego–Kac–Achiezer formulas in terms of realizations of the symbol. J. Func-
     tional Analysis 74, 1, 24–51 (1987).

240. I. Gohberg, T. Kailath, I. Koltracht
     A note on diagonal innovation matrices. IEEE Transactions on Acoustics,
     Speech, and Signal Processing, Vol. ASSP-35, No. 7, 1068–1069, (1987)

241. A. Ben-Artzi, I. Gohberg
     Nonstationary Szego theorem, band sequences and maximum entropy. Inte-
     gral Equations and Operator Theory, 11, 10–27 (1988).

242. R.L. Ellis, I. Gohberg, D.C. Lay
     On two theorems of M.G. Krein concerning polynomials orthogonal on the
     unit circle. Integral Equations and Operator Theory, 11, 87–104 (1988).

243. I. Gohberg, M.A. Kaashoek, F. van Schagen
     Rational contractive and unitary interpolants in realized form. Integral Equa-
     tions and Operator Theory, 11, 105–127 (1988).

244. H. Dym, I. Gohberg
     A new class of contractive interpolants and maximum entropy principles.
     Operator Theory: Advances and Applications, 29, 117–150 (1988).

245. I. Gohberg, M.A. Kaashoek, L. Lerer
     Nodes and realization of rational matrix functions: Minimality theory and ap-
     plications. Operator Theory: Advances and Applications, 29, 181–232 (1988).

246. A. Ben-Artzi, I. Gohberg
     Fredholm properties of band matrices and dichotomy. Operatory Theory:
     Advances and Applications, 32, 37–52 (1988).

247. J.A. Ball, I. Gohberg, L. Rodman
     Realization and interpolation of rational matrix functions. Operator Theory:
     Advances and Applications, 33, 1–72 (1988).

248. I. Gohberg, M.A. Kaashoek, A.C.M. Ran. Interpolation problems for ratio-
     nal matrix functions with incomplete data and Wiener–Hopf factorization.
     Operator Theory: Advances and Applications, 33, 73–108 (1988).

249. I. Gohberg, M.A. Kaashoek
     Regular rational matrix functions with prescribed pole and zero structure.
     Operator Theory: Advances and Applications, 33, 109–122 (1988).

250. D. Alpay, I. Gohberg
     Unitary rational matrix functions. Operator Theory: Advances and Applica-
     tions, 33, 175–222 (1988).

251. I. Gohberg, S. Rubinstein
     Proper contractions and their unitary minimal completions. Operator The-
     ory: Advances and Applications, 33, 223–247 (1988).

252. A. Ben-Artzi, I. Gohberg
     Lower upper factorizations of operators with middle terms. J. Functional
     Analysis 77, 2, 309–325 (1988).

253. I. Gohberg, S. Goldberg
     Factorizations of semi-separable operators along continuous chains of pro-
     jections. Journal of Mathematical Analysis and Applications, 133, 1, 27–43
     (1988).

254. D. Alpay, I. Gohberg
On orthogonal matrix polynomials. Operator Theory: Advances and Applications, 34, 25–46 (1988).

255. A. Ben-Artzi, I. Gohberg
Extension of a theorem of M.G. Krein on orthogonal polynomials for the nonstationary case. Operator Theory: Advances and Applications, 34, 65–78 (1988).

256. I. Gohberg, L. Lerer
Matrix generalizations of M.G. Krein theorems on orthogonal polynomials. Operator Theory: Advances and Applications, 34, 137–202 (1988).

257. I. Gohberg, M.A. Kaashoek
Block Toeplitz operators with rational symbols. Operator Theory: Advances and Applications, 35, 385–440 (1988).

258. R.L. Ellis, I. Gohberg, D.C. Lay
On negative eigenvalues of selfadjoint extensions of band matrices. Linear and Multilinear Algebra, 24, 15–25 (1988).

259. I. Gohberg, M.A. Kaashoek, P. Lancaster
General theory of regular matrix polynomials and band Toeplitz operators. Integral Equations and Operator Theory, 11, 776–882 (1988).

260. I. Gohberg, I. Koltracht
Efficient Algorithm for Toeplitz plus Hankel matrices. Integral Equations and Operator Theory, 12, 136–142 (1989).

261. I. Gohberg, M.A. Kaashoek, H.J. Woerdeman
The band method for positive and strictly contractive extension problems: An alternative version and new applications. Integral Equations and Operator Theory, 12, 343–382 (1989).

262. I. Gohberg, I. Koltracht, P. Lancaster
On the numerical solution of integral equations with piecewise continuous displacement kernels. Integral Equations and Operator Theory, 12, 511–538 (1989).

263. I. Gohberg, T. Shalom
On inversion of square matrices partitioned into non-square blocks. Integral Equations and Operator Theory, 12, 539–566 (1989).

264. A. Ben-Artzi, I. Gohberg
Inertia theorems for nonstationary discrete systems and dichotomy. Linear Algebra and its Applications 120, 95–138 (1989).

265. I. Gohberg, M.A. Kaashoek, H.J. Woerdeman
The band method for positive and contractive extension problems. Journal of Operator Theory 22, 109–155 (1989).

266. J.A. Ball, I. Gohberg
Cascade decompositions of linear systems in terms of realizations. Proceedings of the 28th IEEE Conference on Decision and Control. IEEE Control Systems Society, volume 1, 2–10 (1989).

267. I. Gohberg
Mathematical Tales. Operator Theory: Advances and Applications, 40, 17–56 (1989).

268. I. Gohberg, I. Koltracht
On the inversion of Cauchy Matrices. Signal Processing, Scattering and Operator Theory, and Numerical Methods. Proceedings of the International Symposium MTNS-89, Amsterdam, Vol. III, Birkhäuser, 381–392 (1990).

269. I. Gohberg, T. Shalom
On Bezoutians of nonsquare matrix polynomials and inversion of matrices with nonsquare blocks. Linear Algebra and its Applications, vol. 137/138, 249–323 (1990).

270. I. Gohberg, M.A. Kaashoek, A.C.M. Ran
Regular rational matrix functions with prescribed local zero–pole structure. Linear Algebra and its Applications, vol. 137/138, 387–412 (1990).

271. J.A. Ball, I. Gohberg, L. Rodman
Common minimal multiples and divisors for rational matrix functions. Linear Algebra and its Applications, vol. 137/138, 621–662 (1990).

272. I. Gohberg, B. Reichstein
On classification of normal matrices in an indefinite scalar product. Integral Equations and Operator Theory, 13, 364–394 (1990).

273. J.A. Ball, I. Gohberg, L. Rodman, T. Shalom
On the eigenvalues of matrices with given upper triangular part. Integral Equations and Operator Theory, 13, 488–497 (1990).

274. J.A. Ball, I. Gohberg, L. Rodman
Simultaneous residue interpolation problems for rational matrix functions. Integral Equations and Operator Theory, 13, 611–637 (1990).

275. J.A. Ball, I. Gohberg, L. Rodman
Tangential interpolation problems for rational matrix functions. AMS Proceedings of Symposia in Applied Mathematics 40, 59–86 (1990).

276. D. Alpay, J.A. Ball, I. Gohberg, L. Rodman
Realization and factorization for rational matrix functions with symmetries. Operator Theory: Advances and Applications, 47, 1–60 (1990).

277. A. Ben-Artzi, I. Gohberg, M.A. Kaashoek
Invertibility and dichotomy of singular difference equations. Operator Theory: Advances and Applications, 48, 157–184 (1990).

278. I. Gohberg, M.A. Kaashoek, H.J. Woerdeman
The band method for extension problems and maximum entropy. IMA Volumes in Mathematics and its Applications, Volume 22, 75–94, Springer-Verlag, New York (1990).

279. J.A. Ball, I. Gohberg, L. Rodman
Sensitivity minimization and bitangential Nevanlinna–Pick interpolation in contour integral form. IMA Volumes in Mathematics and its Applications, Volume 23, 3–36, Springer-Verlag, New York (1990).

280. J.A. Ball, I.C. Gohberg, L. Rodman
     Two-sided Lagrange–Sylvester interpolation problems for rational matrix functions. AMS Proceedings of Symposia in Pure Mathematics, volume 51, 17–84, American Mathematical Society (1990).

281. A. Ben-Artzi, I. Gohberg
     Nonstationary inertia theorems, dichotomy, and applications. AMS Proceedings of Symposia in Pure Mathematics, volume 51, 85–96, American Mathematical Society (1990).

282. J. Ball, I. Gohberg, L. Rodman
     Two-sided Nudelman interpolation problem for rational matrix functions. Lecture Notes in Pure and Applied Mathematics. Analysis and Partial Differential Equations, volume 122, 371–416, Marcel Dekker, Inc., New York (1990).

283. I. Gohberg, S. Rubinstein
     Minimal symplectic orbits of rational contractive matrix functions. Integral Equations and Operator Theory, 13, 795–835 (1990).

284. I. Gohberg, S. Goldberg
     Counting negative eigenvalues of a Hilbert–Schmidt operator via sign changes of a determinant. Integral Equations and Operator Theory, 14, 92–104 (1991).

285. A. Ben-Artzi, I. Gohberg
     Band matrices and dichotomy. Operator Theory: Advances and Applications, 50, 137–170 (1991).

286. I. Gohberg, M.A. Kaashoek, A.C.M. Ran
     Matrix polynomials with prescribed zero structure in the finite complex plane. Operator Theory: Advances and Applications, 50, 241–266, (1991).

287. I. Gohberg, M.A. Kaashoek
     The Wiener–Hopf method for the transport equation: a finite dimensional version. Operator Theory: Advances and Applications, 51, 20–33 (1991).

288. I. Gohberg, M.A. Kaashoek, H.J. Woerdeman
     A note on extensions of band matrices with maximal and submaximal invertible blocks. Linear Algebra and its Applications, vol. 150, 157–166 (1991).

289. I. Gohberg, S. Rubinstein
     A classification of upper equivalent matrices. The generic case. Integral Equations and Operator Theory, 14, 533–544 (1991).

290. J.A. Ball, I. Gohberg, L. Rodman
     Boundary Nevanlinna–Pick interpolation for rational matrix functions. Journal of Mathematical Systems, Estimation and Control, Vol. 1, No. 2, 131–164 (1991).

291. A. Ben-Artzi, I. Gohberg
     Dichotomy, discrete Bohl exponents, and spectrum of block weighted shifts. Integral Equations and Operator Theory, 14, 613–677 (1991).

292. J.A. Ball, I. Gohberg, L. Rodman
     The state space method in the study of interpolation of rational matrix functions. Mathematical System Theory, Springer-Verlag, 503–508 (1991).

293. I. Gohberg, M.A. Kaashoek
The State space method for solving singular integral equations. Mathematical System Theory, Springer-Verlag, 509–523 (1991).

294. J.A. Ball, I. Gohberg, L. Rodman
Nehari interpolation problem for rational matrix functions: The Generic case. Lecture Notes in Mathematics, Springer-Verlag, 277–308 (1991).

295. I. Gohberg, M.A. Kaashoek, H.J. Woerdeman
Time variant extension problems of Nehari type and the band method. Lecture Notes in Mathematics, Springer-Verlag, 309–323 (1991).

296. I. Gohberg, M.A. Kaashoek, L. Lerer
A directional partial realization problem. Systems & Control Letters, 17, 305–314 (1991).

297. I. Gohberg, M.A. Kaashoek, H.J. Woerdeman
The band method for several positive extension problems of non–band type. Journal of Operator Theory, 26, 191–218 (1991).

298. I. Gohberg, M.A. Kaashoek, H.J. Woerdeman
A maximum entropy principle in the general framework of the band method. Journal of Functional Analysis, 95, 2, 231–254 (1991).

299. I. Gohberg, B. Reichstein
On H-unitary and block-Toeplitz H-normal operators. Linear and Multilinear Algebra, 30, 17–48 (1991).

300. R.L. Ellis, I. Gohberg
Orthogonal systems related to infinite Hankel matrices. Journal of Functional Analysis, vol. 109, no. 1, 155–198 (1992).

301. I. Gohberg, I. Koltracht, P. Lancaster
Second order parallel algorithms for piecewise smooth displacement kernels. Integral Equations and Operator Theory, 15, 16–29 (1992).

302. I. Gohberg, M.A. Kaashoek, A.C.M. Ran
Factorizations of and extensions to J-unitary rational matrix functions on the unit circle. Integral Equations and Operator Theory, 15, 262–300 (1992).

303. D. Alpay, J. Ball, I. Gohberg, L. Rodman
State space theory of automorphisms of rational matrix functions. Integral Equations and Operator Theory, 15, 349–377 (1992).

304. I. Gohberg, V. Olshevsky
Circulants, displacements and decompositions of matrices. Integral Equations and Operator Theory, 15, 730–743 (1992).

305. I. Gohberg, N. Krupnik
Extension theorems for invertibility symbols in Banach algebras. Integral Equations and Operator Theory, 15, 991–1010 (1992).

306. J.A. Ball, I. Gohberg, M.A. Kaashoek
Nevanlinna–Pick interpolation for time-varying input–output maps: the discrete case. Operator Theory: Advances and Applications, 56, 1–51 (1992).

307. J.A. Ball, I. Gohberg, M.A. Kaashoek
Nevanlinna–Pick interpolation for time-varying input-output maps: the continuous time case. Operator Theory: Advances and Applications, 56, 52–89 (1992).

308. A. Ben-Artzi, I. Gohberg
Dichotomy of systems and invertibility of linear ordinary differential operators. Operator Theory: Advances and Applications, 56, 90–119 (1992).

309. A. Ben-Artzi, I. Gohberg
Inertia theorems for block weighted shifts and applications. Operator Theory: Advances and Applications, 56, 120–152 (1992).

310. I. Gohberg, M.A. Kaashoek, L. Lerer
Minimality and realization of discrete time-varying systems. Operator Theory: Advances and Applications, 56, 261–296 (1992).

311. R.L. Ellis, I. Gohberg, D.C. Lay
Distribution of zeros of matrix-valued continuous analogues of orthogonal polynomials. Operator Theory: Advances and Applications, 58, 26–70 (1992).

312. I. Gohberg, M.A. Kaashoek
The band extension on the real line as a limit of discrete band extensions, II. The entropy principle. Operator Theory: Advances and Applications, 58, 71–92 (1992).

313. J.A. Ball, I. Gohberg, M.A. Kaashoek
Reduction of the abstract four block problem to a Nehari problem. Operator Theory: Advances and Applications, 58, 121–142 (1992).

314. I. Gohberg, M.A. Kaashoek
The band extension on the real line as a limit of discrete band extensions, I. The main limit theorem. Operator Theory: Advances and Applications, 59, 191–220 (1992).

315. I. Gohberg, M.A. Kaashoek
Asymptotic formulas of Szego–Kac–Achiezer type. Asymptotic Analysis 5, 187–220 (1992).

316. J.A. Ball, I. Gohberg, M.A. Kaashoek
Time-varying systems: Nevanlinna–Pick interpolation and sensitivity minimization. Recent Advances in Mathematical Theory of Systems, Control, Networks and Signal Processing I; Proceedings MTNS-1991, Mitra Press, Tokyo, 1992.

317. I. Gohberg, A.C.M. Ran
On pseudo-canonical factorization of rational matrix functions. Indagationes Mathematica, N.S., 4(1), 51–63 (1993).

318. A. Ben-Artzi, I. Gohberg, M.A. Kaashoek
Invertibility and dichotomy of differential operators on a half–line. Journal of Dynamics and Differential Equations, vol. 5, no. 1, 1–36 (1993).

319. A. Ben-Artzi, I. Gohberg
Dichotomies of perturbed time varying systems and the power method. Indiana Univ. Math. J. 42, 699–720 (1993).

320. I. Gohberg, M.A. Kaashoek, H.J. Woerdeman
     The band method for bordered algebras. Operator Theory: Advances and Applications, 62, 85–98 (1993).
321. J.A. Ball, I. Gohberg, M.A. Kaashoek
     Bitangential interpolation for input-output operators of time varying systems: the discrete time case. Operator Theory: Advances and Applications, 64, 33–72 (1993).
322. J.A. Ball, I. Gohberg, L. Rodman
     Two-sided tangential interpolation of real rational matrix functions. Operator Theory: Advances and Applications, 64, 73–102 (1993).
323. I. Gohberg, C. Gu
     On a completion problem for matrices. Operator Theory Advances and Applications, 64, 203–217 (1993).
324. R. L. Ellis, I. Gohberg, D.C. Lay
     Extensions with positive real part. A new version of the abstract band method with applications. Integral Equations and Operator Theory, 16, 360–384 (1993).
325. I. Gohberg, N. Krupnik
     Extension theorems for Fredholm and invertibility symbols. Integral Equations and Operator Theory, 16, 514–529 (1993).
326. I. Gohberg, N. Krupnik, I. Spitkovsky
     Banach algebras of singular integral operators with piecewise continuous coefficients. General contour and weight. Integral Equations and Operator Theory, 17, 322–327 (1993).
327. A. Ben-Artzi, I. Gohberg, M.A. Kaashoek
     A time-varying generalization of the canonical factorization theorem for Toeplitz operators. Indagationes Mathematica, N.S., 4(4), 385–405 (1993).
328. I. Gohberg, I. Koltracht
     Condition numbers for functions of matrices. Applied Numerical Mathematics 12, 107–117 (1993).
329. I. Gohberg, B. Reichstein
     Classification of Block-Toeplitz H-normal operators. Linear and Multilinear Algebra, Vol. 34, 213 245 (1993).
330. I. Gohberg, I. Koltracht
     Mixed, componentwise, and structured condition numbers. SIAM J. Matrix Anal. Appl., Vol. 14, 688–709, July (1993).
331. J.A. Ball, I. Gohberg, L. Rodman
     The structure of flat gain rational matrices that satisfy two-sided interpolation requirements. Systems & Control Letters, 20, 401–412 (1993).
332. I. Gohberg, L. Rodman, T. Shalom, H.J. Woerdeman
     Bounds for eigenvalues and singular values of matrix completions. Linear and Multilinear Algebra, 33, 233–249 (1993).

333. A. Ben-Artzi, I. Gohberg, M.A. Kaashoek
     Exponentially dominated infinite block matrices of finite Kronecker rank.
     Integral Equations and Operator Theory, 18, 30–77 (1994).
334. C. Foias, A. Frazho, I. Gohberg
     Central intertwining lifting, maximum entropy and their permanence. Inte-
     gral Equations and Operator Theory, 18, 166–201 (1994).
335. I. Feldman, I. Gohberg, N. Krupnik
     A method of explicit factorization of matrix functions and applications. In-
     tegral Equations and Operator Theory, 18, 277–302 (1994).
336. I. Gohberg, V. Olshevsky
     Complexity of multiplication with vectors for structured matrices. Linear
     Algebra and its Applications, 202, 163–192 (1994).
337. D. Alpay, J.A. Ball, I. Gohberg, L. Rodman
     J-Unitary preserving automorphisms of rational matrix functions: State space
     theory, interpolation, and factorization. Linear Algebra and its Applications,
     197–198, 531–566 (1994).
338. I. Gohberg, V. Olshevsky
     Fast inversion of Chebyshev–Vandermonde matrices. Numer. Math. 67, 71–92
     (1994).
339. I. Gohberg, M. Hanke, I. Koltracht
     Fast preconditioned conjugate gradient algorithms for Wiener–Hopf integral
     equations. SIAM. J. Numer. Anal., Vol. 31, No. 2, 429–443 (1994).
340. I. Gohberg, N. Krupnik
     Szego–Widom-type limit theorems. Operator Theory: Advances and Appli-
     cations, 71, 105–121 (1994).
341. D. Alpay, J.A. Ball, I. Gohberg, L. Rodman
     The two–sided residue interpolation in the Stieltjes class for matrix functions.
     Linear Algebra and its Applications, 208–209, 485–521 (1994).
342. I. Gohberg, M.A. Kaashoek
     Projection method for block Toeplitz operators with operator-valued sym-
     bols. Operator Theory: Advances and Applications, Vol. 71, 79–104 (1994).
343. A. Ben-Artzi, I. Gohberg
     Orthogonal polynomials over Hilbert modules. Operator Theory: Advances
     and Applications, Vol. 73, 96–126, Birkhäuser Verlag, 1994.
344. I. Gohberg, Y. Zucker
     Left and right factorizations of rational matrix functions. Integral Equations
     and Operator Theory, 19, 216–239 (1994).
345. J.A. Ball, I. Gohberg, M.A. Kaashoek
     Bitangential interpolation for input-output maps of time-varying systems:
     The continuous time case. Integral Equations and Operator Theory, 20, 1–43
     (1994).
346. I. Gohberg, V. Olshevsky
     Fast state space algorithms for matrix Nehari and Nehari–Takagi interpola-
     tion problems. Integral Equations and Operator Theory, 20, 44–83 (1994).

347. D. Alpay, I. Gohberg
Inverse spectral problems for difference operators with rational scattering matrix function. Integral Equations and Operator Theory, 20, 125–170 (1994).

348. J.A. Ball, I. Gohberg, M. Rakowski
Reconstruction of a rational nonsquare matrix function from local data. Integral Equations and Operator Theory, 20, 249–305 (1994).

349. I. Gohberg, H.J. Landau
Prediction and the inverse of Toeplitz matrices. International Series of Numerical Mathematics, Vol. 119, 219–229 (1994).

350. J.A. Ball, I. Gohberg, M.A. Kaashoek
H_∞-control and interpolation for time-varying systems. Akademie Verlag, Mathematical Research, Volume 77, 33–48 (1994).

351. I. Gohberg, V. Olshevsky
Fast algorithm for matrix Nehari problem. Akademie Verlag, Mathematical Research, Volume 79, 687–690 (1994).

352. I. Gohberg, V. Olshevsky
Fast algorithms with preprocessing for matrix-vector multiplication problems. Journal of Complexity, 10 (1994).

353. I. Gohberg
Odessa Reminiscences. Operator Theory: Advances and Applications, 72, xix–xx (1994).

354. J.A. Ball, I. Gohberg, M.A. Kaashoek
Input-output operators of J-unitary time-varying continuous time systems. Operator Theory: Advances and Applications, 75, 57–94 (1995).

355. A. Ben-Artzi, I. Gohberg, M.A. Kaashoek
Discrete nonstationary bounded real lemma in indefinite metrics, the strict contractive case. Operator Theory: Advances and Applications, 80, 49–78 (1995).

356. D. Alpay, I. Gohberg
Inverse spectral problem for differential operators with rational scattering matrix functions. Journal of Differential Equations, Vol. 118, No.1, 1–19 (1995).

357. I. Gohberg, I. Koltracht
Structured condition numbers for linear matrix structures. IMA Volumes in Mathematics and its Applications, Vol. 69, 17–26 (1995).

358. I. Gohberg, I. Koltracht
A fast realization of preconditioned conjugate gradients for Wiener–Hopf integral equations. Applied Mathematics Letters, Vol. 8, No. 6, 65–72 (1995).

359. J.A. Ball, I. Gohberg, M.A. Kaashoek
Two-sided Nudelman for input-output operators of discrete time-varying systems. Integral Equations and Operator Theory, 21, 174–211 (1995).

360. A. Ben-Artzi, I. Gohberg
Inertia theorems for operator pencils and applications. Integral Equations and Operator Theory, 21, 270–318 (1995).

361. I. Feldman, I. Gohberg, N. Krupnik
     On explicit factorization and applications. Integral Equations and Operator
     Theory, 21, 430–459 (1995).
362. R.L. Ellis, I. Gohberg, D.C. Lay
     Infinite analogues of block Toeplitz matrices and related orthogonal functions.
     Integral Equations and Operator Theory, 22, 375–419 (1995).
363. H. Dym, I. Gohberg
     On maximum entropy interpolants and maximum determinant completions
     of associated Pick matrices. Integral Equations and Operator Theory, 23,
     61–88 (1995).
364. V. Boltyanski, I. Gohberg
     Stories about covering and illuminating of convex bodies. Nieuw Archief voor
     Wiskunde, 13, no. 1, 1–26 (1995).
365. I. Gohberg, T. Kailath, V. Olshevsky
     Fast Gaussian elimination with partial pivoting for matrices with displace-
     ment structure. Math. of Computation, 64, 1557–1576 (1995).
366. J.A. Ball, I. Gohberg, M.A. Kaashoek
     A frequency response function for linear, time-varying systems, Math. Control
     Signal 8, 334–351 (1995).
367. I. Gohberg, M.A. Kaashoek, J. Kos
     The asymptotic behaviour of the singular values of matrix powers and appli-
     cations. Linear Algebra and its Applications, 245, 55–76 (1996).
368. A. Ben-Artzi, I. Gohberg
     On contractions in spaces with an indefinite metric: G-norms and spectral
     radii. Integral Equations and Operator Theory, 24, 422–469 (1996).
369. I. Gohberg, Y. Zucker
     On canonical factorization of rational matrix functions. Integral Equations
     and Operator Theory, 25, 73–93 (1996).
370. I. Gohberg, M.A. Kaashoek, J. Kos
     Classification of linear time-varying difference equations under kinematic sim-
     ilarity. Integral Equations and Operator Theory, 25, 445–480 (1996).
371. I. Gohberg, I. Koltracht
     Triangular factors of Cauchy and Vandermonde matrices. Integral Equations
     and Operator Theory, 26, 46–59 (1996).
372. I. Gohberg, S. Goldberg, M. Krupnik
     Traces and determinants of linear operators. Integral Equations and Operator
     Theory, 26, 136–187 (1996).
373. D. Alpay, I. Gohberg
     A relationship between the Nehari and the Caratheodory–Toeplitz extension
     problems. Integral Equations and Operator Theory, 26, 249–272 (1996).
374. C. Foias, A.E. Frazho, I. Gohberg, M.A. Kaashoek
     Discrete time-variant interpolation as classical interpolation with an operator
     argument. Integral Equations and Operator Theory, 26, 371–403 (1996).

375. A. Ben-Artzi, I. Gohberg
     Monotone power method in indefinite metric and inertia theorem for matrices. Linear Algebra Appl., 241–243, 153–166 (1996).
376. I. Gohberg, S. Goldberg
     A simple proof of the Jordan decomposition theorem for matrices. Am. Math. Mon. 103, (2), 157–159 (1996).
377. I. Gohberg, M.A. Kaashoek, L. Lerer
     Factorization of banded lower triangular infinite matrices. Linear Algebra Appl. 247, 347–357 (1996).
378. R.L. Ellis, I. Gohberg
     Distribution of zeros of orthogonal functions related to the Nehari problem. Operator Theory Advances and Applications, 90 (1996).
379. R.L. Ellis, I. Gohberg, D.C. Lay
     On a class of block Toeplitz matrices. Linear Algebra Appl., 241–243, 225–245 (1996).
380. I. Gohberg, S. Goldberg, N. Krupnik
     Hilbert–Carleman and regularized determinants for linear operators. Integral Equations and Operator Theory, 27, 10–47 (1997).
381. Y. Eidelman, I. Gohberg
     Fast inversion algorithms for diagonal plus semiseparable matrices. Integral Equations and Operator Theory, 27, 165–183 (1997).
382. J.A. Ball, I. Gohberg, M.A. Kaashoek
     Nudelman interpolation and the band method. Integral Equations and Operator Theory, 27, 253–284 (1997).
383. I. Gohberg, H.J. Landau
     Prediction for two processes and the Nehari problem. Journal of Fourier Analysis and Applications 3, 43–62 (1997).
384. C. Foias, A.E. Frazho, I. Gohberg, M.A. Kaashoek
     A time-variant version of the commutant lifting theorem and nonstationary interpolation problems. Integral Equations and Operator Theory, 28, 158–190 (1997).
385. C. Foias, A.E. Frazho, I. Gohberg, M.A. Kaashoek
     Parametrization of all solutions of the three chains completion problem. Integral Equations and Operator Theory, 29, 455–490 (1997).
386. Y. Eidelman, I. Gohberg
     Inversion formulas and linear complexity algorithm for diagonal plus semiseparable matrices. Comput. Math. Appl. 33, 4, 69–79 (1997).
387. I. Gohberg, V. Olshevsky
     The fast generalized Parker–Traub algorithm for inversion of Vandermonde and related matrices. J. Complexity 13, (2), 208–234 (1997).
388. H. Bart, I. Gohberg, M.A. Kaashoek
     Wiener–Hopf equations and linear systems. Proceedings of Symposia in Applied Mathematics, Vol. 52 (1997).

389. C. Foias, A.E. Frazho, I. Gohberg, M.A. Kaashoek
     The maximum principle for the three chains completion problem. Integral
     Equations and Operator Theory, 30, 67–82 (1998).
390. D. Alpay, I. Gohberg
     Inverse problem for Sturm–Liouville operators with rationals reflection coef-
     ficient. Integral Equations and Operator Theory, 30, 317–325 (1998).
391. 391. I. Gohberg, M.A. Kaashoek, A.L. Sakhnovich
     Sturm–Liouville systems with rational Weyl functions: Explicit formulas and
     applications. Integral Equations and Operator Theory, 30, 338–377 (1998).
392. Y. Eidelman, I. Gohberg
     A look-ahead block Schur algorithm for diagonal plus semiseparable matrices.
     Computers Math. Applic. Vol. 35, 25–34 (1998).
393. I. Gohberg, M.A. Kaashoek, F. van Schagen
     Operator blocks and quadruples of subspaces: Classification and the eigen-
     value completion. Linear Algebra Appl. 269, 65–89 (1998).
394. I. Gohberg
     Mark Grigorievich Krein: Recollections Integral Equations and Operator The-
     ory, 30, no. 2, 123–134 (1998).
395. I. Gohberg, M.A. Kaashoek, A.L. Sakhnovich
     Pseudo-canonical systems with rational Weyl functions: Explicit formulas and
     applications. J. Diff. Equations 146, (2), 375–398 (1998).
396. R.L. Ellis, I. Gohberg, D.C. Lay
     Singular values of positive pencils and applications. Operator Theory: Ad-
     vances and Applications, 106, 131–146 (1998).
397. A. Dijksma, I. Gohberg Heinz Langer and his work. Operator Theory: Ad-
     vances and Applications, 106, 1–22 (1998).
398. R.L. Ellis, I. Gohberg
     Inversion formulas for infinite generalized Toeplitz matrices. Integral Equa-
     tions and Operator Theory, 32, 29–64 (1998).
399. A. Ben-Artzi, I. Gohberg
     Singular numbers of contractions in spaces with an indefinite metric and
     Yamamoto's theorem. Linear Algebra Applications, 290, 31–48 (1999).
400. Y. Eidelman, I. Gohberg
     On a new class of structured matrices. Integral Equations and Operator The-
     ory, 34, 293–324 (1999).
401. Y. Eidelman, I. Gohberg
     Linear complexity inversion algorithms for a class of structured matrices.
     Integral Equations and Operator Theory, 35, 28–52 (1999).
402. I. Gohberg, M.A. Kaashoek
     State space methods for analysis problems involving rational matrix func-
     tions. Dynamical systems, control, coding computer vision (Padova, 1998),
     93–109, Progress in Systems and Control Theory, Vol. 25, Birkhäuser, 93–109
     (1999).

403. I. Gohberg
Vladimir Maz'ya: Friend and Mathematician. Recollections. Operator Theory: Advances and Applications, 109, 1–5 (1999).

404. I. Gohberg
Mark Grigorevich Krein (A Short Biography). Operator Theory: Advances and Applications, 117, 5–8 (2000).

405. I. Gohberg, M.A. Kaashoek, A.L. Sakhnovich
Canonical systems on the line with rational spectral densities: Explicit formulas. Operator Theory: Differential operators and related topics, Vol. I (Odessa, 1997). Oper. Theory Adv. Appl., 117, 127–139 (2000).

406. I. Feldman, I. Gohberg, N. Krupnik
Convolution equations on finite intervals and factorization of matrix functions. Integral Equations and Operator Theory, 36, No. 2, 201–211 (2000).

407. D. Alpay, I. Gohberg
Connections between the Caratheodory–Toeplitz and the Nehari extension problems: The discrete scalar case. Integral Equations and Operator Theory., 37, 125–142 (2000).

408. I. Gohberg, M.A. Kaashoek, J. Kos
Classification of linear periodic difference equations under periodic or kinematic similarity. SIAM J. Matrix Analysis Appl. 21, 481–507 (1999).

409. I. Gohberg, M.A. Kaashoek, F. van Schagen
Finite section method for linear ordinary differential equations J. Diff. Eqs. 163, No. 2, 312–334 (2000).

410. D. Alpay, I. Gohberg, M.A. Kaashoek, A.L. Sakhnovich
Direct and inverse scattering problem for canonical systems with a strictly pseudo–exponential potential. Math. Nach. 215, 5–31 (2000).

411. I. Gohberg, N. Krupnik
A connection between the determinant and characteristic numbers of an operator pencil. Operator Theory: Advances and Applications, 121, 109–119 (2001).

412. R. Ellis, I. Gohberg
Extensions of matrix valued inner products on modules and the inversion formula for block Toeplitz matrices. Operator Theory: Advances and Applications, 122, 191–227 (2001).

413. Y. Eidelman, I. Gohberg
Fast inversion algorithms for a class of block structured matrices. Contemporary Mathematics, volume 281, 17–38 (2001).

414. I. Gohberg, S. Goldberg, N. Krupnik
Generalization of the determinants for trace-potent linear operators. Integral Equations and Operator Theory, 40, 441–453 (2001).

415. I. Gohberg
Reminiscences of Bela Szőkefalvi–Nagy. Recent Advances in Operator Theory and Related Topics. Operator Theory: Advances and Applications, 127, xiii–xvi (2001).

416. D. Alpay, I. Gohberg
     Inverse problems associated to a canonical differential system. Operator The-
     ory: Advances and Applications, 127, 1–27 (2001).
417. I. Gohberg, M.A. Kaashoek, A.L. Sakhnovich
     Bound states of a canonical system with a pseudo-exponential potential. In-
     tegral Equations and Operator Theory, 40, 268–277 (2001).
418. I. Gohberg
     Peter Lancaster, my friend and co-author. Operator Theory: Advances and
     Applications, 130, 23–27 (2002).
419. I. Gohberg, M.A. Kaashoek, F. van Schagen
     Finite section method for linear ordinary differential equations on the full line.
     Interpolation theory, systems theory and related topics (Tel Aviv/Rehovot,
     1999). Operator Theory: Advances and Applications, 134, 209–224 (2002).
420. I. Gohberg, M.A. Kaashoek, F. van Schagen
     Finite section method for difference equations. Linear operators and matrices.
     Operator Theory: Advances and Applications, 130, 197–207 (2002).
421. Y. Eidelman, I. Gohberg
     A modification of the Dewilde–van der Veen method for inversion of finite
     structured matrices. Linear Algebra Applications, 343–344, 419–450 (2002).
422. I. Gohberg
     On joint work with Harry Dym. Operator Theory: Advances and Applica-
     tions, 134, 25–30 (2002).
423. I. Gohberg, M.A. Kaashoek, F. van Schagen
     Finite section method for linear ordinary differential equations on the full
     line. Operator Theory: Advances and Applications, 134, 200–224 (2002).
424. I. Gohberg, M.A. Kaashoek, F. van Schagen
     Finite section method for linear ordinary differential equations revisited.
     Toeplitz matrices and singular integral equations, Operator Theory: Ad-
     vances and Applications, 135, 183–191 (2001).
425. Y. Eidelman, I. Gohberg
     Algorithms for inversion of diagonal plus semiseparable operator matrices.
     Integral Equations Operator Theory, 44, No. 2, 172–211 (2002).
426. I. Gohberg, M.A. Kaashoek, A.L. Sakhnovich
     Scattering problems for a canonical systems with a pseudo–exponential po-
     tential. Asymptot. Anal., 29, No. 1, 1–38 (2002).
427. I. Gohberg, M.A. Kaashoek, I.M. Spitkovsky
     An overview of matrix factorization theory and operator applications. Fac-
     torization and integrable systems (Faro 2000), Operator Theory: Advances
     and Applications, 141, 1–102 (2003).
428. R.L. Ellis, I. Gohberg
     A generalization of the one-step theorem for matrix polynomials. Integral
     Equations Operator Theory, 47, No. 1, 51–69 (2003).

429. Y. Eidelman, I. Gohberg
     Fast inversion algorithms for a class of structured operator matrices. Linear
     Algebra Appl. 371 153–190 (2003).
430. D. Alpay, I. Gohberg
     A trace formula for canonical differential expressions. J. Funct. Anal. 197,
     No. 2, 489–525 (2003).
431. D. Alpay, I. Gohberg
     Pairs of selfadjoint operators and their invariants. Algebra J. Analiz 16, No.
     1, 70–120 (2004).
432. I. Gohberg, M.A. Kaashoek, F. van Schagen
     On inversion of finite Toeplitz matrices with elements in an algebraic ring.
     Linear Algebra Appl. 385, 381–389 (2004).
433. Y. Eidelman, I. Gohberg
     Direct approach to the band completion problem. Linear Algebra Appl. 385,
     149–185 (2004).
434. I. Feldman, I. Gohberg, N. Krupnik
     An explicit factorization algorithm. Integral Equations Operator Theory, 49,
     no. 2, 149–164 (2004).
435. I. Gohberg, M.A. Kaashoek, F. van Schagen
     On inversion of convolution integral operators on a finite interval. Operator
     Theory: Advances and Applications, 147, 277–285 (2004).
436. I. Gohberg, M.A. Kaashoek
     Our meetings with Erhard Meister. Operator Theory: Advances and Appli-
     cations, 147, 73–75 (2004).
437. Robert L. Ellis, Israel Gohberg
     Seymour Goldberg (1928–2004). In memoriam. Integral equations operator
     theory, 51, no. 1, 1–3 (2005).
438. Eidelman,Y. Gohberg,I. Olshevsky, I.
     The QR iteration method for Hermitian quasiseparable matrices of an arbi-
     trary order. Linear Algebra Appl. 404, 305–324 (2005).
439. Eidelman,Y. Gohberg, I.,Olshevsky, V
     Eigenstructure of order-one-quasiseparable matrices. Three-term and two-
     term recurrence relations. Linear Algebra Appl. 405 1–40 (2005).
440. H. Bart, I. Gohberg, M.A. Kaashoek, A.C.M. Ran
     Schur complements and state space realizations. Linear Algebra Appl. 399,
     203–224 (2005).
441. I. Gohberg, M.A. Kaashoek, A. L. Sakhnovich
     Taylor coefficients of a pseudo-exponential potential and the reflection coef-
     ficient of the corresponding canonical system. Math. Nachr. 278, no. 12–13,
     1579–1590 (2005).
442. Albrecht, Böttcher, Israel Gohberg, Bernd Silbermann
     Georg Heinig (1947–2005). In memoriam. Integral Equations Operator The-
     ory, 53, no 2, 297–300 (2005).

443. I. Gohberg, I. Haimovici, M.A. Kaashoek, L. Lerer
The Bezout integral operator: main property and underlying abstract scheme.
The state space method generalizations and applications. Operator Theory:
Advances and Applications, 161, 225–270 (2006), 225–270.

444. D. Alpay, I. Gohberg
Discrete analogs of canonical systems with pseudo-exponential potential. Inverse problems. Interpolation, Schur functions and moment problems, Operator Theory: Advances and Applications, 165, 31–65 (2006).

445. D. Alpay, I. Gohberg
Discrete analogs of canonical systems with pseudo–exponential potential.
Definitions and formulas for the spectral matrix functions. The state space
method generalizations and applications, Operator Theory: Advances and
Applications, 161, 1–47 (2006).

446. I. Gohberg, M.A. Kaashoek, L. Lerer
The continuous analogue of the resultant and related convolution operators.
The extended field of operator theory, Operator Theory: Advances and Ap-
plications, 171, 107–127 (2007).

447. D.A. Bini, Y. Eidelman, L. Gemignani, I. Gohberg
Fast QR eigenvalue algorithms for Hessenberg matrices which are rank–one
perturbations of unitary matrices. SIAM J. Matrix Anal. Appl. 29, no. 2,
566–585 (2007).

448. D. Alpay, I. Gohberg
Discrete systems and their characteristic spectral functions. Mediterr. J.
Math. 4, no. 1, 1–32 (2007).

449. D. Alpay, I. Gohberg, I.
On generators of quasiseparable finite block matrices. Calcolo 42, no. 3–4,
187–214 (2005).

450. I. Gohberg, M.A. Kaashoek, L. Lerer
Quasi-commutativity of entire matrix functions and the continuous analogue
of the resultant. Modern operator theory and applications. Operator Theory:
Advances and Applications, 170, 101–106 (2007).

451. T. Bella, Y. Eidelman, I. Gohberg, I. Koltracht, V. Olshevsky
A Bjorck–Pereyra-type algorithm for Szego–Vandermonde matrices based on
properties of unitary Hessenberg matrices. Linear Algebra Appl. 420, no. 2–3,
634–647 (2007).

452. Y. Eidelman, L. Gemignani, I. Gohberg
On the fast reduction of a quasiseparable matrix to Hessenberg and tridiag-
onal forms. Linear Algebra Appl. 420, no. 1 (2007).

453. D.A. Bini, Y. Eidelman, L. Gemignani, I. Gohberg
Fast QR eigenvalue algorithms for Hessenberg matrices which are rank-one
perturbations of unitary matrices. SIAM J. Matrix Anal. Appl. 29, no. 2,
566–585 (2007).

454. D. Alpay, I. Gohberg
Discrete systems and their characteristic spectral functions. Mediterr. J. Math. 4, no. 1, 1–32 (2007).

455. I. Gohberg, M.A. Kaashoek, L. Lerer
The continuous analogue of the resultant and related convolution operators. The extended field of operator theory. Operator Theory: Advances and Applications, 171, 107–127 (2007).

456. I. Gohberg, M.A. Kaashoek, L. Lerer
On a class of entire matrix function equations. Linear Algebra Appl. 425, no. 2–3, 434–442 (2007).

457. I. Gohberg, M.A. Kaashoek, L. Lerer
The inverse problem for orthogonal Krein matrix functions. (Russian) Funktsional. Anal. i Prilozhen. 41, no. 2, 44–57, 111 (2007).

458. D.A. Bini, Y. Eidelman, L. Gemignani, I. Gohberg
The unitary completion and QR iterations for a class of structured matrices. Math. Comp. 77, no. 261, 353–378 (2008).

Date: March 2008

List of Graduate Students of Israel Gohberg

With the year and place of the Ph.D. defence.

1. A.S. Markus, 1958 (Kishinev University, Kishinev, with I.A. Itskovich)
2. I.A. Feldman, 1964 (Academy of Sciences, MSSR, Kishinev)
3. L..S. Goldenshtein, 1964 (Academy of Sciences, MSSR, Kishinev).
4. V.N. Vizitei, 1965 (Academy of Sciences, MSSR, Kishinev, with A.S. Markus)
5. M.S. Budianu, 1966 (Academy of Sciences, MSSR, Kishinev)
6. I. Chebotaru, 1966 (Academy of Sciences, MSSR, Kishinev)
7. V.M. Brodskii, 1967 (Moldovan Academy of Sciences, Kishinev)
8. N.Ya. Krupnik, 1967 (Kishinev University, Kishinev, with I.A. Itskovich)
9. V.M. Eni, 1967 (Moldovan Academy of Sciences, Kishinev, with A.S. Markus)
10. V.A. Prigorskii, 1968 (Moldovan Academy of Sciences, Kishinev, with A.S. Markus)
11. M.A. Barkari, 1968 (Moldovan Academy of Sciences, Kishinev)
12. V.I. Paraska, 1968 (Moldovan Academy of Sciences, Kishinev)
13. M.K. Zambitskii, 1969 (Moldovan Academy of Sciences, Kishinev)
14. E.I. Sigal, 1970 (Moldovan Academy of Sciences, Kishinev)
15. Y. Leiterer, 1971 (Kishinev University, Kishinev)
16. R.V. Duduchava, 1971 (Georgian Academy of Sciences, Tbilisi)
17. V.D. Frolov, 1971 (Moldovan Academy of Sciences, Kishinev)
18. Y.M. Spigel, 1972 (Moldovan Academy of Sciences, Kishinev)
19. V.I. Levchenko, 1973 (Moldovan Academy of Sciences, Kishinev)
20. A.A. Sementsul, 1973 (Moldovan Academy of Sciences, Kishinev)
21. G. Heinig, 1974 (Kishinev University, Kishinev)
22. V.A. Zolotarevskii, 1974 (Moldovan Academy of Sciences, Kishinev)
23. L.E. Lerer, 1974 (Moldovan Academy of Sciences, Kishinev, with A.S. Markus)
24. V.P. Soltan, 1974 (Moldovan Academy of Sciences, Kishinev)
25. H.V. Hambartsumyan, 1974 (Armenian Academy of Sciences, Erevan)
26. L. Rodman, 1978 (Tel-Aviv University, Tel-Aviv)
27. S. Levin, 1980 (Weizmann Institut, Rehovot)
28. C.V.M. van der Mee, 1981 (Vrije Universiteit, Amsterdam, with M.A. Kaashoek)
29. I. Koltracht, 1983 (Weizmann Institut, Rehovot)
30. N. Cohen, 1985 (Weizmann Institut, Rehovot)
31. A. Perelson, 1987 (Tel-Aviv University, Tel-Aviv)
32. L. Roozemond, 1987 (Vrije Universiteit, Amsterdam, with M.A. Kaashoek)
33. A. Benartzi, 1988 (Tel-Aviv University, Tel-Aviv)

34. T. Shalom, 1989 (Tel-Aviv University, Tel-Aviv)
35. H. Woerdeman, 1989 (Vrije Universiteit, Amsterdam, with M.A. Kaashoek)
36. S. Rubinstein, 1990 (Tel-Aviv University, Tel-Aviv)
37. R. Vreugdenhil, 1990 (Vrije Universiteit, Amsterdam, with M.A. Kaashoek)
38. Danrun Huang, 1992 (University of Maryland, College Park. MD, with S. Goldberg and M. Michael Boyle)
39. J. Kos, 1995 (Vrije Universiteit, Amsterdam, with M.A. Kaashoek)
40. Ya. Zucker, 1998 (Tel-Aviv University, Tel-Aviv)

Reference

In response to a request by I. C. Gohberg I examined a number of papers written by him while he was a 4th and 5th year student at the Physics and Mathematics Faculty of the Kishinev State University (1950-1951):

1. *On linear equations in Hilbert space.*
2. *On linear equations in normed spaces.*
3. *On linear operators depending analytically on a parameter.*
4. *Some properties of linear equations in Hilbert spaces and their applications.*
5. *On an application of the theory of normed rings to singular integral equations.*

The first three of these papers were published in 1951 in the journal Doklady Akademii Nauk SSSR (Vol. 76, nos. 1 and 4, and Vol. 78, no. 4).

The fourth, on my suggestion, was submitted to the Ukrainskii Matematicheskii Zhurnal, and the fifth one is being submitted to the journal Uspekhi Matematicheskih Nauk.

In these papers investigations of a number of Soviet mathematicians, among them Professor S. G. Mikhlin, Academician N. I. Muskhelishvili (a representative of the Georgian school of mathematics) and Professor S. M. Nikolskii, are refined and deepened.

The results obtained by I. C. Gohberg constitute an important contribution to the theory of one-dimensional singular integral equations. He succeeded in establishing the aforementioned results by employing new methods and tools of functional analysis.

Particularly unexpected and ingenious are the applications that I. C. Gohberg found of the theory of normed rings to problems concerning the study of important classes of singular integral operators.

The research carried out by I. C. Gohberg attests to his outstanding mathematical talent.

Appropriately presented, these investigations should constitute an excellent thesis for the degree of Candidate of Physico-Mathematical Sciences.

Corresponding member of the Academy of Sciences of USSR,
Doctor of Physico-Mathematical Sciences,

Professor M. G. Krein

Editorial comments. This document is a translation into English of a letter in
Russian written by M.G. Krein on July 2, 1951. It has the character of a "to
whom it may concern" letter. When we asked Israel why he requested the letter
he told us the following story.

*In July 1951 I had to defend my "Diploma" work. This had to be done publicly in
front of a Committee appointed by the Minister of Education of Moldavia. For the
master degree one had to do this and to pass two exams. The dean and a few other
teachers were quite anti-Semitic and not friendly to me. I was afraid that they
would make a trick and put me down on my last step at the university. I was not
sure at all. Also they could discover the story with my father. I was very vulnerable.
Nobody in Kishinev would be ready to help me. I was very afraid in this situation. I
decided to insure myself with a reference letter from M.G. It was not clear that he
would be willing to write such a letter without an official request. I went to Odessa
and described him the situation. Krein was on his Dacha (Summer Hous). He asked
me to wait in their garden, went to his study, and in half a hour came to me with
the draft of this reference. He typed it on his famous typewriter, legalized it the next
day in the institution where he worked and gave me two copies. I was extremely
happy with this reference. It was very strong. It gave me some insurance. It helped
me in the defence, it made a very positive helpful impression on the Minister of
Education of Moldavia, and it helped me with receiving an appropriate job.*

Krein's letter is extremely strong indeed. Note that the letter concerns work for
a master thesis, but at the end Krein writes that the material, appropriately
presented, would make an excellent Ph.D. thesis.

Election Moldavian Academy of Sciences

Abstract. At the end of the sixties Israel Gohberg was candidate for promotion to the position of Corresponding Member of the Academy of Sciences of the Moldavian SSR. This required letters of support. Some were sent by telegram, others by ordinary mail. Here some of these telegrams, letters are presented in Engish translation.

Telegram 1

Series II   Kishinev   Academy of Sciences   To: Academician Andrunakevich Novosibirsk 90/0382

The Scientific Council of the Institute of Mathematics of the Siberian Branch of the Academy of Sciences of USSR is unanimously supporting the nomination of Israel Tsudikovich Gohberg to Corresponding Member of the Academy of Sciences of the Soviet Republic of Moldavia.

Director of the Institute of Mathematics of the
Siberian Branch of the Academy of Sciences of USSR

Academician Sobolev

Authenticated copy

Telegram 2

To: Kishinev 1   Lenin Avenue 1

To the Presidium of the Academy of Sciences of the Moldavian SSR

TBL/7   Tbilisi   7/13803   66   3   1710

In recognition of the major contributions of Professor I. C. Gohberg to the elaboration of modern problems of functional analysis, in particular, of the theory of integral equations, we consider it our duty to support his nomination by the Institute of Mathematics and the Computing Center of the Academy of Sciences of

USSR to Corresponding Member of the Academy of Sciences of the Moldavian SSR.

> Academician N. I. Muskhelishvili,
> Academician N. P. Vekua, Academy of Sciences of the Georgian SSR,
> B.V. Khvedelidze, Corresponding Member of the
> Academy of Sciences of the Georgian SSR.

Authenticated copy

Letter of support 1

I consider I. C. Gohberg to be one of the most prominent representatives of functional analysis in the USSR and fully support his promotion to Corresponding Member of the Academy of Sciences of the Moldavian SSR in the specialty "Functional Analysis."

Signed: Academician Vladimir Ivanovich Smirnov

Leningrad,
Kirov Avenue No. 25, Apt. 44

Letter of support 2

September 26, 1969

To the Presidium of the Academy of Sciences of MSSR

I support most energetically the candidacy of the Doctor of Physico-Mathematical Sciences, Professor I. C. Gohberg for the opened vacancy of Corresponding Member of the Academy of Sciences of the Moldavian SSR in the speciality "Functional Analysis."

I am proud that Professor Gohberg is my student and the coauthor of many of our joint works.

Through his research Professor Gohberg has enriched many parts of functional analysis, in the theoretical as well as in the applied directions, and has also found different openings for functional analysis into classical analysis. His name, as a specialist in functional analysis, is widely known, in the USSR as well as abroad. Many of his publications (and there are no less than 90 of them) have been translated into various languages.

Over a relatively short period of time, Professor Gohberg was able to create his own school which does active research in functional analysis within the Institute of Mathematics and Computer Science of the Academy of Sciences of the Moldavian SSR.

I am deeply convinced that Professor I. C. Gohberg will be one of the best orna-
ments of the Academy of Sciences of the Moldavian SSR.

Signed: Corresponding Member of the Academy of Sciences of the Ukrainian SSR,

Doctor of Physico-Mathematical Sciences,

M. G. Krein

Odessa, Arkadya

August 8, 1969

Editorial comments. In the beginning of 1970 Israel Gohberg was successfully
elected to the Academy of Sciences of MSSR, as a corresponding member. In 1974
after his emigration to Israel, he was removed from the list of members of this
Academy. More than 20 years later, in 1996 after the USSR disintegrated and
an independent state of Moldova was created, Israel Gohberg was reinstated as a
corresponding member of the Academy of Sciences of Moldova. This is his status
in the Moldovan Academy till today.

A Review of the Mathematical Work of Israel Gohberg up to 1998

M.A. Kaashoek

Abstract. *A short review is given of the mathematical work of Israel Gohberg, on the occasion of his seventieth birthday.*

Israel Gohberg was born in 1928.[1] He began his mathematical studies around 1950 in Frunze (Kirgizia) and continued in Kishinev (Moldova). He obtained the first doctoral degree in 1954 in Leningrad. In 1959, after a stay of about six years at the Beltsky Pedagogical Institute (where he became head of the mathematics department), he returned to Kishinev as full professor and head of the functional analysis department in the Mathematical Institute of the newly organized Moldavian Academy of Science. In 1964 he obtained the second doctoral degree in Moscow, and in 1970 he was elected a corresponding member of the Moldavian Academy (at that time an unusual honor for a mathematician of Jewish origin). Since 1974 he has been a professor at Tel-Aviv University, where he is the incumbent of the Nathan and Lily Silver chair since 1981, and since 1983 he is an extra-ordinary professor at the Vrije Universiteit in Amsterdam. Among the many honors which have marked his career, we mention his election as a Foreign Member of the Netherlands Academy of Arts and Science in 1985 (in the seat of the late Marc Kac), the Rothschild Prize in Mathematics in 1986, the prestigious Humboldt Research Prize in 1992, and the Hans Schneider Prize in Linear Algebra in 1994. Gohberg has supervised more than 40 doctoral students; many of them are now associate or full professor.

The mathematical work of Gohberg is extensive and influential. He is author or co-author of more than 350 publications and of more than 15 books. Except for two attractive little books on elementary geometry and combinatorial problems, which he wrote together with Boltyansky (1965, 1971), his contributions belong to the fields of Analysis, Operator Theory and Linear Algebra. He is an undisputed leader in the following research areas:

Originally published in *Recent Advances in Operator Theory,* OTAA **124**, Birkhäuser Verlag Basel, 2001, pp. xxvii–xxxii.
[1]For more extended biographical data see also the "Mathematical Tales" and the "Gohberg Micellanea" from OTAA **40**, which are reproduced in Parts I and III of the present book, respectively.

- Integral equations;
- The theory of nonself-adjoint operators;
- Spectral theory and factorization of matrix and operator functions;
- Inversion problems for structured matrices.

His papers and books, which mainly aim at the development of mathematics rather than its applications, are also known and often quoted outside mathematical circles, in particular by (astro)physicists and engineers (from electrical engineering, control theory and system theory).

Gohberg's first research topic was the theory of Fredholm operators and its different generalizations. His first results in this area were published in 1951 in "Doklady" while he was still a student. The main achievements were general theorems about Fredholm operators (perturbation, index and representations), the theorem about the spectra of analytic Fredholm operator valued functions, and the necessary and sufficient condition when a singular integral operator with continuous coefficients is Fredholm. These results were in his master thesis. In this Ph D thesis (which was ready some time in 1952) he made an effective connection between the theory of commutative algebras and that of singular integral operators by showing that the symbol of the operator is the Gelfand transform in an appropriate Calkin algebra. This connection marks the beginning of the modern theory of operator algebras. All these results are of interest even today. But at that time, they were truly pioneer achievements.

On the basis of his innovative work on Fredholm operators, the young Gohberg was accepted as an equal author by the established master of analysis, Mark G. Kreĭn. The collaboration led to many important developments in integral equations, notably the fundamental treatises in "Ushepi Mat. Nauk" (the most prestigious mathematical journal in the former USSR) on Fredholm theory (1957) and on systems of Wiener-Hopf equations (1958). Both articles have been translated into English and became classical; they are often quoted in pure and applied literature. After several more articles, there came the joint work on nonself-adjoint operators which culminated in two world famous books: "Introduction to the theory of linear nonself-adjoint operators" (1965) and "Theory and applications of Volterra operators" (1967). Some of the original and fundamental contributions were: the factorization theory for operators (which is now used in many different areas, such as system theory, probability theory, integral equations, etc.), the theory of singular values for bounded operators (which turned out to be of great importance in interpolation theory of infinite Hankel matrices), the foundation of the theory of operator ideals, and the theory of characteristic operator functions for operators close to unitary (which provides one of the main tools for the classification of operators up to unitary equivalence).

In the mean time, in Kishinev the young Gohberg attracted many strong doctoral candidates, some even from the DDR, who officially or informally became his students and later co-workers. The many joint publications on operator theory, including a number of books, brought a period of fame to the out-of-the way

Institute. Gohberg did fundamental work on factorization of matrix and operator functions which was subsequently (and jointly with Leiterer) expanded into a complete theory, which has now important applications to astrophysics and linear transport theory. In this period he also found a complete description of all algebras of matrix-valued functions of which the elements admit canonical factorization (the so-called decomposing algebras). He developed a new approach to the theory of Wiener-Hopf operators and maximal symmetric operators. These results, which were published in an Uspehi paper, formed the contents for his second doctoral thesis. In the late 60's and the beginning of the 70's he returned to singular integral operators, now for the case when the coefficients have jump discontinuities. He discovered the right description of the symbol of such operators, which surprisingly enough turned out to be in terms of matrix functions even for scalar equations. In this period he also develops numerical methods to solve Wiener-Hopf equations on the half line and on finite interval. The famous Gohberg-Semencul formulas for inversion of finite Toeplitz matrices, which are often used in the electrical engineering literature, were also developed during this time.

A new period in Gohberg's life began with his departure from the Soviet Union, when he started a number of major new research projects, first at Tel-Aviv (where he leads a major group in analysis and operator theory), and somewhat later also in Amsterdam (collaboration with M.A. Kaashoek), in Calgary (collaboration with P. Lancaster), and College Park (collaboration with S. Goldberg, R. Ellis and D.C. Lay). Typical for this new period (and truly remarkable) is his interaction with control and electrical engineers. Inspired by these connections he developed new directions in operator theory, which resulted in new constructive operator theory methods with substantial mathematical contributions to control theory. Three topics are involved.

The first is the state space method. Since the end of the seventies Gohberg and his co-workers have developed systematically a new system theory oriented method to treat problems in analysis and operator theory. This method has its roots in the mathematical system theory of the sixties and is related to the Kalman approach to control problems. It is based on the discovery that analytic functions which are in a natural way related to operators, like the determinant function of a Fredholm integral operator, the symbol of a singular integral operator or the characteristic function of an operator close to unitary, may be viewed as transfer functions of input-output systems, and hence such a function may be analyzed in terms of three or four operators (the coefficients of the corresponding system) which are often much simpler to deal with than the original operator the function came from. For example, for the rational case the symbol of a system of Wiener-Hopf integral equations can be represented in the form

$$W(\lambda) = I + C(\lambda - A)^{-1}B,$$

where A, B, and C are finite matrices, and in this way problems about the original Wiener-Hopf operator can be treated by using linear algebra and matrix theory.

This original approach has proved to be most powerful and is nowadays referred to as the *state space method*. A first main result was a beautiful geometrical principle of factorization for matrix and operator functions which, for example, for the rational case reduces the problem of canonical Wiener-Hopf factorization to a simple matching of finite dimensional spectral subspaces. The state space method proved also to be very effective in the analysis of the zero and pole structure of matrix polynomials and rational matrix functions and associate inverse problems. In this way in a series of articles Gohberg and his associates solved explicitly the matrix-valued versions of the classical interpolation problems of Lagrange-Sylvester, Nevanlinna-Pick, Carathéodory-Toeplitz, Schur-Takagi, boundary Nevanlinna-Pick, and Nehari, and also the more recent bitangential and Nudelman problems. As a by-product, but not less significant, modern control problems, like the sensitivity minimization, model reduction, and robust stabilization, were solved explicitly with the solutions being described in state space form. Gohberg's book with Bart and Kaashoek on "Minimal factorization of matrix and operator functions" (1979), his book on "Matrix polynomials" (1982) with Lancaster and Rodman, and his book on "Interpolation of rational matrix functions" (1990) with Ball and Rodman, which all three employ the state space method systematically, may be described as seminal. The state space method continues to be a of great importance, and is now also used in other branches of analysis (e.g., to derive Szegö-Kac-Achiezer limit formulas or to solve direct and inverse problems for canonical differential systems with rational spectral functions).

Gohberg's interaction with the control and electrical engineers, also led to new completion and extension problems for partially given matrices or operators. In the eighties he developed, first with Dym (1982/1983) and later with Kaashoek and Woerdeman (1988-1992) and Ellis and Lay (1993), the so-called band method, an abstract scheme which allows one to deal with positive and contractive completion problems from one point of view, and which presents a natural strategy to solve such problems by reduction to linear equations. It led to beautiful and easy to handle explicit formulas for the solutions of various completion and interpolation problem. For example, by using this scheme Gohberg and his co-workers solved the operator-valued versions of the Carathéodory problem, the Nehari problem and its four block generalization, and also the time-variant (nonstationary) analogues of these problems. The discovery (in 1990) of the connection between the central completion and maximum entropy solutions was one of Gohberg's of outstanding achievements in this area. Parallel he developed with Kaashoek and Van Schagen (1980-1996) the invariants of various partially given matrices and operators, with remarkable applications to eigenvalue completion problems, to stabilization problems in mathematical system theory and to problems of Wiener-Hopf factorization.

A third major topic, related to the previous two, concerns his original contributions to numerical analysis and numerical linear algebra. They include parallel algorithms (developed jointly with Koltracht and Lancaster 1987/1992) for semi-separable integral operators (which are viewed as input-output operators of systems), and finite section and projection methods for convolution operators. For structured matrices and for their block versions Gohberg developed (1987-1995) fast algorithms, first in joint work with Koltracht, and later also with Kalaith and Olshevsky. His work in this area, which has been tested on numerical experiments, is first class and used by engineers.

In the last fifteen years Gohberg made also significant contributions to the theory of spaces with an indefinite metric (his 1983 book "Matrices and indefinite scalar products", written jointly with Lancaster and Rodman, is the first linear algebra book in which the theory of matrices in indefinite metric spaces is developed systematically), and to the theory of orthogonal polynomials in a number of articles with Alpay (1988), Ellis and Lay (1998/1992-1995), and Lerer (1988). His more recent work includes the development of a time-variant analogue of inertia theorems, of the theory of orthogonal polynomials (both jointly with Ben-Artzi), and of the state space method (jointly with Ben-Artzi and Kaashoek); this orginal work may very well lead to a deep and far reaching nonstationary analogue of the theory of analytic functions on the disc. In the recent years Gohberg also developes further the theory of singular integral equations and writes together with Krupnik two excellent monographs on the one-dimensional theory.

During these years Gohberg also returns a number of times to his earlier research topics, with new ideas and original contributions. His recent work with Krupnik on singular integral equations has already been mentioned. The book on "Factorization of matrix functions and singular integral operators" written jointly with K. Clancey (1981) is an earlier example and his recent analysis of traces and determinants with S. Golberg another.

In addition to all this, Gohberg managed

1. to write three outstanding textbooks "Basic operator theory" (1981, jointly with S. Goldberg) and "Classes of Linear Operators I and II" (1990/1993, jointly with S. Gohberg and M.A. Kaashoek),
2. to found and edit the influential journal "Integral Equations and Operator Theory",
3. to found and edit the series monographs "Operator Theory: Advances and Applications", in which more than 80 volumes have appeared,
4. to create and direct the IWOTA meetings, a series of international workshops on operator theory and its applications, and (5) to organize the biennial Toeplitz Conferences at Tel-Aviv.

Gohberg's mathematical influence is profound and far reaching. His style and fine mathematical taste has attracted and inspired his colleagues and fellow mathematicians. He is a true leader of a mathematical school.

Gohberg's Mathematical Work in the Period 1998–2008

M.A. Kaashoek and L. Lerer

The work of Professor Israel Gohberg is so monumental that no summary can do justice to it. A short review of his work until 1998 is found in the preceding article written by M.A. Kaashoek. Since then the pace of Gohberg's work has not diminished. In the present article a rather short review is given of his work during the last ten years or so.

In this period five new research monographs and numerous research papers have been written by Israel Gohberg jointly with his collaborators. In what follows we concentrate on the five research books, mention briefly the research papers not covered by these books, and at the end we shortly describe two semi-textbooks written during this period.

1. I. Gohberg, M.A. Kaashoek and F. van Schagen, *Partially specified matrices and operators: classification, completion, applications*, Birkhäuser Verlag, Basel, 1995; 333 pp (item 19 in the list of books[1]).

This book explores a new direction in linear algebra and operator theory which emerged in the last couple of decades. The main topics are invariants of partially given matrices and partially given operators, and spectral analysis of their completions. The book centers around two major problems. The first is a problem of classification of partially given matrices modulo transformations that leave invariant the pattern of the given data, and the results here can be seen as a far reaching generalization of the Jordan canonical form. The second problem is a spectral inverse problem referred to as the eigenvalue completion problem. Given a partially given matrix it asks for a description of all possible eigenvalues and their multiplicities of the matrices which one can obtain by filling in the unspecified entries. Both problems are also treated in an infinite dimensional framework. A large part of the book deals with applications to matrix theory, analysis and system theory. Namely, the basic pole assignment problem as well as other stabilization problems in mathematical system theory are viewed as particular cases of

---

[1]See Gohberg's "List of Publications" in the second section of this part

eigenvalue completion. Applications of the general theory in analysis include problems of Wiener-Hopf factorization and of interpolation for matrix polynomials and rational matrix functions. Also applications to the Kronecker structure theory of linear pencils, and to non-everywhere defined operators are presented.

2. C. Foias, A. Frazho, I. Gohberg, and M.A. Kaashoek, *Metric constrained interpolation, commutant lifting and systems*, Birkhäuser Verlag, Basel, 1998; 587 pp (item 20 in the list of books).

This book combines the famous commutant lifting theorem from operator theory and the state space method from systems theory. The aim is to provide a unified approach for solving both stationary and nonstationary interpolation problems with norm constraints on the interpolants. Motivation for interpolation problems (especially their generalizations to matrix- and operator-valued functions and the quest for more practical solution algorithms) has become particularly intense in the last 25 years or so, due to their penetrating role in system theory, particularly, in prediction theory and in H-infinity and H-two control theory. The book provides a grand unified commutant lifting formalism for understanding the conventional time-invariant interpolation theory and the more recent nonstationary version of the theory, along with how the two versions are connected with each other. The authors are able to convert this operator-theoretic formalism to state-space coordinates roughly by simply working with Taylor coefficients. This monograph provides an excellent, general setting for understanding in a unified way the many previous papers on various particular instances of this topic. The book includes the matrix-valued and operator-valued versions of the tangential Nevanlinna-Pick problem, the Hermite-Fejer problem, the Nehari extension problem, the Sarason problem, and the two-sided Nudelman problem (which all originate from complex function theory) as well as the non-stationary analogues of these problems. The main results concern existence of solutions, the explicit construction of the central solutions in state space form, the maximum entropy property of the central solutions, and state space parameterizations of all solutions. Applications to the engineering field of H-infinity control are amply provided.

3. I. Gohberg, S. Goldberg, N. Krupnik, *Traces and determinants of linear operators*. Birkhäuser Verlag, Basel, 2000; 258 pp (item 21 in the list of books).

This book presents a theory of traces and determinants for operators in Banach spaces with special emphasis on applications to Fredholm theory for integral operators. An original new method is developed which allows one to extend the well known results for finite-rank operators to determinants of the type $\det(I + T)$ where T belongs to an embedded subalgebra of the algebra of bounded operators on a Banach space with a stronger norm than the operator norm and satisfying a certain approximation condition. Many examples of such subalgebras are given. The authors also investigate the fascinating interplay between determinants and traces. The very interesting and extensive list of examples contains the classical Hilbert space theory for trace class operators with Lidski's trace theorem and the

Grothendieck's theory of nuclear operators on Banach spaces. The classical form of Fredholm determinants occurring in the expansion of $\det(I+T)$ for integral operators with continuous kernels is included. Regularized determinants, which are non-multiplicative, are also studied and Plemelj-Smithies formulas are derived. New results on determinants and inversion formulas for integral operators with semiseparable kernels (occurring as Green's functions) are included as well.

4. R.L. Ellis, I Gohberg, *Orthogonal systems and convolution operators*, Birkhäuser Verlag, Basel, 2003; 236 pp (item 23 in the list of books).

One of the central features of this book is the interplay between orthogonal polynomials and their generalizations (e.g., continuous analogues) on the one hand, and operator theory, especially the theory of Toeplitz matrices and operators (Fredholm and Wiener-Hopf operators), on the other hand. The far-reaching generalizations of M.G. Krein (and of M.G. Krein and H. Langer) of the classical Szegö theory are presented in full detail. Many recent developments are included (e.g., a continuous infinite analogue of Krein's theorem). A unifying theme is the general problem of orthogonalization with invertible squares in modules over C*-algebras. The book is very nicely written: starting from the most simple examples, an abstract framework is presented that covers many new examples.

5. H. Bart, I. Gohberg, M.A. Kaashoek, A.C.M. Ran, *Factorization of Matrix and Operator Functions. The State Space Method*, Birkhäuser Verlag, Basel 2007, 407 pp (item 25 in the list of books).

The present book deals with various types of factorization problems for matrix and operator functions. The problems originate from, or are motivated by, the theory of non-selfadjoint operators, the theory of matrix polynomials, mathematical systems and control theory, the theory of Riccati equations, inversion of convolution operators, theory of job scheduling in operations research. The book presents a geometric principle of factorization which has its origins in the state space theory of linear input-output systems and in the theory of characteristic operator functions. This principle allows one to deal with various factorizations from one point of view. Covered are canonical factorization, minimal and non-minimal factorizations, pseudo-canonical factorization, degree one factorizations and others. Considerable attention is given to the matter of stability of factorization which in terms of the state space method involves stability of invariant subspaces. A large part of the book is devoted to the theory of factorization into degree one factors and its connection to the combinatorial problem of job scheduling in operations research. This part is completely final dimensional and can be considered as a new advanced chapter of Linear Algebra and its Applications.

Book 5 is the second devoted to the state space factorization theory (a theme that has been dominant in Gohberg's work during the last 35 years). The first appeared in 1979 as the monograph by H. Bart, I. Gohberg and M.A. Kaashoek, *Minimal factorization of matrix and operator functions*, Birkhäuser Verlag, 1976, the first volume in the book series "Operator Theory: Advances and Applications" founded by Gohberg. This second book contains a substantial selection from the first book, in a reorganized and updated form, taking into account a number of new results in state space factorization theory and its applications that have emerged in the period of 30 years after publication of the first book.

Concerning Gohberg's research activities in the last ten years that are not yet presented in monographs, but only in research papers we mention three mayor directions:

- His series of papers (with D. Alpay and with M.A. Kaashoek and A. Sachnovich) on canonical differential systems within the framework of the state space approach. This work is very original and yields nice concrete formulas for solution of the inverse spectral problem, inverse scattering problems and has applications to non-linear PDE's and completion problems.

- His work (with J. Eidelman, V. Olshevsky, I. Koltracht, T. Bella, L. Gemignani) on fast inversion algorithms for diverse classes of structured matrices, which is highly appreciated in the numerical linear algebra community. See, for instance, A bibliography on semiseparable matrices, by R. Vandebril, M. van Barel, G. Golub and N. Mastronardi, in CALCOLO 42, 249-270 (2005).

- His series of papers (with M.A. Kaashoek and L. Lerer) on the common spectrum problems for a class of entire matrix functions of exponential type which eventually lead to a complete solution of the inverse problem for continuous matrix-valued orthogonal Krein's functions which has been open for more than 20 years.

During this period Gohberg has also co-authored two (semi-)text books, one on basic operator theory and one on operators acting on spaces with indefinite inner product. In this period SIAM made a second printing of the book "Invariant Subspaces of Matrices with Applications" (item 13 in the list of books) that originally was published by John Wiley & Sons, in 1986. We conclude this review with a brief description of the two (semi-)textbooks.

I. Gohberg, S. Goldberg, M.A. Kaashoek, *Basic Classes of Linear Operators*. Birkhäuser Verlag, Basel, 2003; pp 423 (item 23 in the list of books).

"Basic Classes of Linear Operators" provides an introduction to functional analysis with an emphasis on the theory of linear operators and its application to differential equations, integral equations, infinite systems of linear equations, approximation theory, and numerical analysis. It contains many sections which one will not find in any standard book. Also in some instances the book brings the reader to the front line of recent research. A part of the book is for undergraduate students, while another part is for graduate students and even for scientists who would like to see a bit more "concrete" functional analysis". The book is designed as a textbook for senior undergraduate and graduate students. It begins with the geometry of Hilbert spaces and proceeds to the theory of linear operators on these spaces.

Included is the spectral theory for compact self adjoins operators with a wide range of applications. Part of the book is devoted to Banach spaces and operators acting on these spaces. The theory is richly illustrated by many examples including Laurent, Toeplitz and Poincaré operators. Presented as a natural continuation of linear algebra, "Basic Classes of Linear Operators" provides a firm foundation in operator theory which is an essential part of mathematical training for students of mathematics, engineering, and other technical sciences. This book is an enriched new version of "Basic Operator Theory" by I. Gohberg and S. Goldberg (item 9 in the list of books). It also serves as an excellent introduction to the two volumes of "Classes of Linear Operators" by the same authors and M.A. Kaashoek (items 15 and 18 in the list of books).

I. Gohberg, Peter Lancaster, Leiba Rodman, *Indefinite linear algebra and applications*. Birkhäuser Verlag, Basel, 2005; pp 357 (item 24 in the list of books).

This book is dedicated to relatively recent results in linear algebra in spaces indefinite inner product generated by an invertible self-adjoint operator H. The basic definitions (of H-adjoint, H-self-adjoint, H-unitary, H-normal operators) carry over immediately from the definite case, but the properties of these operators often turn out to be intricate. A large part of the book deals with the study of canonical forms of pairs of matrices (A, H) of certain classes. The case when A is H-self-adjoint or H-unitary are classified completely and a perturbation theory is developed based on these results. The classification in the case of A-normal operators turns out to be, generally speaking, a "wild" problem. Only the case when A has one negative eigenvalue is settled completely. The book also includes applications to differential and difference equations with symmetries, matrix polynomials and Riccati equations. These applications have been developed in the last fifty years, and all of them are based on linear algebra in spaces with indefinite inner product. The latter forms a new more or less independent branch of linear algebra. This new subject in linear algebra is presented following the lines and principles of a standard linear algebra course. The book has the structure of a graduate text in which chapters on advanced linear algebra form the core. Interesting exercises are presented at the end of each chapter. This together with the many significant applications and accessible style makes it widely useful for engineers, scientists and mathematicians alike.

To conclude, it is a pleasure to mention that Gohberg continues to be actively involved in research, in writing papers and books. We know about at least two books that are close to publication: *Holomorphic operator functions of a single variable. Local/global theory*, jointly with J. Leiterer, and *A state space approach to canonical factorization with applications*, jointly with H. Bart, M.A. Kaashoek and A.C.M. Ran.

The Nathan and Lily Silver Chair

From the left to the right: Nathan and Lily Silver, Israel Gohberg, Helen Friedman and Jack Friedman, President of the Canadian Friends of Tel Aviv University. The picture is taken in Tel-Aviv after the inauguration of the chair in 1981.

Editorial comments
In the Summer of 1980 during his visit to SUNY at Stony Brook, Israel Gohberg received a call from the Rector of Tel-Aviv University, Professor Saul Abarbanel, which passed to him excellent news: the Toronto based family of Nathan and Lily Silver decided to donate a chair in which he will be the incumbent.

To have a chair in Tel-Aviv University is not only very prestigious but it also gives many extra possibilities. It supports partially traveling of the head of the chair, it provides funds to organize invitations of colleagues, organization of seminars, colloquiums and conferences, to buy books and new computer facilities, for typing, telephone and others.

The same year 1980 Israel had the first meeting with Nathan and Lily, and the next year after the Toeplitz Centennial Conference at Tel-Aviv University the Nathan and Lily Silver Chair in Mathematical Analysis and Operator Theory was officially inaugurated. Israel Gohberg became the head of this chair.

The highly respected Canadian Silver Family was based in Toronto, and there was their family business. Three of their four children moved to Israel and served regularly in the Israeli Army. After retirement, Nathan and Lily moved also to Israel (Jerusalem) and their business was taken over by their youngest son Joel Silver. In the course of the years the Silvers and the Gohbergs became friendly, attending family celebrations parties and others.

The Silvers were kind, generous and warm people who never refused any help. Their presence always was very enjoyable. The moral and financial help of the Silvers was very important in the development of Operator Theory in Israel.

Now the Nathan and Lily Silver Family Foundation is run by the children. The parents passed away and the children continue the generous support which started by their parents. After Gohberg's retirement professor Alexander Olevskii was chosen as the new head of the Nathan and Lily Silver chair. It is great that the Nathan and Lily Silver Family Foundation also continues till today to support the work and research of Israel Gohberg.

Part III
Gohberg Miscellanea:
Celebrating the 60th Birthday

This part consists of the Gohberg Miscellanea, written on the occasion of his sixtieth birthday. This biographical text was composed by H. Dym, S. Goldberg, M.A. Kaashoek, and P. Lancaster from reminiscences, notes, letters and speeches prepared by Gohberg's former students, colleagues and friends.

Gohberg Miscellania

This biographical text is composed by the Editors[1] from reminiscences, notes, letters and speeches prepared by Gohberg's former students, colleagues and friends on the occasion of his sixtieth birthday.

Israel Gohberg was born on August 23, 1928 in the town of Tarutino, in the southern part of Bessarabia, then part of Rumania. His father, Cudic (Tsudic), was a printer and his mother, Clara, was a midwife.

Cudic Gohberg was arrested in the Summer of 1940, for reasons which were never made clear, and sent to a labor camp. He was never heard from again. In June 1941, when the Germans moved deep into the Soviet Union, Clara Gohberg fled to Frunze, the capital of Kirgizia in Central Asia, with Israel and his younger sister Fanny. In Frunze, Israel completed his elementary and high school education and then, in 1946, he entered the Pedagogical Institute with plans to switch to engineering as that was the only vocation that he was aware of which would use his already evident mathematical talents. This was a great disappointment to his mother who wanted him to study medicine. His mother's wish to have a doctor in the family was fulfilled by his sister who became a surgeon and, as an added bonus, introduced Israel to his future wife, Bella, who was one of her classmates in medical school.

The Early Years

S. Goldberg (College Park, Maryland, U.S.A.)

Israel Gohberg's life can best be summarized as one filled with outstanding accomplishments, despite the trials and tribulations which he and his family endured. We have seen in his "Mathematical Tales" the obstacles he had to overcome throughout his professional career. I shall now touch upon some of his experiences in his early youth; a detailed account would constitute a very interesting book.

Israel was born in Tarutino, Bessarabia, an August 23, 1928. His was a closely knit, devoted and loving family consisting of his parents and paternal grandparents.

Originally published in *The Gohberg Anniversary Collection*, Operator Theory: Advances and Applications **40** (1989), 58–80.
[1] Meant are the Editors of the *The Gohberg Anniversary Collection*, that is, H. Dym, S. Goldberg, M.A. Kaashoek, and P. Lancaster.

His sister Feia was born 5 years later. Israel's mother Clara worked as a nurse-midwife and his father owned a one-man print shop dealing with announcements and invitations. The Gohbergs were a typical Jewish family in Tarutino.

On Israel's twelfth birthday the Soviet secret police arrested his father without any formal charges. He was dragged away and tried by three officials (Troika) and sentenced to eight years at hard labor in a Gulag in Siberia. The so called trial took place without the presence of a defense witness or a defense lawyer. He was never seen by the family again. Clara was now the sole support of her two children. After pleading with various authorities and writing many letters, she learned years later that her husband perished in the labor camp.

In June 1941, the German army invaded the Soviet Union forcing the Gohbergs to flee for their lives. This was the start of the family's trek East from village to village seeking food and shelter which was to last throughout the war. The journey often took place in carts and open trains in bitter cold, accompanied by snow and rain. In order to survive, the Gohbergs worked as farm hands, with Israel also mending shoes to obtain additional food. During this time, Clara made sure that her children attended school and kept up with their home work.

Some of the schools which Israel attended did not offer an environment conducive to studies. Many of the students were much older than Israel and were waiting to be drafted into the Army. Nevertheless, Israel worked hard and did exceedingly well in his classes. In fact, he was later awarded scholarships. In applying for these grants, he was always concerned that the authorities would learn of his father's demise in a Gulag and would deny Israel any support.

After more than twenty years, Clara received a terse note that her husband was posthumously declared "free of guilt". No apology or compensation was offered to the family.

The experiences of the Gohberg's which, in small part, I've described above, were told to me by Clara. She was an exceptional person whose courage, devotion and judgment enabled the children to survive some terrible times. Her endeavors motivated her two children to excel in their professions – Feia became a prominent pediatric surgeon who received a "woman of the year" award from a major Israeli magazine.

Israel was awarded a Stalin Fellowship after his first year of study at the Frunze Pedagogical Institute. Two years later, in 1948, he was encouraged by one of his professors to transfer to Kishinev University. Israel's mother and sister joined him in Kishinev in 1954. By this time Israel was already working with M.G. Krein.

Reminiscences

M.G. Krein (Odessa, U.S.S.R.)

I have had a number of students I feel proud of. The list includes M.A. Krasnosel-ski, H.K. Langer, M.S. Livsic, M.A. Naimark and V.P. Potapov. In this brilliant company a conspicuous place is occupied by Israel Gohberg.

I first met Israel Gohberg in 1950. I was then on the faculty of the Odessa Marine Engineering Institute. I gave a seminar talk on (as far as I can remember) the inverse problem for the oscillating string with "beads" and noticed the face of an unfamiliar young man in the audience. After the talk he came up and asked me whether I could spare some time for him. This young man was Israel Gohberg. I invited him to come to my house the same evening. He came and showed me some unpublished manuscripts and asked me to suggest possible continuations. This work was done under the supervision of Docent Itskovich of Kishinev University. After perusing the manuscripts I concluded that they ought to be published. (It turned out that one had already been presented to "Doklady of the USSR Academy of Sciences" with the help of S.M. Nikolskii, who at that time was not yet an academician.) I suggested to Israel Gohberg that he study the paper of I.M. Gelfand, D.A. Raikov and G.E. Silov on normed rings which had appeared just before the war in "Uspehi". I had a reprint of that paper and asked I. Gohberg to look for it in one of my bookcases. I remember well that at this very moment, with his face turned to the bookcase and his back to me, he said that he would be happy to work under my supervision.

I cannot now remember exactly whether he was a graduate student or an undergraduate student at that time (however, it seems possible that he was never in "aspiranture", the regular graduate program). I do remember having sent a personal letter to the Education Minister of Moldavia about a work assignment for I. G. after his graduation. In particular, I wrote that his diploma work could serve as a base for a Ph.D. dissertation and requested that he be placed in an environment favorable for research. As a result of this letter, he was assigned to the Beltsy Pedagogical Institute, and later he moved to Kishinev, the capital of Moldavia.

Despite a huge teaching load, he managed to find opportunities (and in this, I must mention that he was helped by the administration) to travel to Odessa to see me. His research advanced successfully and fruitfully.

At one of our meetings (in 1955) I suggested that we write a joint paper for "Uspehi". This marked the beginning of a long and fruitful collaboration in a broad field of problems which I look back on with a sense of joy and satisfaction. Israel Gohberg's energy, his easy accessibility and cheerfulness, quickly made him a favorite of both the Odessa circle of mathematicians and of my whole family. His visits invariably brought joy and excitement.

In these days of I. G.'s birthday celebration I transmit to him my affection and respect and extend to him my best wishes for many years of vitality, health and creativity.

For want of a better place, Israel used to work on the dining room table in his home, filling reams of paper with calculations and mathematical symbols. His sister recalls how at the end of one particularly long day he picked up the accumulated stack of papers, handed them to his mother and said "you see all this, you can put it in the garbage". This caused his family concern. It was especially worrisome because it was difficult for them to understand what he and Krein were doing. "We ask ourselves questions and then try to answer them", was the best explanation that Israel could manage. "Then why don't you ask yourselves easier questions?" was his mother's immediate commonsense reply.

Enough hard questions were answered to insure Israel's rapid advancement on the Russian academic ladder. In relatively short order he was appointed Head of the Department of Functional Analysis in the Institute of Mathematics at the Moldavian Academy of Science. In addition to this full-time position, he also held a half-time teaching position at Kishinev University. Advanced students at the University often participated in seminars at the Academy. Leonid Lerer, then a third-year student at the University, attended a Seminar on s-numbers which was given by Israel Gohberg at the Moldavian Academy in 1962–63. Leonid recalls that he and the other students were completely magnetized by these lectures. They left feeling that they had been exposed to one of the most exciting chapters of modern mathematics; many of them later became his students.

Letter of A. Markus, N. Krupnik and I. Feldman (Kishinev, U.S.S.R.)

Dear Israel Cudicovich,

We have it from an absolutely reliable source that on the 23rd of August 1988 you will be 60 years old. You were nearly half as young when in Kishinev a small group of mathematicians started to work under your leadership. The fifteen years during which we worked together in Kishinev were very fruitful years, and it is with mixed feelings of joy and sadness that we, your first students, recall those years. Your boundless energy, your cheerfulness and your limitless interest in mathematics, were to some extent passed on to us. During the following fourteen years you were far from us, but this did not prevent you from obtaining an entire series of first class results, to publish a number of papers and monographs, become the laureate of the Rothschild Prize, and to travel extensively all over the world (unfortunately we cannot say the same for ourselves).

Continuing in the arithmetic line of this letter, and basing ourselves on Jewish tradition, we wish you a healthy, happy, creative, active life till 120. Hearty greetings to Bella Jakovlevna, your children and grandchildren. Please convey our wishes for much success to the participants of the conference, which by good luck coincided with your anniversary.

Affectionately,

Kishinev, 25 July, 1988   A. Markus, N. Krupnik and I. Feldman

Speech of R. Duduchava (Tbilisi, U.S.S.R.), delivered at the banquet

I am authorized to transfer to you greetings from Professor I. Simonenko from Rostov. He wrote me that all the mathematicians in Rostov like you as much as before and they think that you are just on a long business trip abroad.

Ten days ago I was in Kishinev and met there your former collaborators, Naum Krupnik, Israel Feldman and Alexander Markus. This meeting was not occasional of course. I must underline that the friendly atmosphere, which you leave among your pupils and collaborators in the Sovjet Union, is preserved and we have very tight and friendly relations. They (Krupnik, Feldman and Markus) also asked me to transfer to you their best wishes on the occasion of your 60th birthday. They regret very much that they cannot express their feelings personally.

In conclusion I would like to say several words in my own name as well.

I am very grateful to you as a person and as a mathematician. When I came to Kishinev 20 years ago I was an inexperienced young man. What I know now and what I am now, is mostly due to you. I will never forget those two and a half years in Kishinev, which were among the best in my life. You and your nice family supported me, even when I had difficulties in my private life.

I wish you a long and happy life so that you can still do a lot of kind and necessary work in this world.

My Years in Kishinev

G. Heinig (Karl-Marx-Stadt, G.D.R.)

The years from 1971 to 1974 I spent at Kishinev State University for Ph. D. studies. These years were very important for me, from the view-point of my scientific career they were surely the most important in my life. In Kishinev I found a fertile mathematical climate, excellent teachers and good friends. Most of what I know how to do in mathematics I learned there.

The man who taught me this subject was Professor Gohberg. He embodied for me and the other students all the qualities that a mathematician should possess. We appreciated him for his superlative mathematical research activity and for his outstanding results but also for his abilities as an expositor and teacher of modern mathematics. Still more striking was his permanent care for the personal matters of his students. On holidays he used to invite his students to his house. There they got an unforgettable impression of the excellent cuisine of the Gohberg women. As I got to know, Professor Gohberg is not only an outstanding adviser but also an ideal co-worker. "Mathematics should be done like football" – these are his words. The enormous number of co-authors of his papers shows that he is an excellent player and game organizer.

There is no doubt that at the time I stayed in Kishinev the Gohberg team was one of the most active research groups in the world in die Fields of operator theory and integral equations. Many issues of the Kishinev journal "Mathematicheskie Issledovanija" of that time are still important now. Active members of the group were

A. Markus, N. Krupnik, I. Feldman, A. Semencul and others. The seminar of die group became a model for me how to run a real research seminar. Many outstanding mathematicians such as V. Macaev, B. Mityagin, H. Langer, S. Prössdorf, R. Duduchava gave talks in the seminar regularly. Especially I remember with pleasure the sometimes controversial but always productive discussions between Professor Gohberg and A. Markus.

What I also learned in Kishinev is that serious mathematics does not exclude humor. Professor Gohberg was a master in joke telling. I wish that he will retain this humor for many years. I wish him further successful research, many good students and, last but not least, good health.

In 1964 Seymour Goldberg visited Kishinev. He came to discuss the book that he was writing an unbounded operators. He spent a few days discussing mathematics with Israel with the aid of an interpreter and a little Yiddish. This was Israel's first meeting with an American mathematician. Other contacts with Western mathematicians were made in 1970 when Israel was permitted to participate in the Tihany Conference in Hungary which was organized by B. Sz.-Nagy. There he met L.A. Coburn, J.D. Pincus, M.A. Kaashoek, J.W. Helton, P.R. Halmos, F.F. Bonsall, Chandler Davis and P. Masani for the first time, and renewed his acquaintanceship with Ciprian Foias and Heinz Langer.

First Meeting

S. Goldberg (College Park, Maryland, U.S.A.)

I first met the Gohbergs in Kishinev in April, 1964. The family then consisted of Clara, Israel, his wife Bella, and their daughters Zvia and Yanina. The purpose of my visit was to discuss with Israel the manuscript of my book "Unbounded Linear Operators". I knew of Israel's seminal work in operator theory and was sure he would offer valuable suggestions.

Israel greeted me with a "bear hug" and invited me to have dinner with his family at their apartment. I was accompanied by Israel's colleagues – I.A. Felduran, A.S. Markus, – and an interpreter who was also an mathematician. We had a wonderful dinner which was prepared by Bella and Clara. When introduced to Clara, I had the impression that she was a very frightened woman. I tried to converse with her in Yiddish, which she pretended not to understand. It was obvious to me that she knew the language since Israel spoke it fluently. Years later, Clara told me (in Yiddish) that she was very worried that, by entertaining an American, the family would suffer dire consequences. This is certainly understandable in view of the tragic loss of her husband during the Stalin reign of terror.

The next day, a reception was held in my honor at Kishinev University to enable me to meet administrators and various faculty members. I was informed that I was the first mathematician from the West to visit Kishinev on an official basis. They all were very friendly and wanted to know about life in America; in particular, how university faculty fared. When I informed them that my teaching load was six hours, they were surprised that I meant six hours per week, not per day. This was substantially less than their teaching loads. Many eyebrows were raised when I announced that my trip to various cities in the Soviet Union was sponsored by the U.S. Air Force. Later I learned that the authorities wanted to know whether I was really a mathematician visiting for professional reasons or an agent on some secret mission. Recall that this was about the time that a Yale professor, visiting the U.S.S.R, was accused of being a spy.

After several more days in Kishinev, I bid a sad farewell to my new friends. We did not believe that we would ever meet again. Who would have thought that ten years later the Gohbergs would join us for dinner in our home in Maryland. To this day, Israel shakes his head in disbelief.

What amazes me is that in spite of the hardships encountered by Israel, he remains a warm, kind and gregarious human being. He is an inspiration and a delight to all those who know him.

The Moldavian Academy was a forty-minute walk from Kishinev University. Leonid Lerer recalls that he often walked with Israel from the University to the Academy and that during these walks many problems, mathematical and other, were thrashed out. On one of these walks Leonid revealed that he was thinking of applying for permission to immigrate to Israel. Gohberg encouraged him in this even though it could well have been a cause of future embarrassment for him. Leonid applied and received permission some four months later. Israel followed suit in November 1973. However, his application was rejected and the Gohbergs went through a very difficult period. The combined pressure applied by many foreign friends and colleagues, who were familiar with and sympathetic to the case, finally led to reversal. In June 1974 Israel received permission to leave. Late in July the Gohbergs left Kishinev by train for Vienna enroute to Israel. Household furnishings and "approved books" were sent ahead. Unfortunately many reprints and unpublished manuscripts had to be left behind; only a small portion of this material arrived later in Israel with the help of the Dutch embassy, which was and still is representing Israel in the U.S.S.R.

From Two Sides

L. Lerer (Haifa, Israel)

My initial acquaintance with Israel Gohberg came during my student years at Kishinev. Thanks to his lucid and captivating lectures, seminars and informal discussions, I was introduced to a wide range of exciting mathematical theories and problems. Since then Gohberg has been my "guiding star", and his judgement and support have played a very important role for me.

Gohberg arrived in Israel full of ideal for far-reaching projects. Despite a long exhausting journey of several days, while still an the way from Ben-Gurion Airport to an absorption center in Tel-Aviv, he told me about some of his plans. Very soon we started to work together.

Our first joint works gave me a unique opportunity to learn at close range from a great mathematician. One cannot help being impressed by Gohberg's unerring instinct for what is important and potentially fruitful, and by his striking talent for analyzing a complicated problem via a transparent and motivating special case. The high standards he applies to himself and to his co-workers, and his diciplined work habits taken together with his unimpeachable honesty, kindness and a very special sense of humor, create a wonderful school for every young mathematician working with Gohberg.

I had the privilege of going through this school at a very important stage of my development. At that time we often worked at Gohberg's apartment where a warm friendly atmosphere reigned thanks to his wife Bella Jakovlevna, his late mother "babushka" Clara, and his lovely daughters Zvia and Janina.

The above remarks, as incomplete as they are, indicate that working with Gohberg is an enjoyable and exciting experience. Fortunately, our collaboration continues and many projects are in progress. From the bottom of my heart I wish Israel Gohberg many healthy years of enjoying his great love – Mathematics.

Israel the mathematician received offers from most of the universities in Israel the state. He chose Tel-Aviv University because he felt that he could play a useful role in shaping the growth of its Mathematics Department, which was then still young and unformed.

Israel Gohberg at Tel-Aviv University

D. Amir (Tel-Aviv, Israel)

Those who know Izia Gohberg and are familiar with his wonderful sense of humor will certainly understand why it seems natural to me to begin my Gohberg recollections with a Jewish joke.

A young man, not very rich and not too clever, comes to a matchmaker to ask for a bride. The matchmaker says: "I have got something very special for you: She is from a very good family, rich, young and beautiful too." The man says: "This seems perfect to me; but, how come such a wonderful girl is still unmarried – doesn't she have any faults?" The matchmaker answers: "To tell you the truth – she has one small problem: Once a year she becomes insane for one day." The young man reflects for a while and says: "One day a year is not that bad. I am willing to marry her. In fact, I would like to marry her as soon as possible. Why don't we go there at once?." The matchmaker calms him down: "You can't do it now ; we have to wait till the day she becomes insane and then she may consent to marry you..."

When we received Gohberg's C.V. in 1974, just after he applied to emigrate from the USSR, naturally we became very enthusiastic. It was obvious that here was a distinguished mathematician, highly esteemed and holding a high position in the USSR in spite of his Jewishness. He was very rich, in publications of course. I believe his publication list contained 137 works at that time, including several books. He was known to be an outstanding teacher, with 25 Ph.D. students to his credit. Moreover, he was even willing to marry us, i.e., to join Tel-Aviv University. But there was still some tiny problem. Mother Russia did not approve of the marriage. His application was refused with the excuse that, if allowed to emigrate, "he would take out with him the ideas of his colleagues..."

Nevertheless, in June 1974, The Mother's consent finally being acquired, we received a telegram from Kishinev telling us the good news. On July 28th he arrived in Vienna and on the next day – he was finally in Israel. Two days later he was already studying in an Ulpan (an intensive Hebrew course for newcomers) and shortly thereafter he accepted a position as a Professor of Mathematics at Tel-Aviv.University. This, besides showing how eager we were to absorb him, shows also how fast Izia adapts to new conditions, like a cat always manages to fall on its feet... In fact, he adapted so fast that in the same fall he already went, for the first time in his life, to visit the real West, i.e., the U.S.

I was serving then as the Chairman of the Department of Pure Mathematics. I still remember how he came to my room and handed me an invitation which he had received from Stony Brook and an application to go there for a short visit. I looked at the papers and said: "That's O.K. You may go." He remained seated, waiting for something else to be said by me – some "only" or "but..." He just could not believe that this was all the procedure needed when you want to go abroad... Anyhow, as you all know very well, he learnt this, too, quite fast.

There's this joke about the guy who, when told that his wife is being unfaithful to him, and that with more than one man, answers as follows: "My late father, may he rest in peace, was a successful merchant. Before he died he said to me: My son, this is the advice I can give you: It is much better to share a good business with several people, then to be the single owner of a bad one..." Thus, we had to share Izia with the Weizmann Institute, Stony Brook, College Park, Calgary, and Amsterdam... Still Gohberg is a wonderful business investment, bearing Tel-Aviv University's name, and making his home here in Israel.

And what a home! Most of us have experienced, more than once, the proverbial "Russian hospitality" in that warm home. His charming wife, Bella, the Doctor, helped by their daughters, and sometimes even by his sister, made evenings at the Gohberg's unforgettable experiences. In the first years there was "Babushka" too, Izia's wonderful mother, who kept an eye on each of us to make sure that we are all well fed.

Gohberg turned out, of course, to be as popular in Tel-Aviv, as he had been in Kishinev, and as he is everywhere. As a teacher he is strongly beseiged by the students. When I taught the introductory course in Hilbert Spaces in the second semester, while he gave the same course in the first semester, the distribution of students was about 5:1 in his favour, and others didn't score better... He holds a local record in the number of graduate students, and many of them can be seen queuing in the corridor to his office. As everybody knows, he is very diligent, and is an ideal partner for research and publication. His current list of publications holds about 300 papers, among them a dozen or more books, and he is still only

60 years young! On top of that, one has to add his other activities: Founding
and editing the journal "Integral Equations and Operator Theory" and the book
series "Operator Theory: Advances and Applications", running the Israel seminar
on Operator Theory, and initiating and organizing the biannual Otto Toeplitz
Memorial Lectures etc.

Some of this unbelievably intensive activity should probably be attributed to the
very special Gohberg personality, some of the ingredients of which are a basic
optimism and a rare sense of humor, which did not let him down during the worst
political or physical crises and enabled him to recover and bloom again each time.
Thus it seems appropriate to conclude with a classical traditional Jewish story.

A traveller walked in the desert. After ten days in which he met no city, no inn,
no tree, no water and no living soul, he found a tree standing on a sweet water
spring. He sat down, cooled himself under the thick shadow of the beautiful tree,
ate from the tree's tasty fruit, drank the water and rested. When he had to leave
he said: "Oh tree, what blessing can I wish you? You are sitting on water, you are
beautiful, your shadow is thick and your fruit is plentiful and delicious! The only
blessing I can add to all that is – LET THERE BE MANY MORE LIKE YOU
WITH US!

Letter of B. Trachtenbrot (Tel-Aviv, Israel)

Dear Israel,

I hereby join your many friends, colleagues and students in greeting you on the
occasion of your anniversary.

As early as the fifties you had made a name for yourself as one of the most brilliant
mathematicians of your generation. Even though my area of expertise is far from
your field of research, I always followed with great interest the development of your
scientific career. There were personal reasons also for my interest and admiration.
After all, we are what in Russian is called "Zemliaki" (fellow countrymen), and I
was very impressed and moved by your contribution towards scientific development
in the Bessarabia region, in which we were both of us born, and from which I
departed many years ago.

You were the first of the senior academy members who reached and realized the
difficult and courageous decision to make aliya to Israel. In this way you encouraged
me as well as others to work towards aliya.

You succeeded in raising in the old-new homeland a young generation of students
as well as continuing in your excellent research. You achieved this thanks to your
superlative talents and to your never-ending conscientious work.

Best wishes to you, your family, your students and your admirers.

Much success in your life's endeavor.

Sincerely,

Boris Trachtenbrot and wife Bertha

*Shortly after his arrival in Israel, Ronald Douglas and Joel Pincus invited him
to visit Stony Brook for the Fall semester. Israel requested permission from his*

Department Chairman to accept the Invitation. Permission was promptly granted. Israel, unaccustomed as he was to the freedom of the West, was sure that the Department Chairman had not understood the question properly. He therefore repeated it twice more, each time receiving assurances that it was okay to go, before he was ready to believe that he could really go. In mid October 1974 Israel made his first trip to the United States.

At Stony Brook arrangements had been made for room and board with David and Barbara Ebin, an orthodox Jewish family, because it was believed that M.G. Krein was orthodox, and therefore that Israel must be so too. This was based on the observation that Krein kept his head covered and was very careful about what he ate. These conditions, although necessary, where not sufficient. Nevertheless the boarding arrangements were very satisfactory to all parties concerned, and exposed Israel to customs which he had been isolated from in the Soviet Union and brought him new friends. Stony Brook offered him a full time position , but Israel decided to stay permanently in Tel-Aviv.

In January and February of 1975, Israel was invited to the University of Maryland at College Park, by Seymour Goldberg. At Maryland he met Rien Kaashoek, who was visiting David Lay. They showed him their work. There were some strong differences of opinion. Prompted by Rien, Israel presented a couple of informal lectures on promising directions for future research and on open problems. One of these, the problem of linearizing analytic Operator functions, caught Rien and David's fancy and led to a program for joint research. A paper of Gohberg, Kaashoek and Lay was completed in the Spring of 1976. More importantly, however, this chance meeting marked the beginning of the Dutch connection which was to blossom and expand in subsequent years.

The Dutch Connection

M.A. Kaashoek (Amsterdam, The Netherlands)

Israel Gohberg's first visit to Amsterdam took place in December 1975. We had met before in 1970 at the Hben Space Operators and Operator Algebras conference in Tihany and, much more intensively, at College Park, Maryland in the beginning of 1975. The latter meeting, although unplanned, turned out to be most important. At that time Israel was in College Park for two months, as a guest of Seymour Goldberg, and lectured on various topics from his work with M.G. Krein, N. Krupnik, and J. Leiterer. I was in College Park for the Spring semester and planned to work with David Lay on problems related to papers of Gohberg-Sigal, not knowing that Israel would be in town. Israel was quite critical about a paper Harm Bart, David Lay and I had written, and for me our first meetings in College Park can be best described in terms which are sometimes used in the Dutch weather forecast: heavy Storms and critical conditions at the dikes. But the winds were wann and the atmosphere was friendly. Before Israel left College Park at the end of February 1975 the first main results for a Gohberg-Kaashoek-Lay paper on linearization and factorization existed, there was a set of notes (never published, but still worth doing so) of two stimulating lectures of Israel about open problems on operators and operator-valued functions, there were plans for further collaboration, and I had the permission of my chairman to invite Israel to Amsterdam for the Autumn semester of 1976. Israel's visit to Amsterdam in December 1975 was just the first in a long chain.

In the Autumn semester of 1976 Israel developed a great deal of activities in Amsterdam. He gave an advanced course on singular integral equations and Toeplitz operators and ran a wonderful seminar on matrix polynomials, including the fresh and new Gohberg-Lancaster-Rodman results. Israel, Harm Bart and I were planning to work on related problems for rational matrix functions using the approach of the theory of characteristic operator functions. Harm and I were introduced to the latter subject in one long afternoon session, where Israel explained to us the main ideas of the theory by browsing through the Brodskii book. After the session both of us felt that we had known the subject all our lives. Such experiences we had many times. That semester, in lectures, seminars and informal discussions, Israel introduced the operator theory group in Amsterdam to the beautiful way of mathematical thinking in (what I shall call for simplicity) the Gohberg-Krein school, and we learned much about the great mathematical achievements in Odessa and Kishinev. We also experienced the stimulating and joyful working atmosphere that Israel knows how to create, which involves many hours of intensive mathematical discussions (larded with relaxing stories and jokes) and also non-mathematical affairs as mushroom dinners, bicycle rides and family visits. In short it was a great period.

At the end of the 1976 Autumn semester it was decided that each year Israel would come to Amsterdam for two or three short periods, each of a couple of weeks. The contract would be for three years, but for each short period a formal appointment was needed and each time all forms would have to be filled in. In the Dutch context, with its well-developed tax and social security system, the number of forms is not negligible. Hence each period started with a couple of hours of filling in forms and answering questions which are not meant for persons who were born in Rumania, emigrated from the U.S.S.R to Israel and came to Amsterdam to do mathematics. Soon Israel's file at the personnel department of the Vrije Universiteit exceeded the place reserved for it and he knew more about the Dutch working conditions than most of us. In 1983 the three year contract was replaced by an appointment as an extra-ordinary professor. We appreciate the appointment and miss the forms.

By now Israel is well established in Amsterdam, and the results of his visits are clearly visible. During the past years many new ideas emerged, new students were attracted and joined the group, new connections were made, with the mathematical system theory people and the electrical engineers, new books appeared, dissertations were written and many papers published.

Israel's activities in Amsterdam are not restricted to the Vrije Universiteit. He has developed contacts all over the country. Also he has brought many mathematicians to Amsterdam. Several have become good friends and co-workers. His many mathematical achievements are well recognized and highly appreciated in the Netherlands. In 1985 her majesty the Queen appointed him a foreign member of the Royal Dutch Academy, and, as one may guess from the size of the country, it is more difficult to become a foreign member than an ordinary one. I wish that in Amsterdam we may enjoy his visits for many years to come.

Speech of L. Frank (Nijmegen, The Netherlands), delivered during the conference

The Gohberg-Krein theory of convolution operators an the half-line is one of those fundamental contributions to Operator Theory which is extensively used in many different Fields of Pure and Applied Mathematics.

Coercive Singular Perturbations is one illustration of the important role that this dass of operators plays in Singular Perturbation Theory. Therein, in order to reduce a coercive singular perturbation to a regular one in a constructive way, one needs the symbolic calculus with a small parameter (modulo operators with small norms), whose version without parameter (modulo smoothing operators) was developed earlier by Boutet de Monvel. The core of Boutet de Monvel's calculus and its version with a small parameter is the Gohberg-Krein theory mentioned above. Thus, not only as an admirer, but also as a user of the fundamental results by I. Gohberg and M.G. Krein, I consider it a special privilege to have been invited to give a talk at the conference honoring the 60th birthday of Israel Gohberg.

I first saw Professor Gohberg at the Moscow State University at one of the sessions of Gelfand's much celebrated Monday seminar. At that time, Israel Gohberg was not yet a corresponding member of the Moldavian Academy of Sciences. I met Israel Gohberg for the second time in 1974 at the Hebrew University of Jerusalem after his alya (immigration) to Israel. At that time he was no longer on the list of the members of the Moldavian Academy of Sciences. The last time I saw Israel Gohberg (before coming to the Conference honoring his 60th birthday) was in Nijmegen, at my home University, where he gave a talk on determinants and traces for some classes of linear operators. At this time he was already a foreign member of the Dutch Royal Academy of Sciences. Quite a way to go along and further up to the International Conference honoring the 60th birthday of Israel Gohberg.

Albert Einstein said once: "If my Relativity Theory is true, Germany will claim that I am German and France will say that I am a Citizen of the World. However, if my Relativity Theory turns out to be wrong, then France will claim that I am German and Germany will say that I am Jewish."

It is my belief that in the case of Israel Gohberg each country where he feels at home, is happy to claim Israel Gohberg's belonging to its scientific elite, even if he happens to be only a part-time Dutchman, Canadian and American, being a full-time Israeli.

From Maryland, Israel also made short excursions to Toronto and Calgary at the invitation of Chandler Davis and Peter Lancaster, respectively. The visit to Calgary was to prove particularly fruitful.

Gohberg in Canada

P. Lancaster (Calgary, Alberta, Canada)

I was surprised and pleased to receive a letter from M.G. Krein in 1968 requesting a copy of the monograph on "Lambda-Matrices and Vibrating Systems" that grew out of my doctoral thesis. It showed that my work had attracted some attention in the U.S.S.R.. Subsequently, I got to know the Gohberg-Krein book on non-selfadjoint operators, as well as several important papers that had appeared in translation by Gohberg and Krein, Gohberg and Markus, and Gohberg and Sigal. So I was very glad to hear from Chandler Davis in 1974 that Israel Gohberg had emigrated, and would be able to visit North America in early 1975.

As a result, he first came to Calgary in February of 1975. This was not a long visit, but enough to begin to identify problem areas where we might work together. But that was not our only common ground. It was a cold month with plenty of snow, and we still remember his delight in our Cottage in the mountains and in the clear winter weather. Since then he has enjoyed our wilderness in most seasons of the year and, along the way, we have joined in some most enjoyable mathematics.

We met again in the following year while I was on sabbatical leave in Scotland. After this visit and my later visit to Tel Aviv, the shape of our first researches on matrix polynomials began to emerge. I probably met Israel's Ph.D. Student, Leiba Rodman, at about that time who, under Israel's guidance, was already producing very interesting results on analytic matrix functions. So began a very productive and enjoyable three-way collaboration. Since those early years we have been able to meet, for shorter or longer periods, in Tel-Aviv, or Calgary, or elsewhere, on an annual basis.

Since we first met I have been fascinated by Israel's powers of gentle persuasion. His extraordinary talent in stimulating mathematical investigations is very much in evidence in these volumes. But he never needs to explicitly persuade anyone to join with him on a mathematical adventure. One needs only to show enthusiasm for a problem that he considers interesting and worth-while, and there is a good chance that something interesting and worth-while will actually be produced; after some intense effort and several revisions, to be sure. He seems always to be able to bring out the best mathematical talents in those around him, whether they are beginners, engineers, pure or applied mathematicians.

During these years of joint activity warm, and very rewarding relationships have evolved between our three families. It is a privilege, not only to have worked with a genius of our time, and with Leiba Rodman, but also to have shared many happy times, and a few sad times, with their families.

I am very glad to have had the opportunity to help in providing a sixtieth birthday present in the form of the Calgary conference and these volumes. In these pages, and at Calgary in August of 1988, we have been fortunate to bring together Israel's immediate family, and many members of his extended family of collaborators and admirers representing his career both in the U.S.S.R. and in the West.

At the end of February 1975, upon returning to Tel-Aviv, Israel found Leiba Rodman on his doorstep. Leiba wanted Israel to be his Ph.D. advisor. Israel agreed and suggested Matrix Polynomials as a research topic. By the Spring of 1976 they had their first results. This, as Leiba writes, was the easy part. Writing it in acceptable form took longer – "he would make me rewrite drafts many times. Although it was a privilege to work with a great master in creating mathematics, it was often frustrating due to his high demands and expectations. There was a point in my Ph.D. work (after being told to rewrite a draft for the eighth time) when I seriously contemplated quitting. Why I did not quit, I do not know... I do know that doing a Ph.D. with Gohberg means going through an outstanding school of mathematics." A paper of Peter Lancaster was found to be relevant to Leiba's and Israel's work on matrix polynomials. Peter who was then on sabbatical in Scotland, came to visit Tel-Aviv in the Spring of 1976. This led to the first Gohberg-Lancaster-Rodman publication and marked the beginning of another successful partnership.

In the Spring of 1975 Israel also started to lecture in Tel-Aviv University using the English language which he had picked up during his visits to the United States and Canada, lecturing in Hebrew later. The same semester he also accepted a half-time position at the Weizmann Institute of Science. This led in due course to a long and fruitful collaboration with Harry Dym.

Israel at the Weizmann Institute

H. Dym (Rehovot, Israel)

I first met Israel Gohberg in the Spring of 1975 when he joined the Weizmann Institute as a half-time faculty member. That Spring he delivered a series of lectures on Topics in Operator Theory. They were outstanding. He had a gift for focusing on the essence, of presenting just enough to give the flavor and the main results without overwhelming the listener with technical details. It all seemed so clear, so natural and so elegant. It was only later, upon reviewing the lecture that one realized how deep many of the results were.

These virtuoso performances were to be repeated many times in the years to come on a number of different topics ranging from highlights of his books with Krein and with Feldman to the newly developing theory of matrix polynomials.

In the early days at Weizmann our contacts were limited. I listened to his lectures, and asked him an occasional question connected with some problems I was working on at the time, which, as it happened, were based on an Operator theoretic interpretation of Szegö's formula. Israel used to come to the Institute twice a week. On one of these days he would lecture. He spent most of the rest of the time talking with his newly acquired student, Sonia Levin (daughter of the well-known refusenik Alexander Lerner). Still this was not enough to keep a man of his vitality fully occupied. He used to sit in an office two doors down the hall from mine. His door was always open and as I passed by from time to time, I could not help feeling that it was a tremendous waste just to leave him to his own devices and, more than that, I sensed that he too would welcome some active intervention with the "natives".

In the Spring of 1976 I approached him and expressed an interest in working with him. He was receptive. We didn't get much done in the remaining few weeks of that semester, but contact had been initiated.

In the Fall semester, Israel returned from Amsterdam with a question on extensions which was to mark the real beginning of our collaboration. Our meetings were somewhat sporadic in the beginning. Israel had many invitations and I had a less exotic invitation to do a couple of months of military service. Nevertheless we had amassed some results before I left for a partial sabbatical in May of 1978. That Summer, while visiting Tom Kailath, I stumbled across Burg's Ph.D. thesis. If memory serves me correctly, I just happened upon it in the xerox room. This was to prove to be a wonderful find. The search for analogues of Burg's maximum entropy principle in the context of a variety of different extension problems was to turn out to be a major influence on much of our subsequent work. We continued to collect results throughout the coming academic year.

Some weeks later, in the Summer of 1979, we met again in the Van Gogh museum in Amsterdam on a Sunday afternoon. It was one of those grey days which the Dutch excel in; rain just around the corner and never quite making it beyond a sprinkle here and there – perfect museum weather. I came in from Delft with the family and met Israel at the museum. Irene and the boys went to look at the paintings and Israel and I seated ourselves on a couple of comfortable armchairs to formulate strategy. A fat loose-leaf of notes was placed on a nearby table amidst a stack of art magazines and catalogues. They were picked up from time to time by passing browsers, but they were always replaced very quickly. It was a fruitful meeting which led to some nice refinements of our accumulated results and agreement on the general lines of time separate papers, all of which appeared not so long thereafter. Curiously enough, the one which we thought to be just a side issue in the general context of our developing machinery, seems to have attracted the most interest.

I learned a great deal working with Israel. It was equivalent to serving an apprenticeship with a master craftsman. Even more than that, it was a living connection with the great Russian school of Functional Analysis of Achiezer and Krein. Israel was an ideal teacher and colleague. He seemed to remember everything that he had ever worked on, where to find it, and how to explain it. Moreover, he was always patient, always cheerful, and always, always optimistic. When Israel first met my youngest son Michael, he asked him "How many children are there in your class?." "Forty-one" was the reply. "Wonderful" said Israel, "so many friends".

For many years it was Israel's habit to come to the Weizmann Institute early on Sunday mornings. One morning he came very late. It turned out that he had had a traffic accident enroute. Another car had hit his car in the rear. As accidents go, it was relatively minor, but still the expense and inconvenience was far from negligible. After settling in, he called his wife to tell her what happened. "But why did he do that to you" was her immediate wifely response. "Bellachka", said Israel, ever so gently, "this question you have to put to him, not to me".

Israel spent the Fall Semester of 1975 at Stony Brook, where he gave a graduate course on Fredholm operators and an undergraduate course on linear algebra. That semester he met Kelvin Clancey who was also visiting Stony Brook while on sabbatical leave from Athens, Georgia. This was the beginning of their joint work, which continued later with visits of Israel to Athens and produced a number of nice papers and a book on factorization of matrix functions and Singular integral operators.

By the Fall of 1976 a number of collaborations were in operation. Permutations of these, supplemented by students and colleagues of the original partners, generated many new collaborations. Moreover, the roster of partners kept growing as Gohberg shuttled between Israel, Amsterdam, the United States and Canada.

In 1978 the first issue of "Integral Equations and Operator Theory", a new journal under his editorship, appeared. In 1979 it was followed by the first volume in the accompanying series of books: "Operator Theory: Advances and Applications" (also under his editorship). Both of these ventures succeeded because of his optimism, enthusiasm and hard work. With characteristic forethought, Israel had enough collaborations running by this time to sustain the Journal over its initial critical period. In 1979 the first Toeplitz Lectures Series were set up. Peter Lax and Ciprian Foias, the latter then a recent refugee from Rumania, were the first guest lecturers. One of the other highlights of that event was a party in Israel's house catered by the Gohberg women: Mother, wife, sister and two daughters, whose combined energies and talents were truly awesome. His wife and sister took a few days off from their medical careers to cook and bake. It was a spectacular party that was surpassed only by the Toeplitz Memorial Conference Party which was held at his home two years later.

This Calgary conference is testimony to the success of all these activities. Hugo Woerdeman, perhaps the youngest participant, spoke for us all when he observed: "Professor Gohberg is still so young too! He could sit back and relax and Look back on a wonderful career as a mathematician. But he still does all this travelling and is still so active. I am glad he is doing all this and I hope that he continues because I learned a great deal from him and hope to learn a lot more".

Letter of B. Khvedelidze U.S.S.R.), read at the banquet

Dear Israel !

With my whole heart I would like to congratulate you on the occasion of your sixtieth birthday.

Thanks to the organizers for inviting me to the conference, dedicated to this event. I regret very much that the doctors at the last moment advised me to cancel this trip which I was so much looking forward to make.

Since I lost the opportunity to join your pupils, collaborators and colleagues during the celebration in Calgary, I decided to express in writing my admiration of you as a person and as a mathematician.

Your fundamental investigations in many topics of functional analysis certainly influenced their development and determined their current substance. These investigations brought you worldwide fame. I appreciate especially your result on integral equations with singular kernels, which completed the investigations in this field.

I must also express my gratitude for educating my student R. Duduchava, who became an experienced expert in integral equations under your guidance.

My colleagues and I remember well the nice meetings with you here in Tbilisi and in Kishinev. During these meetings you impressed me as an honest, kind and sensitive person. I pray to God that we may have more scientists like you.

Unfortunately we haven't seen each other for a long time and it's a pity for me to miss the opportunity to correct this situation.

Due to the last positive changes in our lives I hope I will be able to invite you very soon as a cherished guest of honor.

Sincerely yours,

Tbilisi, 15 August, 1988   Professor Boris Khvedlidze

Speech of M.M. Drzjabashian (Yerevan, U.S.S.R.), delivered at the banquet

Dear colleagues and friends,

We are gathered together here today to mark with honor the 60th birthday of one of the outstanding mathematicians of our time, Professor Israel Cudikovich Gohberg.

All of us heartily greet and congratulate you, dear celebrant, your wonderful family, your daughters and wife, on this festive day. Wherever you may be, and wherever your creative activity and your organizational ability develops, you serve the gratifying and noble task of the development of the science which all of us here love - mathematics – the science to which you devoted all your energy and talent.

I think it would not be out of place here to recall what is perhaps known to only a few of you, that from ancient times the people of Armenia had an interest in the science of mathematics. Even in the 11th century a then famous scholar and political figure in Armenia, Gregory Magister, occupied himself with mathematics. A visit to the Matenadaran, in Yerevan, will serve to convince oneself of this fact. This is where die ancient manuscripts of die Armenian people are kept, and where eleven pages of an old translation of die "Elements" of Euclid in the Armenian language was miraculously kept safe from the Mongol-Seljuq barbarians. Because of these barbarians, it was only after a very long period of time that Armenian mathematics started to develop once more, in fact only recently, during the last four or five decades.

The people of Armenia worked the land, and they were particularly successful at tending vineyards and producing wine. Exactly 60 years ago, in this sunny land, there was an exceptionally good grape harvest, and from this harvest of Armenian grapes it was decided to produce a very good cognac, which is now exactly 60 years old. Allow me on this day to present to you as a souvenir, a bottle of this Armenian cognac which is precisely 60 years old. Allow me also to convey my heartfelt wish to you and your family that you should open this 60 year old bottle of cognac in good health on the day of your one hundredth birthday, with your grandchildren and your greatgrandchildren, gathered around you.

The present notes form only a partial biography. Mathematics is hardly discussed, and many important events and meetings are not touched upon. These topics remain for a next version. Moreover, it is our hope that many new developments will take place and that new sections will have to be added later.

The Editors

Part IV
Celebrating the 70th Birthday

This part contains the texts of the speeches that were given at the conference dinner of the IWOTA meeting in Groningen, in the context of a pre-celebration of Israel Gohberg's 70th birthday later in 1998. The order of the texts corresponds to the way the speeches were presented during the dinner.

Speeches Given at the Conference Dinner in Celebration of Israel Gohberg's 70th Birthday

At the conference dinner a number of speeches were given, as a pre-celebration of Israel Gohberg's 70th birthday later that year. The texts here below have been prepared by Harm Bart, Harry Dym, Bill Helton, Heinz Langer, Alek Markus, Cora Sadosky, and Hugo Woerdeman. The order of the texts corresponds to the way the speeches were presented at the dinner.

1. To You, Izea!, by Alek Markus

I have known Israel Gohberg most of my life - since 1951. And for most of my life I have been proud of being his first disciple. I don't mean the best, but the first I was.

In 1959 we began to work together, for almost 15 years he was my boss and a very good boss, I have to say. Unfortunately I was not a very good subordinate, as I understand now. Whenever he asked me to do something, I immediately asked him if it was worth doing at all. He never lost patience with me and very soon I realized that he was almost always right and I was nearly always wrong. He realized that too, because he always told me: "You should never argue with me, because I am old and wise." He did seem old to me then - he was already 31 in 1959. But about 10 years before the time described, Israel was not quite so old and, I suspect, not quite so wise.

To illustrate the last point I would like to tell you a story. There were two characters in the story - one Israel Gohberg, the name of the other one will come up later. The events in the story are connected with elections to the Soviet Parliament. Those who lived in the former Soviet Union know that the Soviet Parliament was nothing like a parliament and the elections were nothing like elections. There was always one candidate for every vacant position and the result of the election was always 99.99% "for".

On the morning of some election day Israel came to his polling station and took a ballot paper with the name of the only candidate ... Joseph Stalin! In accordance with his previous firm decision Israel went into the booth, which itself

Originally published in *Operator Theory: Advances and Applications* **124** (2001), xiii–xxii.

was a very unusual and risky action, and crossed out the hated name. He then carefully folded the ballot paper and put the paper into the ballot box. Next morning he opened the newspaper. He was eager to know how many people were as brave as him. What he read was: "The result of the voting for Joseph Vissarionovich Stalin was one hundred percent." To make sure that the figures were not rounded up the newspaper said: "All the voters as one person voted comrade Stalin their deputy."

It seems to me it was the last time Israel acted naively. But he also acted bravely and that was not the last time in his life. Many years later Israel was one of the first to decide to leave the Soviet Union. He needed all his courage for this step. And this step proved to be very wise. To you, Izea! To your courage! To your wisdom! To your health!

2. Dear Israel, by Hugo Woerdeman

First of all I would like to wish you a happy 70th birthday. One thing that has always amazed me, and still does, is your endless energy. Travelling all year round to do mathematics in all corners of the world. As I said ten years ago: you seem so young!

At the time I was the youngest participant at the conference in Calgary honoring your 60th birthday, and while I am no longer the youngest participant you are still as young as ever. During your travels you inspire many mathematicians, including myself, to work on challenging problems. You build bridges between different projects and make connections with other disciplines, most notably engineering. I have profited from your energy and am still profiting from it. Under the guidance of both Rien and you I have been provided with an excellent foundation on which I am still building. Sure, you were demanding at times, though you would probably argue you were not demanding enough! You have taught me to pursue problems that are versatile and applicable. Further, I admire your true artistry in explaining hard problems and complicated notions in a way that makes them seem easy and transparent without compromising the mathematics. Israel, I am grateful for the education and inspiration you have given me and look forward to future interaction with you, as I know that you will remain active for a long time to come.

Thank you!

3. Dear Israel, by Heinz Langer

While Hugo Woerdeman, who spoke before me, is one of the youngest professors among your students here, I belong to those who have known you for a very long time. I do not remember if it was our first meeting, but in December 1961, you were at the age of 33 and I was just 26, you gave a lecture at Krein's seminar at the Odessa Civil Engineering Institute. I am still keeping my notes of this lecture.

It was about your abstract factorization theorem in rings. You considered two elements such that one is a one-sided inverse of the other, and then you proved the existence of a factorization for certain functions on the circle. After your lecture, as usually, Krein made some remarks and said "You made something out of nothing", meaning that you have built up a beautiful theory starting from only very few assumptions.

I think, in some sense, these words of Mark Grigor'evitsch are also true for your activities in general. Wherever you have worked, you started something new, you created a school and influenced people very much by your knowledge and enthusiasm, under whatsoever conditions. First in Kishinev, then in Tel Aviv and other places in Israel, in Amsterdam, Maryland, Calgary etc. You started to issue new journals, first one in Kishinev, and now the internationally leading journal "Integral Equations and Operator Theory", which is your journal, and you started a new series of monographs and books in which until now more than 100 volumes have appeared. Your activities and your mathematics, which evolved in almost 400 articles and about 20 books, had a tremendous influence on the development of Operator theory in the whole world. And last but not least you also initiated the IWOTA conferences, of which we enjoy right now already the 10th very much.

And almost at the age of 70, your enthusiasm hardly seems to diminish. For all this we are very grateful to you!

Let me also mention something more personal. Some people consider family like relations between mathematicians, the teacher is the father (or mother) and so on. So did Krein sometimes, and he proved that Euler was our grandgrand grandfather. From this point of view we are brothers. Of course, you are the older brother, or the big brother, whom I always really admired and who was always watching me. I do not mean this in the ironic or dark sense of Orwell. I mean that you always took great personal interest in me and in general in other people, in your colleagues and friends. In particular, and this is also one of my personal experiences with you, after you went to the West, you tried to support those who were left in the East in all possible ways, including really practical help. Maybe it is hard to imagine for somebody who never lived in the East what it meant to have such a trustworthy friend.

So, in conclusion, I would like to thank you very much for everything you did for me personally and for us as mathematicians and friends. I wish you, Bella and your family all the best! I hope you will be inspiring and watching us for many years to come!

4. Wishing Professor Gohberg a Happy 70th Birthday and Many More to Come!, by Cora Sadosky

We are together to honor our friend Israel Gohberg, who has taught us so much, not only through his mathematics but through his way of being. Those who spoke before already mentioned some of the special qualities of our honoree: great mathematical talent, a formidable capacity for collaboration, enduring generosity and loyalty towards friends, extraordinary organizational skills as well as an enormous energy to put them into practice. I want to add a few words on a most remarkable trait of Israel Gohberg: his inclusiveness.

The conjunction of all of Gohberg's talents have yielded many realizations. Among them, the writing of a collection of seminal books on Operator Theory, the editorial leadership of a main international journal and of an outstanding series of books, the gathering of groups around the world to work in interdisciplinary problems, and the organization of many international meetings, quite especially, the IWOTAs.

The IWOTAs, as well as his journal, Integral Equations and Operator Theory, bear Israel's trademark: inclusiveness. These meetings running every two years in conjunction with the MTNS meetings of engineers and mathematicians are a prime example of the Gohberg vision. A place to discuss mathematics with people coming from different strands of operator theory, in the broadest sense, from applied linear algebra, functional and harmonic analysis, complex function theory, differential equations, mathematical physics, and their applications to systems theory and H^1 control. At IWOTA we learned what was going on far beyond our sometimes initially narrow range of specialization, and we had the opportunity to interact with senior mathematicians as if they were our peers. Beyond that, Israel always succeeds in bringing together all sorts of people: not just from different mathematical backgrounds, different schools, and different levels of maturity, but from many countries, including those in the peripheries. And to do so he had to work hard, since enacting inclusiveness requires resources and cooperation. Thanks to Israel's vocation for inclusiveness, the IWOTA meetings are a true intellectual melting pot. He helps mathematicians from all over to become acquainted with the tumultuous, exhilarating world of Operator Theory as he sees it: inclusive. It was Israel who made this international community of mathematical interests possible, and this feat is the result of his clear and deep appreciation of the universe of mathematics, that is, as himself, open, unrestricted, non-sectarian, and multifaceted.

Thank you, Israel, and mazel tov on your young seventy years!

5. Rehovot Days Redux, by Harry Dym

Israel the mathematician, arrived in Israel the State towards the end of July 1974. Shortly thereafter he was offered a position in the Department of Pure Mathematics (as it was then called) of the Weizmann Institute of Science, as well as at a number of Israeli Universities. Israel accepted a full time position at Tel Aviv University. The Institute then offered him a half time position in the hope that he would eventually leave Tel Aviv and come to the Weizmann on a full time basis. He never did. However, he continued on a half time basis at the Institute for almost ten years. This period came to an end when the Institute went through a financial crisis and cut back on all part time positions.

Israel's presence at the Institute was like a breath of fresh air for me and my students and the enlightened few who were interested in Operator Theory and its applications. Even before Israel arrived I was an ardent admirer of the two wonderful Gohberg–Krein books. The first one was a marvellous source of information for some problems on trace formulas connected to a generalization of the Szegö formula that I was working on. The second was useful for a problem I had worked on with Naftali Kravitsky, my first doctoral student. The problem was to recover a string from its principal spectral function when that spectral function is a small perturbation of the principal spectral function of a known string. Gohberg–Krein factorization along a chain of orthoprojectors enters in an essential way in the resolution of this problem. It was mind boggling to have one of the principal architects of this theory show up in the flesh and on top of it to be so friendly and personable.

As it happened, Naftali was also an immigrant from the Soviet Union. He arrived in Israel in 1971 and, although he had already written a dissertation for the degree of "Kandidat" under the supervision of V.B. Lidskii at Moscow University in 1968, he was advised by a well meaning official who did not understand the Soviet system to get a "real" PhD. Presumably, the unfamiliar term "Kandidat" contributed to the confusion. Naftali was a wonderful student and a wonderful lecturer. He went on afterwards to make important contributions to the theory of commuting nonselfadjoint operators. Unfortunately, he passed away prematurely in August 1998 and never obtained a full measure of the recognition that he so richly deserved.

During Israel's Rehovot period he would generally come to the Institute twice a week: a full day on Sunday and a longish afternoon later in the week. On Sundays Israel would usually lecture from 11 a.m. to 1 p.m. Then, after lunch, we would work together for a couple of hours, take a tea break and then continue on until 6:30 or so. The pattern on the second afternoon was more or less the same. During breaks between work our conversation would drift into other topics. I was always keen to hear about M.G. Krein and, on a different tack, about how one coped with life under communism. Humor helped. Two of my favorite jokes stem from this period:

A high government official comes to a communal farm to see how the poultry division is being run. He approaches one of the local workers and asks him what he feeds his chickens. "Wheat", was the reply. "What, you feed your chickens wheat when we do not have enough wheat to bake bread for the citizens of our great country. Off to Siberia!" The official then approached a second worker, Ivan, with the same question. "Chocolate", Ivan replied, "I feed my chickens chocolate." "What, you feed your chickens chocolate when we do not have sweets to give to our children. Off to Siberia!" The pattern continued this way until finally the official asked a little old Jewish man, Yaakov, the same question. "Feed them", replied

Yaakov. "I don't feed them! I give them a ruble and tell them to buy what they want."

A second Story from that period concerns a bakery that is selling bread for two rubles a loaf. A customer purchases a loaf and, after he has it safely under his arm, says to the baker. "Tell me, why do you charge two rubles a loaf when the bakery down the street only charges one ruble?" "Ah", says the baker, "that's because they don't have any bread. If I didn't have any bread, then I would also charge only one ruble."

As our friendship grew, Israel shared some of his plans with me and in fact incorporated me into them. Thus, I became involved with the journal Integral Equations and Operator Theory and a conference organizer for the second (of what was to become) IWOTA (International Workshop on Operator Theory and its Applications). It is hard to be a passive bystander when Israel is involved.

His enthusiasm and optimism are infectious. I do, however, remember a luncheon meeting at Tel Aviv University with Bill Helton, Tom Kailath, Rien Kaashoek (and possibly others) in which we managed to dissuade him from starting a second journal, dedicated to engineering applications of Operator Theory. I believe that although initially disappointed at our lack of enthusiasm for this project, Israel was ultimately grateful not to have this extra responsibility to contend with.

One of the dominant aspects of Israel's personality is optimism. It manifests itself in a number of different ways: enthusiasm, vitality, generosity of spirit. I never heard Israel deprecate other mathematicians. By and large Mathematics is not a particularly sociable activity. There are collaborations and a limited contacts with the few others who are working on the same set of problems. This changed when Israel arrived on the scene. In short order he established a network of collaborators and colleagues that extended over large parts of Europe, the United States and Canada, with branch offices in Amsterdam, Calgary, College Park, Rehovot, Tel Aviv, to mention a few of the longer lasting ones. In addition to the Journal, he began the OT book series (which at last count had more than 107 volumes), initiated periodic workshops (which evolved into the IWOTA), started the biannual Toeplitz Lecture Series at Tel Aviv University and was a principal organizer of a number of Oberwolfach meetings.

The Oberwolfach meeting on Operator Theory in the Fall of 1976 was akin to a family reunion of distant relatives who had been heard of but not seen (at least by many of us). It was Israel's first visit to Oberwolfach and also his first visit to a Michelin three star restaurant in nearby Strasbourg. The participants that I remember included Harm Bart, Kevin Clancey, Lewis Coburn, Chandler Davis, Ron Douglas, Paul Fuhrmann, Bernhard Gramsch, Bill Helton, Rien Kaashoek, Leonid Lerer, Erhard Meister, Reinhard Mennicken, Joel Pincus, plus, of course, many others.

The first of the Toeplitz Lecture series featured Peter Lax and Ciprian Foias as the main speakers. There was a party in their honor at the Gohberg home in Ra'anana catered by three generations of Gohberg women: Israel's mother Clara, his wife Bella, his sister Fanny, and his two daughters Zvia and Yanina. Bella and Fanny, both of whom were practicing physicians, took three days off from work to help cook and bake. The apartment was wall to wall people and food. It was an amazing event, especially considering the fact that Israel was a relatively recent immigrant from the Soviet Union (that was none trivial in those days; you can read about it in OT40) and that Ciprian had even more recently made a dramatic escape from Romania via Finland. (In fact, Peter Lax also had an "interesting" history, having made the passage from Hungary to New York in December 1941.)

I would like to dose with two stories about Israel that I reported on earlier in the first of the two volumes dedicated to Israel's sixtieth birthday and will undoubtedly wish to repeat for his eightieth because, though short in length, they are rich in content.

When Israel first met my youngest son Michael, he asked him "How many children are there in your class?" "Forty one", was the reply. "Wonderful", said Israel, "so many friends".

One Sunday morning Israel came to the Institute much later than usual. It turned out that he had had a traffic accident enroute. Another car had hit his car from the rear. As accidents go, it was relatively minor, but still the expense and inconvenience was far from negligible. After settling in, he called his wife to tell her what happened. "But why did he do this to you?" was her immediate wifely response. "Belochka", said Israel, ever so gently, "this question you must put to him, not to me".

6. Wishing Israel Gohberg a Very Happy 70th Birthday, by Bill Helton

Rien has been asking everyone to be quiet, so the speakers can be heard. This is not actually necessary while I am speaking, because I have brought a large collection of slides which I did not get time to finish at my talk this afternoon, and if the noise becomes great I can just show them to you.

Toasting Israel Gohberg is a great pleasure. At a personal level because of his warmth and friendship. At a scientific level because of the many wonderful things he has done which effected me.

I first met Professor Gohberg at a conference sponsored by Sz-Nagy in Hungary in 1970. Several of us here met for the first time at that conference. I had just gotten out of graduate school and this was the first conference where I was invited to give a talk. I was very very nervous and it must have showed, because after the talk the famous Professor Gohberg patted me on the back and gestured that the talk had gone well. This made me feel much better, but soon I came to realize that Professor Gohberg's reassurance was firmly based on the fact that he did not speak any English. Throughout the conference though we did communicate some in German and Israel spent most of his time trying to phrase jokes in German so simple that I could understand them. At the time Israel worked in pure mathematics and he had done many important things with a number of collaborators. It was a great thrill as a beginning mathematician to meet him.

When Israel came to the West he took up engineering mathematics with great energy and had a remarkable impact on the subject. At that time there were just a few of us in the West who worked on the boundary between Operator Theory and linear systems theory. On one side was Paul Fuhrmann and I, and on the other side (always) is Patrick De Wilde, who is sitting over there in the corner. Soon the landscape changed. Israel and collaborators like Rien trained a whole generation of very strong young mathematicians who extended and greatly united engineering systems theory. The operator side of the field became lively and stimulating with many people to talk to. In less than a decade Israel built on his strong base in pure mathematics to become proficient in applied mathematics.

Israel has created much more than many theorems as his personal achievement. He has created a whole culture. Please let us toast a remarkable person, with important achievements, whose great warmth makes him a friend who is cherished by us all.

7. Total Quality Management, by Harm Bart

Ladies and gentlemen, dear friends, it is a pleasure to say a few words in anticipation of Israel Gohberg's seventieth birthday later this summer. In 1988 Israel's sixtieth birthday was celebrated with an impressive conference in Calgary. In the ten years that since then have passed, my professional life changed considerably. The change is – to use a metaphor – That from a mathematical butterfly living in a world of scientific beauty to an administrative caterpillar running a faculty in financial distress.

Why to bring this up in a short speech that should focus on Israel Gohberg and not on myself? There is a simple reason, lying in the fact that the deanship that I hold is not in a Mathematics Department. It is in a School of Economics, including a large section of Business Economics, a Dutch variant of what Americans call Business Administration. This gives new insights, some of which can be used to enliven this dinner.

One colleague from Business Administration recently told me his view on the difference between the Mathematical Sciences and the Behavioral Sciences, including Business Administration. Mathematics, he said, is like mining. You dig coal or precious stones and bring them to the surface. There they can lie an indeterminate time and still keep their value and usefulness. Business Administration, on the other hand, is like fishing. What you catch is valuable also, but you should not let it lie for too long. Otherwise it will spoil and become a nuisance.

Our friend Israel Gohberg certainly is a miner. He brought up diamonds where others would only expect coal. Precious stones that gladdened the hearts of many mathematicians and will continue to do so in the future.

There is a part of Business Administration where one studies developments and trends. Sometimes a certain development is not so much a trend but a fad. Recently, in my faculty a thesis was produced with the title "Beyond Fads". Such an expression certainly applies to Israel Gohberg. He is a trendsetter, leading his followers into mines full of precious stones. In this, he is usually gentle. Like Einstein, who wrote in 1943 to junior high school student Barbara Wilson: "Do not worry about your difficulties in mathematics; I can assure you that mine are still greater."

But Israel can also be demanding! This brings me to the third issue from my present background: "Total Quality Management", often abbreviated as TQM.

Indeed, Israel practiced TQM long before the notion was invented in modern management theories. Those that have worked with him know his relentless drive for perfection. They will appreciate the following small anecdote.

A singer gives a concert. After performing a beautiful aria, the audience applauds and shouts: "again, again". She is very pleased and repeats the aria. After she finishes, the same thing happens. Elated she sings the piece for the third time, thinking that this will be enough, and that she can look back at a very satisfactory performance. But once more, the audience shouts: "again, again". This makes her uneasy and hesitatingly she asks: "Again?" Then a heavy voice booms out: "Yes! Again, again and again! Until you do it right."

Back to real life. For me it was a privilege to work with Israel. I am grateful for the many things he taught me, especially for his lessons in "mathematical taste". But also for the warm personal contacts in which our families were involved too. Where of course – I also wish to mention emphatically Israel's wife, Bella. She played an important role in all of this!

Thank you very much.

Part V
About Colleagues and Friends

This part presents sixteen articles that were written or co-authored by Israel Gohberg himself. Some of those have the character of an *In Memoriam*, paying tribute to a dear colleague who has passed away. Others are *Recollections, Reminiscenses* or *Reviews* that highlight the work and personality of a friend celebrating a special occasion. These documents taken together give a fascinating and sometimes moving insight in the human factors that have influenced the development of the field.

Opening Address
Toeplitz Memorial Conference

I. Gohberg

Ladies and Gentlemen,

It is a great pleasure to welcome you all to this Toeplitz Memorial Conference. Many of you have crossed thousands of miles to be with us and we appreciate it very much. We have representatives from North America, Europe and Asia who have gathered here to honour the memory and the 100th anniversary of the birth of Otto Toeplitz, an outstanding scholar, teacher and personality.

Otto Toeplitz was one of the founders of operator theory. Many concepts, notions and theories can be traced back to him. He understood the importance of what today are called Toeplitz matrices and Toeplitz operators. A quick glance at the programme of this conference will illustrate just how wide have spread these matrices and operators. Numerical range, normal operators and many other notions were initiated by Otto Toeplitz. I was told by experts that in the day to day operations in the oil industry is used an algorithm which can be traced back to the original investigations of Otto Toeplitz.

A few words about his books. They are of such quality and cover such wide areas that they cannot be overlooked. The first was a book together with Hellinger. This is a volume of the mathematical encyclopedia which appeared in 1929. It was a real encyclopedia for operator theory of that time. The next book was written together with Rademacher and is a popular book on numbers and figures. This is a true masterpiece. It has been translated into many languages and almost every high school or university student interested in mathematics has read it. The third book is a text book which appeared after the death of Otto Toeplitz and was edited by G. Köthe. It is called "Calculus: A Genetic Approach" and is a course in calculus with an extremely fascinating historical background, which, at least in my opinion, should be required reading for every teacher of calculus.

As a scholar and teacher Toeplitz had a great deal of influence in Germany. All this activity was cut short in 1935 during the dark period of the Nazis, when he was dismissed from his post at the university because he was a Jew. During the period

Originally published in *Operator Theory: Advances and Applications* **4**, Birkhäuser Verlag, Basel, 1982, pp. 13–15.

1935-1939 he helped young Jewish students to find places to continue their studies in the United States. He devoted his full time and all his energy to organizing the education of Jewish children and representing the Jewish community. All these activities were often both tragic and hopeless. Only in 1939 did his colleagues convince him that his life was in great danger and he finally left Germany for Palestine, already a broken man. He worked as advisor to the Rector of the Hebrew University in Jerusalem for a short period before his death in 1940.

Otto Toeplitz dedicated most of his life to science and mathematical education in Germany, and Germany is the country in which he suffered so much. I think that it is very appropriate that this conference honouring the memory of Otto Toeplitz is a joint venture with the German Mathematical Society. I think that Israel is the most appropriate place for this conference to be held because it was here that he found his home and Israel is the home of his children and grandchildren.

I would like to thank the German Mathematical Society, the co-organizers, Professors Bernhard Gramsch and Heinz König, for their efforts and cooperation, and also the German Federal Ministry for Development and Technology for their support. We are grateful to the Embassy of the Federal Republic of Germany in Israel for their interest.

The conference enjoyed strong support here in Israel from the Council for Research and Development and from all levels of the Tel Aviv University, starting with the President, Rector, Dean, Chairman, and including all the secretarial staff.

I would also like to thank the Department of Mathematics, Weizmann Institute of Science, and the Institute of Mathematics of the Hebrew University, Jerusalem for their hospitality.

It is a pleasure to have with us today the family of Otto Toeplitz. His son, Dr. Uri Toeplitz and his daughter, Mrs. Chava Wohl, together with their families, along with other relatives. I would like very much to thank Dr. Uri Toeplitz for his advice, cooperation and help in the preparation of this conference, including the musical part.

I would like to say a few words about the organization of the conference itself. It was first planned as a conference whose main aim was to speak about the influence of Toeplitz and the state of modern operator theory. While working on it, the topics expanded and it grew by itself and crossed the borders of the initial plan, even when we restricted it as much as possible. We tried to avoid parallel sessions and in this we succeeded, but we had to pay for it with a very heavy schedule. We did not follow any special principle in the order of the lectures. Maybe only one - that everyone should find something of interest between any two breaks. As you will see we did not specify a special period for questions and breaks between lectures. We would like to ask lecturers to save from the time of their lecture three to five minutes for this purpose.

Again, let me welcome you all to the conference and wish you all a fruitful, pleasant and enjoyable stay in Israel.

David Milman (1912–1982)

I. Gohberg, M.S. Livšic and I. Piatetski-Shapiro

On 12th July, 1982, in Tel Aviv, Professor David Milman passed away. He was one of the famous pioneers in functional analysis and he is remembered for his outstanding contributions.

David Milman was born in 1912, the Saturday before the Jewish Passover, in a small town called Chichelnik near Vinitza, Ukraine, now USSR. For the sake of "simplicity", the local registrar recorded his birthday as January 1, 1913. D. Milman began his education in Heder, where he received both religious and secular instruction and, until age 14, continued reading in Hebrew both original and translated works (amazingly, these included the works of Shakespeare). At the same time he also attended a Ukrainian school, and from age 14 a chemistry college (called the Chemistry Professional School). There he met Israel Gelfand and together they became interested in mathematics. Unfortunately he was forced to leave the Chemistry Professional School because his social origins were viewed unfavourably by the new regime: his mother's father was the rabbi of Chichelnik and his father owned a shoe store before the Russian revolution.

In 1929 his family moved to Odessa. Once there, David by chance came across an article in a popular weekly magazine, "Ogonyok", about Fermat's Last Theorem. Soon afterwards he met Ivan Timchenko, Professor of Mathematics at Odessa State University, who helped him find an error in his "proof" of Fermat's theorem, but suggested that he should apply to the university. David was accepted one month later (1931) as a second year mathematics student at Odessa State University. He graduated in 1934. Again, because of his social origins, the authorities did not allow him to proceed with his mathematical education until 1937, when he became a graduate student of M.G. Krein. He received his Ph.D. in 1939. These were the boom years of functional analysis. M.G. Krein was leading one of the first seminars in functional analysis in the USSR (Odessa), and David Milman was one of the active participants in the seminar. Over various periods of time the participants in this seminar included V.A. Artemenko, M.S. Brodskii, F.R. Gantmacher, V.R. Gantmacher, B.Ja. Levin, M.S. Livsic, M.A. Naimark, V.P. Potapov, M.A. Ruttman, V.L. Smul'jan. Many contributions which now form a

Originally published in *Integral Equations and Operator Theory* **9** (1986), 1–7.

part of the basis of functional analysis were produced and discussed by the members of this seminar. The topics were varied, starting with geometry in Banach spaces and abstract operator theory (linear and non-linear), and up to applied problems in mechanics and engineering and their connections. David Milman is remembered by members of the seminar as a man who was a constant source of new ideas. It is recalled how the Brodskii/Milman theorem on the center of a convex set was proved. Milman and Brodskii worked during the summer vacation whilst living in the Dachas which are on the outskirts of Odessa in the resort area on the Black Sea coast. Every morning David Milman would visit Brodskii and bring with him a new idea of the proof. They would sit together and discuss the proof until they found that the idea they were working on would not work. This procedure was repeated for twenty-nine consecutive days, until on the thirtieth day, the thirtieth idea went through.

Unfortunately, soon after World War II this seminar and the school of functional analysis at Odessa University, came to an end because M.G. Krein was dismissed and his students had to leave. The seminar was very important for David Milman as a forum in which he could try out his ideas.

David Milman was very devoted to mathematics, research and education, and this dedication knew no bounds, in either space or time. On one occasion he came to M.G. Krein after midnight to discuss a fresh, and for him, very important mathematical idea. He could not understand why the hour should be a factor preventing immediate discussion. Even Krein's protest that he had to catch an early train the next morning, had no effect on Milman.

In 1938, David married Nemo Tsudikova, a physics major. Together they raised three sons, all of whom are mathematicians. Vitali is a professor of mathematics at Tel Aviv University; he continues to work on functional analysis, and especially on geometry of Banach spaces. Peter is a professor at Toronto University, and Vladimir, who is still in Russia, works in computer science. From 1939 to 1945 David worked as an Associate Professor (Dozent) at the Polytechnic Institute. He joined the Odessa Institute of Communications in 1945 and worked there until 1974. He played an important role in revising the mathematics program at this institution and he served as chairman of the Mathematics Department from 1949–1966. In 1974 he immigrated to Israel and was immediately offered a professorship at Tel Aviv University. He arrived speaking Hebrew fluently and lectured in Hebrew from the very beginning until his untimely passing eight years later.

The first publications of D. Milman (1938–1939) deal with criteria for reflexivity of Banach spaces. He proved that a Banach space E is reflexive if it is uniformly convex; i.e. for every $\varepsilon > 0$ there exists $\delta = \delta(\varepsilon) > 0$ such that $\frac{1}{2}\|x + y\| > 1 - \delta$, $\|x\| = \|y\| = 1$, implies $\|x - y\| < \varepsilon$. This theorem was not only the first result in the Geometric theory of Banach spaces, the active study of which continues to the present, but also an inspiration for other results to follow. He also established the following necessary and sufficient condition for reflexivity; the unit ball is weakly compact and every nested collection of nonempty closed convex subsets of the unit sphere has a nonempty intersection.

Later, David Milman [8], Vitold Smul'jan [Mat.Sb. 5(47) (1939), 317–328 (Russian) MR 1, 335], W.F. Eberlein [Proc. Nat. Acad. Sci. U.S.A. 33 (1947), 51–53. MR 9, 42; MR 10, 855], and also David Milman together with Vitali Milman [23, 25, 26] found even stronger criteria for reflexivity of Banach spaces.

In his work with B.Ja. Levin [4], David Milman proved that the only closed subspaces of both the Banach space of functions of bounded variation and the space of continuous functions on $[0,1]$ are finite dimensional subspaces.

The same year (1940) he published with M.G. Krein the well-known Krein–Milman theorem [3]. By definition, extreme points of a convex set K are the points that do not belong to interior of any interval in K. In a finite dimensional space every point of K is a centre of mass distributed among the extreme points of K. The famous Krein–Milman theorem says: *Every bounded regularly convex subset K of a dual to a Banach space is the regularly convex hull of its extreme points.*

This theorem entered all text books in functional analysis in the following form due to Bourbaki: *a compact convex set in a separable locally convex space is the closed convex hull of its extreme points.* It was a starting point of the "Method of extremal points and centres" developed by D. Milman in the 1940's and early 50's in [8,10,13,17,19], in his Russian second doctoral thesis (1951) and later in a monograph [31]. This theory has been further developed by G. Choquet, V. Klee and others (see [31]). The Krein–Milman theorem and "the method of extreme points and centres" have important applications. Let us mention a few of them:

The theorem of I.M. Gelfand and D.A. Raikov on the existence of complete systems of irreducible unitary representations of locally compact groups [Mat. Sb. 13 (55) (1943), 301–316; Amer. Math. Soc. Transl. (2) 36 (1964), 1–15. MR 6, 147];

M.G. Krein's representation theorems for positive-definite functions [Amer. Math. Soc. Transl. (2) 34 (1963), 69–164. MR 14, 480 and MR 12, 719];

M.A. Naimark's "Rings with involution" [Uspechi Math. Nauk 3:5 (27), (1948), 52–145] and I.M. Gelfand and M.A. Naimark's "Normed rings with involution and their representations" [Izv. Akad. Nauk SSSR Ser. Mat. 12 (1948), 445–480 MR 10, 199] (see also D, Milman [8, 17]);

D. Milman's theorem on the existence of invariant measure in dynamical systems defined by functionals [12].

Inspired by the theory of normed rings and the method of extreme points, D. Milman [8,9] introduced the notion of "T-boundary". The T-boundary of a topological space Q with a fixed set R of continuous functions on Q is the set T of all points $q_0 \in Q$ such that for every open neighbourhood $U(q_0)$ of q_0, there exists a function $f \in R$ such that

$$\sup_{q \in U(q_0)} f(q) > \sup_{q \in Q - U(q_0)} f(q).$$

In case of complex-valued functions on Q one can consider $\mathrm{Re}\, f(q)$ or $\|f(q)\|$ for $f \in R$. For example, for a convex regularly closed subset Q of a dual to a Banach space with the R being linear functionals, the T-boundary is the closure of the extreme points of Q; for a family R of harmonic (or analytic functions on a domain

Q in the complex plane, the T-boundary is the usual boundary of Q; family R of analytic functions of several variables (in a domain $Q \in \mathbb{C}^n$), the T-boundary is the skeleton of Q . In the separable case, the T-boundary is the closure of the set of all points $q^0 \in Q$ such that for at least one $f \in R$ and for all $q \neq q_0$, $f(q_0) > f(q)$. D. Milman proved that every function $f \in R$ admits an integral representation

$$f(q_0) = \int_T f(q) d\partial(I_q, q_0)$$

with measure $\partial(I_q, q0)$ supported on T and determined by q_0 uniquely in the separable case.

In the important paper [14] written by D. Milman jointly with M. Krein and M. Krasnoselskii, the gap between two subspaces of a Banach space was introduced and studied. Theorems have been proved which establish equality of dimensions for subspaces with a small gap. These results are then used for the study of defect numbers of linear operators in Banach spaces, angular operators and others. This paper played an important role later in the theory of Fredholm operators.

Among other results, D. Milman [7] extended I. Shafarevich's criterion for the existence of a norm on a commutative topological field to the case of commutative topological rings with unit. In particular this implies a characterization of dense subrings of complete topological rings of continuous functions on a compact set. Jointly with Brodskii, D. Milman [9] introduced the notion of a normal structure of a convex set and proved that if K is a convex, w-compact set with a normal structure, then there is a common fixed point for the set of all surjective isometries of K (moreover, they have proved this theorem for a larger class of distance nondecreasing mappings).

He introduced the notion of a normal cone K in a Banach space E; i.e., there exists $\delta > 0$ such that if $x, y \in K$ and $\|x\| = \|y\| = 1$ then $\|x + y\| > \delta$. It turned out to be very useful: By the two main theorems of the theory of cones in Banach spaces due to M.G. Krein, normality of the cone is equivalent to the condition that for every $f \in E^*$ there exist f_1 and $f_2 = f - f_1$ from K , and it also implies the existence of an isomorphism i of E with a subspace of continuous functions on a compact set, which identifies K with the nonnegative functions in $i(E)$.

D. Milman also found an extension theorem for sublinear functionals [24,30] which implies the three fundamental Banach theorems. In Israel he extended this result to the case of subadditive functionals [32]. (Note that on $\mathbb{R}^1$ the set of all symmetric sublinear functions p is one dimensional, namely $p(t) = \lambda \cdot |t|$ for $\lambda \geq 0$, while the set of all symmetric subadditive functions p has infinite dimension, containing all $p(t) = |\sin \omega \cdot t|$ for $\omega > 0$.) Numerous theorems of functional analysis are special cases of Theorem B of [32]. Also in Israel (1979–81), D. Milman introduced the notion of the central subset of an open domain G in $\mathbb{R}^n$ and studied its structure [33, 34, 35]. In the case that the boundary $Q = \partial G$ is a smooth submanifold of $\mathbb{R}^n$, the central subset of G is R. Thom's cut locus of Q, see [35]. These results of D. Milman have already found recognition in singularity theory (see, for example, the introductions to J.N. Mather's "Distance from a submanifold in Euclidean space", Proc. of Symposia in Pure Math. vol. 40 (1983), 199–215 and Y. Yomdin's "On the local structure of a generic central set", Compositio Math. 43 (1981), 225–238). His last manuscript [36], which is to appear in this issue, is a continuation of his earlier work in [20, 27].

David Milman had a philosophical inclination and a tendency towards simple, general and fruitful ideas. He was an extraordinary lecturer, devoting much time and effort to education. He also was a kind and sympathetic man who always offered help to those in need.

D.P. Milman – List of Publications

[1] On some criteria for the regularity of spaces of the type (B). *Doklady Akad. NAUK SSSR (N.S.)* **20** (1938), 243–246 (Russian).

[2] On some properties of regular spaces. *Doklady Akad. NAUK SSSR* **22** (1939), 394–398.

[3] (with M.G. Krein) On extreme points of regular convex sets. *Studia Math.* **9** (1940), 133–138; *Math. Rev.* **3**, 90.

[4] (with B.J. Levin) On subspaces of C, consisting of functions of bounded variation. *KHARKOV NOTES of Math. Soc.* (4), **16** (1940), 102–105.

[5] (with M.G. Krein and M.A. Rutman) A note on bases in Banach space. *Comm. Inst. Sci. Math. Mec. Univ. Kharkoff (Zapiski Inst. Math. Mech.)* (4) **16** (1940), 106–110 (Russian, English summary); *Math. Rev.* **3**, 49. (1942).

[6] Sur une classification des points du spectre d'un operateur Lineaire. *Doklady Akad. NAUK SSSR (N.S.)* **33** (1941), 279–287.

[7] Norms in topological rings. *Doklady Akad. NAUK SSSR* **47** (1945), 166–168.

[8] Characteristics of extremal points of regularly convex sets. *Doklady Akad. NAUK SSSR* **57** (1947) 110–122 (Russian); *Math. Rev.* **9**, 192.

[9] (with M.S. Brodskii) On the center of a convex set. *Doklady Akad. NAUK SSSR (N.S.)* **59** (1948), 837–840 (Russian); *Math. Rev.* **9** (1948), 448.

[10] Accessible points of a functional compact set. *Doklady Akad. NAUK SSSR* **59** no. 6 (1948), 1045–1048 (Russian); *Math. Rev.* **9** (1948), 449.

[11] Isometry and extremal points. *Doklady Akad. NAUK SSSR* (N5) **59** (1948), 1241–1244 (Russian); *Math. Rev.* **9** (1948), 516.

[12] Dynamical systems defined by functionals and invariant measures on them. *Doklady Akad. NAUK SSSR* **59** (1948), 1397–1398 (Russian); *Math. Rev.* **9** (1948), 449.

[13] (with M.A. Rutman) On more precise theorem about com- pleteness of the system of extremal points of a regularly convex set. *Doklady Akad. NAUK SSSR* **60** (1948), 25–27 (Russian); *Math. Rev.* **9** (1948), 448.

[14] (with M.G. Krein and M.A. Krasnoselski) On the defect numbers of linear operators in Banach space and some geometric questions. *Sbornik Trudov Inst. Acad. NAUK Uk. SSR* **11** (1948), 97–112 (Russian).

[15] Extremal points and centers of convex bicompacts. *Uspehi Math. Nauk (N.S.)* **4** no. 5 (33) (1949), 179–181 (Russian); *Math. Rev.* **11** (1950), 117.

[16] Multimetric spaces. Analysis of the invariant subsets of a multinormed bicompact space under a semigroup of nonincreasing operators. *Doklady Akad. NAUK SSSR* **67** (1949), 27–30 (Russian); *Math. Rev.* **11** (1950), 117.

[17] On the theory of rings with involution. *Doklady Akad. NAUK SSSR* **76** (1951), 349–352. (Russian); *Math. Rev.* **12**, 618.

[18] The facial structure of a convex bicompact space and integral decompositions of means. *Doklady Akad. NAUK SSSR* **83** (1952), 357–360 (Russian); *Math. Rev.* **13**, 848.

[19] On integral representations of functions of several variables. *Doklady Akad. NAUK SSSR* **87** (1952), 9–10 (Russian); *Math. Rev.* **14**, 546.

[20] Some theorems of non-linear functional analysis and their application in the theory of local groups. *Scientific Communications, Uspehi Mat. Nauk* **12** no. 1 (73) (1957), 222–226; *Amer. Mat. Soc. Trans.* (2) , V16, 437–442.

[21] (with V.A. Breskin, and A.Ju. Lev) On contractions of the frequency spectrum of binary informations with small probability of one of the states. *Izv. Vyzov., Radiothechnik* **4** (1958), 466–473.

[22] Some theorems of non linear functional analysis and their applications to the theory of local groups. *TRUDY of the 3. Soviet Math. Conqress* **4** (1959), 58–59.

[23] (with V.D. Milman) Some geometric properties of non- reflexive spaces. *Doklady Akad. NAUK SSSR* **152** (1963), 52–54; *Soviet Math. Dokl.* **4** (1963), 1250–1252; *Math. Rev.* **28**, 1478.

[24] Separators of non linear functionals and their linear extensions. *Izv. Akad. NAUK SSSR Ser. Mst.* **27** (1963), 1189; English transl.: *Amer. Math. Soc. Transl.* **54** no. 2 (1966), 153; *Math. Rev.* **29**, 498.

[25] (with V.D. Milman) Some properties of non-reflexive Banach spaces. *Mat. Sb. (N.S.)* **65** (107) (1964), 486–497; *Math. Rev.* **30**, 1383.

[26] (with V.D. Milman) The geometry of nested families with empty intersection. The structure of the unit sphere of non reflexive space. *Mat. Sb. (N.S.)* **66** (188) (1965), 109–118; *Math. Rev.* **31**, 2596; *Amer. Math. Soc. Trans.* (2) **85**, 233–243.

[27] Formulation and solution of a general boundary value problem in Operator theory as an aspect of funtional analysis problems of Cauchy and Dirichlet type. *Doklady AN SSSR* **161** no. 6 (1965); English transl.: *Soviet Math. Mar–April*, 1965 **6** N2, 592–595.

[28] (with A.Ju. Lev, and V.D. Milman) Convolution of information in the classical probabilistic scheme. *Problemy Peredaei Informa* **2** no. 2 (1966), 29–38; *Math. Rev.* **34**, 8877.

[29] Sublinear envelope of functionals; Applications. Theses of short communications of the ICM, Section V, Moscow (1966), 65.

[30] Sublinear continuations of functionals, *Doklady Akad NAUK SSSR* **186** no. 2 (1969); English transl.: *Soviet Math. Pohl* **10** no. 3 (1969).

[31] Facial characterization of convex sets; Extremal elements. *Trudy Moscow Math. Obsc.* **22** (1970); English transl.: *Trans. Moscow Math. Soc.* **22** (1970), 69–139.

[32] On the set of minimal extensions of a subadditive functional. Topics in functional analysis. *Advances in mathematics supplementary studies* **3** (1978).

[33] Eine geometrische Ungleichung und ihre Anwendung. In: *General Inequalities 2*, ISNM **47**, Birkhäuser, 1980.

[34] The central function of the boundary of a domain and its differentiable properties. *Journal of Geometry* **14** no. 2 (1980).

[35] (with Z. Waksman) On topological properties of the central set of a bounded domain in R^m. *Journal of Geometry* **15** no. 1 (1981).

[36] The abstract linear boundary problem (a non group approach). *Integral Equations and Operator Theory* **9** (1986), 11–30.

Biography of M.S. Livšic

Harry Dym, Israel Gohberg, Naftaly Kravitsky

Moshe Livšic (Mikhail Samuilovich Livšic) was born on the 4th of July, 1917 in the small town of Pokotilova near Uman, a province of Kiev, in the Ukraine (according to the 1897 census, Pokotilova numbered 3030 citizens, including 1670 Jews). When he was four years old, his family moved to Odessa where his father held the position of associate professor of mathematics in an academic institute. His father's influence on Moshe, an only child, was very great, and until today he has fond memories of the Yiddish songs and Jewish prayers sung by his father, who before his mathematical career, was a cantor at the synagogue.

Moshe Livšic's father often spoke about the great mathematicians who were active at that time in Odessa: N.G. Chebataryov, an outstanding algebraist; V.F. Kagan, an outstanding expert in geometry, especially non-Euclidean; Yu. I. Timchenko, an expert in mechanics; M.G. Krein, who was at that time a young postgraduate student; and S.O. Shatunovsky. They were all close friends of Moshe's father.

S.O. Shatunovsky was an especially intimate friend of the elder Livšic. Shatunovsky was an excellent lecturer and an extraordinary person, well known to many of the townsfolk. Even today the older generation still recall the professor whose fascinating lectures in mathematics were attended by hundreds of students from all the faculties of the university.

In the West relatively little is known about the work of S.O. Shatunovsky (1859–1929). He was one of the first representatives of constructive mathematics, and a pioneer in intuitionist logic and modern algebra. He devoted much time to the law of the excluded third, and in 1901 he was the first to indicate that the formal transfer of this law to infinite sets is not obvious. In his dissertation, completed in 1917, S.O. Shatunovsky constructed a foundation of algebra and in particular of Galois theory as a theory of congruences with respect to the functional moduli introduced by Cauchy, without invoking the law of the excluded third to infinite sets. S.O. Shatunovsky found an original generalization of the limit notion. Without using limits, he defined the volume of a polyhedron as a certain invariant. Independently of Hilbert, and approximately at the same time (1897–1898) he laid

Originally published in *Operator Theory: Advances and Applications* **29** (1988), 6–15.

the axiomatic foundation of the theory of areas. In 1910, together with V.F. Kagan, S.O. Shatunovsky founded a mathematical publishing house in Odessa. This publishhing house, "Matesis", played a significant role in mathematical education and the popularization of mathematics in Russia. "Matesis" published translations of European classics in mathematics and a number of textbooks, including the well-known book of Dedekind on the theory of irrational numbers. In 1923 "Matesis" published S.O. Shatunovsky's book "Introduction to Analysis" which contained his lectures on the subject. "Matesis" continued operating until some time after the October revolution.

In 1931 Moshe graduated from school, which at that time comprised seven grades. At school he became friendly with Israel M. Glazman, and their friendship continued intermittently until the tragic death of this outstanding mathematician an May 30th, 1968. At the age of sixteen, the two friends worked out a plan for their future education which culminated in a deep study of philosophy (Kant, Hegel, and the works of a number of English and French philosophers). It was clear to them that the study of philosophy in the 20th century had no value without a fundamental knowledge of the natural sciences and so they planned to study physics and chemistry first. However, since a study of the natural sciences was impossible without a fundamental knowledge of mathematics, the two friends reached the logically inevitable conclusion that they should begin by studying mathematics. Many years later M.S. Livšic lightheartedly summed up the youngsters' program: "I succeeded in thoroughly studying some fields in mathematics. I succeeded less in studying some fields in physics. There was no time for philosophy." Nevertheless a tendency to a philosophical understanding of scientific results remained.

As a youth Moshe Livšic was attracted to the new technology of radio, and he dreamed of becoming a radio technician. After graduating from school he entered the radio class at the Technical College for Communication in Odessa. However, in 1933, he interrupted these studies to enroll in the newly created Department of Physics and Mathematics at the Odessa State University.

During the first years at the university mathematics was taught by M.G. Krein, F.R. Gantmakher (who shortly thereafter left for Moscow), and M.A. Naimark, who was a graduate student, assistant and, subsequently, a colleague of M.G. Krein. Later, when Moshe Livšic was already a third year student, B.Ya. Levin, a prominent specialist in the theory of analytic functions, joined the faculty.

Those who influenced Moshe Livšic most during his university years were M.G. Krein and B.Ya. Levin. He especially remembers B.Ya. Levin's course on theory of functions of a complex variable, and M.G. Krein's courses on integral equations and on the differential equations of mathematical physics, both from the viewpoint of functional analysis. These courses were on the frontiers of research of that time.

Among his fellow students and friends, Moshe Livšic remembers especially his schoolfriend I.M. Glazman. He also remembers V. Smushkovich, a very talented mathematician and a handsome young man (lost in the Second World War), and A.P. Artyomenko, the most talented student of M.G. Krein (and much older than the others). Artyomenko obtained important results in the theory of Hermitian positive-definite functions. He and M.S. Livšic had a warm relationship. After World War II, A.P. Artyomenko disappeared and all attempts to find him were of no avail. A younger student who entered the university a year after Moshe Livšic was V.P. Potapov, who later also became an eminent mathematician. The two, having common scientific interests, became close friends.

Among the older students whom Moshe Livšic remembers were V.L. Shmulyan, D.P. Milman, and M.A. Rutman, all of whom became graduate students of M.G. Krein and later outstanding mathematicians. Along with Moshe Livšic they were active participants in the seminar an functional analysis. This seminar and its participants became famous in the mathematical world where they were known as the Odessa School of functional analysis, with M.G. Krein as its undisputed leader. This school had a great impact an the development of functional analysis both in the USSR and abroad. V.L. Shmulyan perished at the front in the Second World War. His achievements were highly appreciated and were later continued by other mathematicians. From the front Shmulyan wrote letters with mathematical results, and M.S. Livšic corresponded with him until his death.

M.G. Krein devoted much of his time to his students. He could often be seen taking a stroll an Deribassovskaya Street (the main street of Odessa), in the company of his students, deep in discussions of a mathematical, philosophical and ethical nature. M.S. Livšic and I.M. Glazman admired and loved M.G. Krein. They often visited him at his home where discussions would continue long into the night. These were fruitful and pleasurable hours, which made life more vital and interesting. The students were preoccupied with the role of abstract and classical analysis and M.G. Krein impressed them with his balanced attitude towards both of these streams. This attitude was reflected in his work and served as a source of inspiration.

From B.Ya. Levin Moshe Livšic learned a love of the theory of analytic functions. He used the ideas, techniques and methods of analytic functions theory throughout all of his research. B.Ya. Levin made friends easily with his students, and they regularly visited him at his home. He was a bachelor, and in his small room, thick with tobacco smoke, lively discussions an literature, philosophy, mathematics, and politics took place. M.S. Livšic's first research paper, published in "Mathematicheskii Sbornik", was a joint paper with B.Ya. Levin.

M.S. Livšic began his research work at the end of his student period. His initial scientific interests were formed under the influence of his teachers, M.G. Krein and B.Ya. Levin. After graduating in 1938 from the Odessa University he continued his studies in the same university as a Ph.D. student under the supervision of M.G. Krein.

During those years M.G. Krein was involved in research related to the moment problem, and B.Ya. Levin was occupied with the theory of quasianalytic functions. M.S. Livšic's masters thesis and his research works published during the following three years were all devoted to these topics. However, soon afterwards he abandoned them both. He became interested in operator theory to which he came directly from the moment problem. He often recalls the following episode which had some influence on him. It was a remark made by M.G. Krein during the defense of a Ph.D. thesis (one of the first in the Odessa University) to the effect that the person who could construct a spectral theory of nonselfadjont operators had not yet been born. Moshe had read the famous book of M.H. Stone on operator theory, and the papers of J. von Neumann in which the theory of unbounded operators was exposed for the first time. The lecture series of N.I. Akhiezer "Infinite Jacobi matrices and the moment problem" also sparked this interest. In this lecture series the connection between Jacobi matrices, the moment problem and general operator theory was traced out in a clear and lucid way. The first edition of these lectures appeared in Kharkov in mimeograph form; later they were published in Uspekhi Mat. Nauk, V. 9, 1941. In the same issue of this journal also appeared, for the first time in the USSR, a review paper on operator theory by A.Y. Plessner. The work of N.I. Akhiezer and A.Y. Plessner also had a great influence on M.S. Livšic.

During World War II, M.S. Livšic was evacuated, together with the University from Odessa, first to Maikop and then in 1942, when the Nazis approached the town, to Bairam-Ali in Turkmenia. Because of his bad eyesight he was not conscripted. In Maikop, in 1942, M.S. Livšic defended his Ph.D. thesis which was devoted to applications of Hermitian operator theory to the generalized moment problem. His opponents were M.G. Krein and F.R. Gantmakher.

In 1945, whilst in Bairam-Ali, Moshe Livšic completed the dissertation for his second doctoral degree. It comprised two parts. The first part was a continuation of his previous research and showed the connection between the extension problem for symmetric commuting operators and the extension problem for Hermitian positive-definite functions. These results were never published. In the second part he generalized the extension theory of J. von Neumann. He introduced a certain matrix valued analytic function which characterized the nonselfadjoint extensions of a symmetric operator with finite coefficiency indices, and named it the characteristic function of a nonselfadjoint operator. This function has a number of remarkable properties. In particular the nonselfadjoint operator may be recovered from it up to unitary equivalence. Later the notion of characteristic function was introduced also for bounded operators.

In May 1945 at the age of 28, just three years after his first Ph.D. degree, he successfully defended his second dissertation in the Steklov Mathematical Institute of the USSR Academy of Sciences in Moscow. His opponents were S. Banach, I.M. Gelfand, M.A. Naimark and A.Y. Plessner.

During the postwar period the situation in Odessa changed considerably. The first indications were felt by Moshe when his name was dropped by the administration from the list of faculty members returning to Odessa to work in the University because, in the words of the vice-rector, "he was not suited to represent the Ukranian culture". Instead he was sent to work at a relatively minor institution in Kirovograd, also in the Ukraine. In 1944 M.G. Krein was dismissed from Odessa University and his students also had to leave. Thus, the famous school of functional analysis which had existed at this university for many years, came to an end. Shortly after B. Ya. Levin left for Kharkov. In 1945 Moshe Livšic returned to Odessa to head the Mathematics Department of the Odessa Hydrometerological Institute (a second rate academic institute with no mathematics majors), where he remained until 1957. He found his part time work in the Odessa Pedagogical Institute more interesting. Here he collaborated with M.S. Brodskii and P. Potapov, who at that time proved an important theorem which generalized Blaschke product decompositions to the case of analytic matrix valued functions. This theorem arose from the problem of factoring the characteristic function into elementary factors. M.S. Livšic used Potapov's theorem to construct triangular models for nonselfadjoint operators. M.S. Brodskii developed the theory of nonselfadjoint operators, the theory of characteristic functions and triangular models, using geometric methods.

During the period 1955–57 Moshe Livšic became interested in physical interpretations of his theory. He published a number of papers in which it was shown that, under certain conditions the characteristic matrix function coincides with the Heisenberg scattering matrix. M.S. Livšic's students at that time included L.A. Sakhnovich, A.V. Kuzhel and B.R. Mukminov. Their research was also related to the theory of nonselfadjoint operators and was an extension of the M.S. Livšic theory and his philosophy. The M.S. Livšic theory was already recognized at this time as a major contribution to functional analysis. It was utilized by many researchers and had great impact. The monographs of I. Gohberg with M.G. Krein, and B.Sz.-Nagy with C. Foias, may serve as examples.

In 1957 Livšic moved with his family to Kharkov, where until 1962 he was head of the mathematics department of the Kharkov Mining Institute. He then joined the department of mathematical physics in the Kharkov State University, where N.I. Akhiezer was chairman of the department. Here he continued working an the physical interpretation of the theory of nonselfadjoint operators and developed his theory of "open systems" which are physical systems which interact with the environment. These investigations are described in two monographs: "Operators, oscillations, waves. Open systems" and "Theory of operator colligations in Hilbert spaces" (the latter jointly with A.A. Yantzevich), both of which are published in English translation. E.R. Tsekanovski, L.L. Waksman, A.G. Rutkas, A.A.Yantzevich, and V.K. Dubovoi were among the students of M.S. Livšic in Kharkov.

In 1975 M.S. Livšic moved with his family to Tbilisi, where he worked for three years in the Institute of Agricultural Machines. The reason behind this move was his desire to immigrate to Israel and he knew that for various reasons it would be impossible for him to do so whilst in Kharkov. It was at that time that Livšic first became interested in extending the theory of one nonselfadjoint operator to the case of two or more commuting nonselfadjoint operators. During this period he obtained his first results in this direction, which included in particular a generalization of the Cayley-Hamilton theorem to the case of two commuting operators with finite dimensional non-Hermitian components.

In 1978 M.S. Livšic arrived with his family in Israel. He settled in Beersheva and began working at the Ben Gurion University of the Negev. It was the third time in his life that he actively engaged in building up a new school. There he started working with N. Kravitsky an developing a theory of several commuting operators. It turned out that the theory of pairs of commuting operators is closely connected with a problem of algebraic geometry of obtaining a determinantal representation of algebraic curves. Certain elementary transformations which allow one to find all the possible non-equivalent determinantal representations of a given algebraic curve if one such representation is given were recently found by V. Vinnikov, Moshe's most recent student. In all of the investigations of this period an active part was taken by N. Kravitsky. H. Gauchman, another colleague at the Ben Gurion University, recently generalized a number of these results to the case of Hilbert bundles.

From his early days at Beersheva, Moshe organized and successfully ran an active seminar an operator theory and systems. Since 1985 Moshe holds the David and Helena Zlotowski Chair in Operator Theory and Systems.

Moshe has always been very active and dedicated to his work. His recent breakthroughs in the theory of characteristic functions for several commuting operators indicate that in spite of his seventy years, mathematically Moshe is still a young man. He is loved by his friends, colleagues and students for his wonderful qualities of integrity, honour and benevolence, and respected as a great mathematician.

Odessa Reminiscences

I. Gohberg

I remember how during one of my visits to Odessa in the mid fifties M.G. Krein and myself were out walking when we met V.P. Potapov with his wife. Mark Gregorievich introduced us. I already knew about V.P. Potapov and his achievements from Krein who considered him to be one of his most brilliant students of whom he was proud. Krein told me about Potapov's second doctoral thesis which he had defended not long before in the Steklov Institute in Moscow. The thesis was about multiplicative decomposition of contractive matrix functions and he considered the results to be fundamental, innovative and opening a new area of research. At that time Potapov was the Dean of the Faculty of Mathematics and Physics of the Odessa Pedagogical Institute and M.G. spoke of him with great respect.

The problem of multiplicative decompositions of contractive matrix functions came from the theory of nonselfadjoint operators and was stated by M. Livsic (a colleague and friend of V.P.). Later M. Livsic used the results of Potapov in the first version of his theory of characteristic functions, the analysis of invariant subspaces of nonselfadjoint operators and their triangular forms. Today these results and their far reaching generalizations are obtained by purely operator theoretic methods. In general Potapov's research very often started with problems appearing in operator theory. Operator theory was used also to predict the results and the formulas for the solutions. At the same time Potapov would usually present his results as pure complex analysis and without operator theory in the final versions of his work. Operator theory had an essential influence an Potapov's work. In its turn his work influenced important research in operator theory and its applications.

In the late forties a wave of antisemitism swept through the Soviet Union and especially Ukraine. M.G. Krein was accused of Jewish nationalism. As usual in such cases no official accusation actually was made, but there were obvious signs that such an accusation was included in the secret files kept in the offices of the Communist Party in Kiev. This accusation was based on the fact that among the first generation of M.G. Krein's Ph.D. students (A.B. Artemenko, M.S. Livsic, D.P. Milman, M.A. Naimark, V.P. Potapov, M.A. Rutman, V.L. Shmulyan) the

Originally published in *Operator Theory: Advances and Applications* **72** (1994), xxiii–xxv.

majority were Jewish. The passage of time proved that M.G. had much more objective reasons for his choice of students. All of them made outstanding contributions to mathematics, but this was of no importance to the party headquarters. These files were often responsible for the fact that M.G. was rejected many times as a candidate for prizes or promotions proposed by most prestigious mathematical organizations. More than that, these files would be used at any time to harm him and his followers, and the outcome could even be a jail or gulag sentence. During these years one of M.G. Krein's students defended his doctoral thesis in Kharkov. There he was officially asked if Krein was still going free.

Among the first generation of M.G. Krein's students only A.B. Artemenko and V.P. Potapov were not Jewish. The former disappeared after the end of World War II, so only Potapov remained. His presence was very important and those in M.G.'s circle understood this very well. I am sure that Potapov also realized the responsibility this carried. Starting with the late forties Krein was forbidden to take on any Ph.D. students who were Jewish. This type of restriction existed in general but V.P. Potapov fought it and was able to overcome it in many cases. D.Z. Arov, L.A. Saknovich and Yu.P. Ginzburg were all Ph.D. students of Potapov during these most difficult years of antisemitism in the Soviet Union. More than that he was very protective of them. Probably no Jew could do this. I can remember the following nontrivial case.

In 1958 an All Union Conference on Functional Analysis was held in Odessa. M.G. Krein was the chairman of the organizing committee, of which V. Potapov was a member. The conference was a success and many outstanding experts from different cities and republics participated. During the organization it turned out that the number of talks to be given by Jews was very high. Since this was an extremely dangerous situation, which could lead to serious consequences for Krein, he suggested imposing some restrictions, in particular not to include in the programme a talk to be given by a young man, Yu.P. Ginzburg, at that time one of V. Potapov's doctoral students. M.G. thought this would not harm the young man and there would be other opportunities for him to give his talk. When Potapov learned about this intention he was furious and threatened to boycott the conference if the talk of Yu. Ginzburg was not included. Ginzburg's talk was included and in this instance everything went smoothly.

I came to know Potapov better some time later when I was trying to arrange for him to move to Kishinev to work in the Moldavian Academy of Sciences. During these days I understood his importance, both as a mathematician and as a personality. I knew he was interested in making this move because at that time he was going through a major change in his personal life (the first of many such changes), and he wanted to move to Kishinev with his new wife. They visited Kishinev and discussed the matter with the authorities of the Academy. I was quite optimistic. To my regret this did not work out and the invitation fell through. Till today the reasons for this are not clear. I can only speculate.

Let me finish with a little anecdote. A colleague was visiting Potapov one day. Potapov offered his visitor a drink, but the latter declined, saying that he had just recently had some heart trouble. Potapov insisted: "The drink will be very good for you. It widens the blood vessels and helps the heart." The colleague replied that after you stop drinking the blood vessels would shrink again and make things even worse. "You are a clever man, but you come to the wrong conclusions. Is it not clear what you have to do – just don't stop." This was Vladimir Petrovich Potapov.

Opening Address

Israel Gohberg

Dear Colleagues and Dear Friends
Ladies and Gentlemen

I am thankful for this opportunity to open the conference on the occasion of the sixtieth birthday of M.A. Kaashoek, the founder and head of the Analysis and Operator Theory group in Amsterdam at the Vrije Universiteit. He and his group have made important contributions to traditional operator theory as well as developing new areas. I was closely related to this group and helped to build and nurture it. Together we established and developed connections between operator theory, integral equations and systems theory. Those connections allowed us to enrich considerably all theories involved and to build a method, based on state space techniques, for solving explicitly problems in Analysis.

Over a period of 20 years the group developed into an important world renowned School of Operator Theory and its Applications. Here were organized many summer schools, and national and international conferences. The members of this School published numerous monographs and important papers, and several outstanding young experts were born and bred in this School.

I have had the pleasure to work together with Rien for more than twenty years. Five books (with H. Bart, S. Goldberg, F. van Schagen, C. Foias and A. Frazho) were written as well as many papers. One book with Harm Bart and Andre Ran is in preparation. Together we educated many graduate students. I will always remember with joy and satisfaction those weeks, and weekends, when we worked together. Usually the weekends included, as well as attractive mathematics, also interesting excursions and bicycle tours in Netherlands and enjoyable dinners with the family. The parties organized by the hospitable Kaashoek family were always remarkable for the excellent food and very warm and friendly atmosphere. I hope that this wonderful time is not over and we will be able to continue to enjoy this warm hospitality for many years.

Originally published in "The M.A. Kaashoek Anniversary Volume," *Operator Theory: Advances and Applications* **122** (2001), xxxi–xxxiii.

By the way we have very different mentalities by descent and education, but during the twenty years of joint work there were never any arguments between us. As can be expected, sometimes our opinions differed, but always the difference was resolved before reaching the dangerous level of an argument. Each time this required convincing discussions, flexibility and compromise. As a result this led to a better understanding and decisions that satisfied both of us.

I would like to mention that Rien is a wonderful teacher, friend and coauthor. Rien can listen to others and can express precisely and persistently his wishes. Discussions with him are always pleasant. He is not only very talented and capable, but also very reliable, organized, kind and generous. I learned many good things from him. Today, on the occasion of his sixtieth birthday, I would like to congratulate Rien on his outstanding achievements and to thank him heartily for this wonderful lasting collaboration and friendship. I would also like to congratulate today the faculteit of Wiskunde and Informatics on its excellent group in Analysis. In my opinion it is certainly a day of celebration for the entire Vrije Universiteit.

Rien Kaashoek obtained an excellent mathematical education in Leiden University in the school of one of the famous pioneers of functional analysis, Professor A.C. Zaanen. Rien's assignment was in new territory, unbounded operator theory. He successfully started his research in a new area for this school. At the end of the sixties he started to build up his group at the Vrije Universiteit.

I first met Rien at a conference in Tihany, Hungary, in 1970. We could hardly communicate; at that time I did not speak English and he did not speak Russian. The next time we met was after I immigrated to Israel. Our meeting was in College Park, Maryland, where he was spending his sabbatical and I was visiting as a guest of Seymour Goldberg. This was at the beginning of 1975. Soon we were already looking for an area of joint interest. We continued our work in Amsterdam at the end of 1975 and in the spring and autumn of 1976, and finished this first period with three important things: two large papers and a nice bicycle tour to the Keukenhof. The first paper, jointly with David Lay, was in a sense in the area of our old interests, and the second, with Harm Bart, was in a new area. It was mostly on matrix polynomials and used the approach of characteristic functions. This was the beginning of connections with the interests of the Odessa School of M.G. Krein and with mathematical systems theory and control from Electrical Engineering. Our first paper was soon published in the American Journal of Functional Analysis; the second was rejected by a respected journal. Unfortunately the reviewer did not make the slightest effort to understand the paper or to understand the framework in which it was written. This paper was a new beginning, and we decided that it would be fitting to publish it in the first issue of the new journal, Integral Equations and Operator Theory. Finally, the Keukenhof astonished me by its rich beauty of tulips.

After this beginning we met many times for extended periods in Amsterdam and Tel-Aviv. I was visiting Amsterdam on a regular basis more than three months per year and Rien spent a large part of his sabbaticals in Tel-Aviv. We published many papers in many different areas, all the time extending our fields of interest. Let me mention some of them: factorization of matrix-functions and operator-functions and its applications to integral and differential equations, homogeneous and nonhomogeneous interpolation problems for matrix-functions and applications to problems in control, completion problems for matrices and applications, applications of the state space method to different problems in analysis, inverse spectral problems and applications to nonlinear equations, different time dependent prob lems in analysis, and the band method.

One of the important features of our work was its closeness to electrical engineering (especially systems theory). We were able to speak to the theoretically oriented engineers and to understand their mathematical problems. We listened to many talks by engineers and participated in their conferences. We were active in the steering committee of MTNS, and one of the big meetings of MTNS was organized at the VU (co-chaired by M.A. Kaashoek, A.C.M. Ran and J.H. van Schuppen). We certainly did not become engineers, we remained mathematicians, but without any doubt all of this influenced our work and gave us a beautiful feeling of having next to us the motivations and applications.

I was planning to have a second part to my talk. The second part was to have been devoted entirely to the paper which was rejected, and I had planned to present in detail the ideas contained in this paper and their developments. Unfortunately I had to undergo an urgent and difficult operation, and I have had to postpone my plans for another time. For the same reason I am unable to participate in this conference in which I was so keen to participate. I regret very much having to miss this conference and meetings with close friends.

I am very happy to open the conference dedicated to the sixtieth birthday of my close friend and coauthor Marinus Adriaan Kaashoek. Dear guests, welcome to Amsterdam and have a nice time in this lovely city. Special thanks to my friend Seymour Goldberg for presenting this address to the conference.

Mark Grigorievich Krein: Recollections

I. Gohberg

1. Curriculum Vitae

Mark G. Krein was born into a Jewish family of modest means in Kiev on April 3, 1907. His father was a lumber merchant. As a youngster he exhibited a talent for mathematics. At the age of 14 he was already attending research seminars. He never obtained an undergraduate degree. In 1924 he ran away from home to Odessa and in 1926 he was accepted for doctoral studies by N.G. Chebotarev at Odessa University. He completed the degree requirements in 1929.

M.G. Krein was an excellent and enthusiastic teacher. He attracted many students. In the thirties he created at Odessa University one of the strongest centers of functional analysis throughout the world. His interests included matrices and integral equations, geometry of Banach spaces, moment problems spectral theory of linear operators, extension problems and applications. Many of the results of his outstanding students of this period (A.B. Artemenko, M.S. Livsic, D.P. Milman, M.A. Naimark, V.P. Potapov, M.A. Rutman, V.L. Shmuljan), as well as joint results together with his friends and colleagues (N.I. Achiezer, F.R. Gantmacher), are now characterized as classical and appear in textbooks on functional analysis.

During World War II, from 1941 to 1944, M.G. Krein held the Chair of Theoretical Mechanics at the Knibyshev (now Samara) Industrial Institute. M.G. preferred this position over the Chair in Mathematics because he thought that in technical institute of higher education it offered more interesting work and a wider set of possibilities and responsibilities. In 1944 he returned to Odessa. Soon he and his closest friend, B.Ja. Levin, were dismissed from Odessa University. This step was a direct result of the antisemitic Communist Party policy and corruption of the administration of the university (by the way the letter of dismissal was handed to M.G. on his birthday). Also their students had to leave. This was the end of the famous center of Functional Analysis at Odessa University.

The administration of Odessa University replaced M.G. Krein by more "reliable" mathematicians, such as N.A. Lednev, and N.I. Gavrilov. The former became famous for his Marxist critique of Einstein's Theory of Relativity, while the latter claimed publicly that he had resolved a number of the outstanding open problems

Originally published in *Integral Equations and Operator Theory* **30** (1998), 123–134.

in Mathematics, including the Riemann hypothesis. Each time elementary mistakes were found, and he tried to force the acceptance of his arguments under the pressure of the Communist Party administration. Further comments, I think are superfluous.

From 1944 to 1952 M.G. Krein held a part-time position as head of the Department of Functional Analysis and Algebra at the Mathematical Institute of the Ukrainian Academy of Science in Kiev. He was dismissed from this post in 1952. The official reason given was that he was not a permanent resident of Kiev. The real reason is easy to guess, it happened soon after the tragedy of the Jewish physicians.

In the period 1944–1954 M.G. was a professor at the Chair of Theoretical Mechanics at the Odessa Marine Engineering Institute. For reasons that are still unclear he did not try to extend his contract with this institution, instead he moved to a less prestigious institute – Odessa Civil Engineering Institute. Here he held the Chair of Theoretical Mechanics till his retirement. During the last period of his life he was a consultant to the Institute of Physical Chemistry of the Ukrainian Academy of Sciences.

2. Achievements

One of the most eminent mathematicians of our time, Mark Grigorievich Krein is the author of more than 270 papers and monographs of unsurpassed breadth and quality. His work opened up new areas of mathematics and greatly enriched the more traditional ones. He educated dozens of brilliant students in the USSR and inspired the work of many mathematicians, engineers and physicists all over the world.

A list of themes where M.G. Krein's research was fundamental, and in many cases even determined the future of the field, includes: oscillating (totally positive) kernel functions and matrices; problem of moments, orthogonal polynomials, and approximation theory; cones and regular convex sets in Banach spaces; the theory of gaps between subspaces and operators in spaces with two norms; the extension theory of Hermitian operators, continuation of Hermitian positive definite functions and helical arcs, entire operators; integral operators, direct and inverse spectral problems for nonhomogeneous strings and Sturm–Liouville equations; trace formula and scattering theory; method of directing functionals; stability theories for differential equations; Wiener–Hopf, Toeplitz and singular integral operators; operator theory in spaces with indefinite metric, indefinite extension problems; nonselfadjoint operators, characteristic operator-functions and triangular models; perturbation and Fredholm theories; interpolation and factorization theories; prediction theory for stationary stochastic processes; problems in elasticity theory, and ship waves and water resistance.

A profound intrinsic unity is characteristic of Krein's work. Interlacing of general abstract and geometric ideas with concrete and analytical results and applications are also characteristic of his work.

Krein was a very fine pedagogue and lecturer. He would always share his new ideas and plans with his students and colleagues. He was known for his scientific generosity and enthusiasm, as well as his kindness and attention to young mathematicians.

In 1982 M.G. Krein was awarded the prestigious international Wolf Prize in Mathematics in Jerusalem. The citation to this prize reads in part as follows: "His work is the culmination of the noble line of research begun by Chebyshev, Stieltjes, S. Bernstein and Markov and continued by F. Riesz, Banach and Szegö. Krein brought the full force of mathematical analysis to bear on problems of function theory, operator theory, probability and mathematical physics. His contributions led to important developments in the applications of mathematics to different fields ranging from theoretical mechanics to electrical engineering. His style in mathematics and his personal leadership and integrity have set standards of excellence."

Among his honorary awards, he was elected corresponding member of the Ukrainian Academy, 1939; honorary member of the American Academy of Arts and Sciences, 1968; Foreign Member of the National Academy of Sciences of the United States of America, 1979. He was also awarded the State prize of Ukraine, 1988.

3. Our First Meeting

I met Mark Grigorievich Krein in the autumn of 1950. M.G. Krein was already then a world renowned mathematician with a list of publications containing more than 75 very important papers and books. He was considered to be one of the strongest experts in functional analysis and its applications. His main position at that time was in the Marine Engineering Institute in Odessa at the Chair of Theoretical Mechanics.

I was at that time a student at the Kishinev University, in my fifth and final year. Our meeting was at my initiative, and my aim was to discuss with him my first results in operator theory. They were about Fredholm theory and index. After I met him he invited me to listen to his seminar talk about extremal distributions of the mass of a string and inverse problems. The talk was a masterpiece and made a great impression on me. He invited me to his home in the evening and we had detailed discussions on mathematics, my future education and other subjects.

He drew my attention to two topics, Toeplitz operators and normed commutative rings. He presented the topics in such an interesting and attractive way that I already fell in love with them, and planned to spend a lot of time studying them. During that evening M.G. told me that his first impression was that I am not Jewish, and he had planned to take me on as a graduate student. He explained to me that antisemitism is very strong and that the administration is not allowing acceptance of Jewish graduate students. He also told me that he had recently been dismissed from Odessa University, which had become very antisemitic, and relations towards him were personally hostile and extended also to his family and his friends. M.G. invited me to write to him and to visit him. After this conversation it became clear to me that my dream of becoming his graduate student would never be realized.

It was difficult to imagine then that I would one day be privileged to have M.G. Krein as a teacher, coauthor and friend. Joint work with M.G. Krein was an outstanding university in the best and most friendly atmosphere. I will always remember M.G. Krein with gratitude, affection and admiration.

4. Joint Work

For more than twelve years I worked intensively with M.G.. During those years we wrote twenty papers and two books. A third book was in our plans, but we never finished it. For the remaining twelve years M.G. was mostly busy with other projects and our joint work was less intensive. At other times I remember him to be busy with large joint projects with V.M. Adamjan and D.Z. Arov; and with each of the following colleagues: M.Sh. Birman, Yu.L. Daletsky, I.S. Iokhvidov, V.A. Jakubovich, I.S. Kac, H. Langer, G.Ya. Liubarsky, A.A. Nudelman, Yu.L. Shmuljan, and I. Spitkovsky.

Over a long period of time I spent two or three months a year in Odessa for joint work. Most of the time we worked in Krein's apartment; a small room there got the official title of Izea's office. At the beginning of our joint work we worked late into the night. In fact we often stopped working in the early hours of the morning, and that would cost me a ruble to get the porter of the house to open the door which was already locked for the night. Later we changed our hours of work for more convenient ones. Some of the time was during my official vacation when we would work on M.G.'s dacha on the shores of the Black Sea. On occasions we would even allow ourselves an hour off for a swim.

Mark Grigorievich was extremely demanding of himself and of everyone working with him. If, by a total change in the writing of a paper or a chapter he could improve it by even a small fraction, he would not think twice. It was never a question of whether it was worthwhile or not. Papers which were almost ready would lie for years until they attained perfection. He maintained that the writing of a paper is no less scientifically important than proving the result. As a rule the polishing process leads to new proofs, new connections and new mathematics. At the beginning of our collaboration M.G. himself wrote the formulas in by hand in the first copy of the paper, he did not trust me enough to let me do this. When I asked if this was so, he answered that he likes to do it himself because this gives him an opportunity to think again over the proofs and the whole material.

The writing of the joint books took up more and more time. Each time my wife would ask, "Is it finished?" and each time I would reply that a little more needs to be done. It is still told in Odessa, now as a joke, though at the time it was true, how L.A. Sakhnovich met me and asked, "How is the book going?" "Well, it is 85 percent ready," I replied. "Then why do you look so sad? That is wonderful." "Yes," I answered, "but if you had asked me yesterday I would have said it was 95 percent ready." Our students and friends made fun of us; I.S. Iokhvidov even wrote a little poem on the eve of the new year in 1963 (thanks to C. Davis for translating it into English).

> Around the festive table, all our friends
> Have come to mark our new book's publication.
> The fresh and shiny volume in their hands,
> They offer Izia and me congratulations.
> The long awaited hour is here at last.
> The sourest sceptic sees he was mistaken,
> And, smiling, comes to cheer us like the rest,
> And I'm so delighted,... I awaken.

When making selections or discussing expositions or proofs we sometimes had differences of opinion, and this often led to an argument. In general M.G. was a fighter and could make himself very convincing. But he would also listen attentively to my arguments. During such discussions we sometimes exchanged positions; I would adopt his position and he mine. But we always ended up on a cheerful note with a common point, jokes often being used to illustrate arguments.

After we wrote the first joint paper I suggested that his name appear before mine on the manuscript. My argument was based on the fact that this would be to the benefit of the paper, which would be more appreciated and attract more attention with his name first. M.G. refused, maintaining that he would not write joint work with a person who could not be an equal partner, and he always used the alphabetical order of names in joint papers. Any change may cause speculations in the minds of the reader. After a while I drew his attention to a paper by "M.G. Krein and M.A. Krasnoselsky". He laughed and remarked that the difference is in the third letter. M.G. considered joint work as the work of a team and after deciding that we would work together on certain topics thereafter we would not distinguish who did what in these topics. M.G. would compare joint work with a team of sportsmen, especially in a long term collaboration. Each player passes the ball on to another player to the best of his ability, and it is impossible to delineate which pass was the most important. These principles worked very well and during twenty four years of our collaboration we never had any misunderstanding.

Our joint work usually started with plans, discussions, preparations of ingredients for proofs and changes, changes, changes. In such discussions appeared also the basis and topic for a new paper. Then the joint work was realized in different ways. Sometimes one of us would write the first draft of the manuscript and the second would then polish the manuscript and make additions. This process would continue and the manuscript would change hands many times until it was considered ready for publication. Sometimes each of us would prepare a part of the manuscript and we would put the parts together at a very late stage. Sometimes we would write the manuscript together, discussing the presentation on the spot. Usually the final step in the process of preparation of the manuscript was its joint reading. We continued thinking about a paper even after it was considered ready for publication. Many times this led to surprising results and to many last minute changes.

Often we continued our mathematical discussions during walks in town. Part of the walk would take us along the main street – Deribasovskaya. It was especially pleasant on a sunny day. The street was very crowded and usually M.G. met many people he knew. Sometimes he would stop for a short conversation about the latest Odessa news. Next we would visit the main bookstore. M.G. would check the newly arrived books in mathematics, mechanics, physics and astro-physics. He usually made an order for a book in advance and his copy would be waiting for him. Our next stop would be the office supplies store. Here he would buy paper, ink, pens, pencils, glue and other accessories which we used in our work. Later I found out that M.G. liked to meet a young saleswoman in this store. She was really beautiful. She had a biblical look, with dark hair and black eyes. She always welcomed M.G. with a modest smile. We would also visit a second-hand store in the neighborhood. M.G. was looking for a typewriter with Latin letters. He planned to replace his old one (in the usual stores one could buy only typewriters with Russian lettering). He never found one. Sometimes we would also visit a hardware store. M.G. liked to buy new original tools for the household. He greatly admired the ingenuity of such tools.

M.G. and his family were very kind to me. This wonderful family always insisted that I eat with them. Once M.G. introduced me to a nice looking girl who was a friend of his daughter. Much later I learned that M.G. thought we would suit each other and that we would make the perfect couple. At the time I did not understand his intentions and I probably disappointed him.

Working with M.G. was a joy. I was never a formal student of his, but for me he was much more than an ordinary instructor. When I started to work with M.G. I was warned by a friend that I would get little credit for this joint work. I had also heard many tales about what a difficult person M.G. Krein was, but during the twenty four years in which we worked together I found Krein to be otherwise. He set very high standards for himself and I found his demands on others to be fair and reasonable.

5. Seminars and Lectures

For many years M.G. Krein ran a mathematical seminar at the Scientists House in Odessa. This house used to be an old palace, but since the revolution it served as a kind of club for scientists. The rooms with high ceilings were very luxurious, covered with expensive marble and wood paneling. Missing was only a good blackboard. The meetings were always in a very pleasant atmosphere. M.G. would usually open with an interesting introduction to the subject under discussion, and from this introduction one could feel how rich and wide was his experience. During the talk he would ask important questions and sometimes add remarks with an historical perspective, or other approaches and proofs. As a rule he would present details, closing the talk with illuminating comments. All was done in a very friendly manner and with a great respect for the lecturer and audience. At the end of such seminars I felt myself enriched. M.G. would leave the Scientists House for his home accompanied by a large group of his admirers.

Outstanding mathematicians participated in these seminars, some of them members of the famous older generation of M.G.'s students and colleagues, and also the younger generation. Among them was V.M. Adamjan, D.Z. Arov, M.L. Brodsky, Ju. P. Ginsburg, I.S. Iokhvidov, V.A.Javrjan, I.S. Kac, K.R. Kovalenko, H. Langer, F.E. Melik-Adamjan, S.M. Mkhitarjan, B.R. Mukminov, A.A. Nudelman, I. Ovcharenko, G.Ja. Popov, Sh.N. Saakjan, L.A. Sakhnovich, P.A. Shwartsman, Ju.L. Schmulyan, V.G. Sizov, and I. Spitkovsky. M.G. Krein also regularly held other interesting seminars on a smaller scale in the institutions where he was working. For instance he ran a seminar in Hydrodynamics in the Marine Engineering Institute. A.A. Kostiukov, V.G. Sizov and Yu.L. Vorob'ev were participants in this seminar. In Kiev were Yu.M. Berezansky, B.I. Korenblum, M.A. Krasnoselsky, and S.G. Krein (M.G.'s younger brother), in Kuibyshev, G.Ya. Liubarsky, A.V. Strauss and ON. Svirsky.

I recollect M.G.'s seminars and lectures as one of the best I ever heard. His lectures and talks were always well prepared with historical remarks and examples. Almost every year M.G. gave courses of lectures for experts and Ph.D. students. Usually these lectures were based on his recent results. Many of them were even unpublished. His lectures on the theory of entire operators (prepared for publication by the Gorbachuks) will appear only now in print, in Birkhäuser Verlag. Lectures on the theory of cones were circulating among experts for many years and were never published. Lectures on the theory of the string exist only in rough handwritten notes. A manuscript of a book on group representations (written together with M.S. Brodsky) is not finished.

As a matter of fact, almost no one from Odessa University attended M.G. Krein's seminars and lectures. They probably did not even realize how much they were missing.

6. Travel Abroad

M.G. Krein never traveled abroad. The reason for this was that he would not have been granted an exit visa had he applied. An exit visa for a trip abroad was considered by the authorities as a special privilege given only to the very "reliable". He received many invitations, then he would apply for an exit visa, entailing the filling in of thick official forms. He would then get the inevitable negative reply, or no reply at all. He even developed a sort of allergy to the filling in of forms, and during his last years it was very difficult to get him to undertake such a task, even on his wife's insistence. He already knew it would be of no use, so why bother.

In one case was he given permission to travel abroad. That was in 1970 to attend a conference in Tihany on the Balaton Lake in Hungary. It probably worked out this time because he used a private invitation which did not have to go through the high official channels. But this time he could not use the visa because precisely at that time there was an epidemic of cholera in Odessa and no one was allowed to leave Odessa. I took part in this conference and at the request of M.G. I gave Professor B. Sz.-Nagy (the organizer of the conference) regards from M.G. and told him the reason why M.G. could not come. Professor Sz.-Nagy smiled and answered, "So it's now called cholera, is it?" In the West people were already used to the various reasons that were invented to justify the absence of M.G. Krein. But this was the only time that the reason given was the true reason.

M.G. liked very much to meet and to receive foreign colleagues. However those meetings always brought repercussions. He would have to answer unpleasant questions put to him by KGB agents (in the best case).

The small technical institutions that he worked for did not have the official facilities to receive foreigners and to organize such visits (usually they existed only in large universities or research institutes). For instance, in the case of the one year visit of H. Langer an unusual solution was found. Langer was well established at Odessa University. There was his residence and there he received his salary, but scientifically he was connected only with M.G., with his seminars and his School.

Very often Mark Grigorievich would avoid a meeting with a colleague in order to prevent the inevitable repercussions and diffculties. In order to avoid similar difficulties and problems with the administration he did not send preprints and reprints abroad. Unfortunately foreign colleagues did not always understand these sad facts of real life in the USSR.

7. The Residence

M.G. Krein loved Odessa. He was well acquainted with the history of the town and even the history of the main streets. He liked especially the satirical books written about Odessa. Often he read out the funny bits for his guests.

All his mature life (with a relatively short exception during World War Two) the Kreins lived in Odessa in an old apartment house on 14 Artema Str., App. 6. This was the apartment in which his wife and her parents lived previously. The apartment building, which had been built at the beginning of this century, was luxurious. It had an elevator, heating, bathrooms, and every apartment had many large high ceilinged rooms, and separate quarters for servants. The owner of the building used to send a cleaner every week to polish the parquet flooring and to take care of other small things. In general he wanted to make sure that the house would be suitably maintained as such a building deserves.

After the revolution normal life in such apartments became extremely difficult and in many cases impossible. First of all the houses were nationalized and the newly appointed caretakers were more involved in watching the occupants and reporting on their doings than in maintaining the building. In each apartment new families were installed. Very often each family had only one room, sharing the use of the kitchen and other facilities. Very soon the elevator and heating broke down and the water failed to reach the upper floors. Everyone knew what was going on in each of the neighboring rooms and what was being served for dinner in each room. Situations arose which were described with great humor in Russian literature of the post revolution period. All of this was the lot also of the Krein's apartment. There were times when this apartment accommodated six families (according to the number of rooms). Sometimes running water did not reach the apartment (certainly not in the summer). Hot water they never had. Each morning there was a queue for the toilet. For many years heating was a serious problem. I remember one occasion during lunch or dinner while a lively discussion was taking place one of the members of the family made a sign, and the topic under discussion was automatically changed. The sign meant that someone from another room was passing by their room on the way to the bathroom or to the telephone and their discussion could be overheard. People from other rooms would often call in on the Kreins during a visit by colleagues.

Television was introduced to the Krein family relatively late. The family would watch together different programmes. They liked theater, opera, classical music, as well as other programmes. The Kreins liked the summer when they moved to the dacha on the shores of the Black Sea (in Arcadia) without the television.

Mark Grigorievich was happily married to Raisa L'vovna Romen. She was, an expert in naval architecture and worked in the Marine Engineering Institute. They had only one child. Their daughter, Irma Maxkovna Krein (Kozdoba), has a Ph.D. in Philology and is an expert in cybernetics. They also had a grandson Aleosha and and a great grandson, also called Mark.

8. Battling Hostilities

In general, M.G. Krein was a fair, very amiable and kind person. However, all of his life he battled against mediocrity. After the Second World War he had to contend with hostile elements which fought fiercely against him using the officially supported antisemitism which was rife in the Ukraine, and especially so in Odessa. He was accused of Jewish nationalism, presumably for having had too many Jewish students before the War. This accusation was certainly entered into his classified file and was held against him all of his life. Presumably, it played a significant role in his two dismissals which were mentioned earlier. He was not allowed to have Jewish students and was deprived of a university base. All attempts on the part of various societies, academies and individuals in the Soviet Union to gain for him some measure of the official recognition which he so richly deserved, were unsuccessful. Remarkably enough, he was never elected as a full member of the Ukranian Academy of Sciences. Their standards must have been very high. Worse than that, there were times when his friends feared that he was in serious danger of arrest. In 1948, in an article in a local Odessa newspaper, M.G. Krein was called a "rootless cosmopolitan". He was accused of too often quoting foreign mathematicians and following too much their ideas and ignoring the achievements of Russian and Soviet mathematicians. The official establishment considered this to be a crime. This accusation was supported publicly by a professor of the University in Astronomy. The latter based his conclusions on the famous paper of M.G. Krein and M.A. Rutman (Linear operators leaving invariant a cone in Banach space, Uspekhi Mat. Nauk 3, No. 1 (1948), 3–98; English translation: Amer. Math. Soc. Transl. (1), 10 (1962), 199–325). I guess he could understand only the general introduction to this paper and the list of references; remarkable that he did not need to know any more in order to make the accusation. This was the usual way of starting many of the campaigns against Jewish scientists, writers and cultural activists which often led to arrests. M.G.was lucky, he escaped without an arrest.

M.G. Krein gave a plenary talk at the Mathematical Congress held in Moscow in 1966. It was a very deep talk with a detailed analysis of the state of affairs in operator theory and its applications. Special attention was paid to the theory of nonselfadjoint operators. In this talk M.G. introduced, for convenient references, a mathematician Gokr (as a shortened form of Gohberg–Krein). My friend V.G. Boltjansky was working in the Press Committee of the Congress and mentioned this talk and Gokr in the public press. After the Congress he had serious problems with the mathematical authorities. He was accused of passing to the press information which had a "wrong political orientation".

In the second half of the sixties the situation worsened. In particular two very important and influential committees became almost openly antisemitic. I have in mind the Committee of Experts (Chairmen V.A. Il'in, V.S. Vladimirov) which dealt with approval of degrees in mathematics in USSR, and the Publication Committee (chairman L.S. Pontrjagin) which oversaw all publications in mathematics in USSR. Approval of the degree of "Doctor of Sciences" was absolutely refused for Jews, and projects for books by Jewish authors were rejected. Laughable reasons were given to justify all of this.

M.G. Krein responded to these hostile surroundings in the only way open to him: by deep research and hard work. He and many of his students were protected by virtue of his outstanding achievements. In retrospect, it seems clear that he won this very difficult struggle. Firstly, he was able to devote all his life to mathematics (teaching and research): the work he loved so much. Secondly, he was able to spend most of his life in Odessa, a town which he had always regarded with love and affection (some of his friends thought that his life would have been much easier in Moscow or Leningrad). Thirdly: he was always the leader of a strong and dedicated group of colleagues and followers who loved and respected him. (This group existed almost on a private basis: holding many of its meetings in his house, or at the Scientists Club.) Fourthly: he had a great impact on the development of mathematics and its applications throughout the world. Even though he was never allowed to travel abroad: his brilliant work knew no borders.

In spite of all the difficulties that surrounded him: M.G. was cheerful and optimistic. He liked to tell stories and jokes, some of which he invented himself. At one time he went to the rector of his institute and asked if there was any danger that he could be accused of Armenian nationalism since he had four graduate students who were Armenians. The rector did not understand the joke and tried, in all seriousness, to explain that in this case there was no danger.

9. Epilogue

This fight took a heavy toll on his health, and towards the end of his life he suffered from depression. This condition worsened after the tragic loss within one year of his wife, and his only grandson, Alcosha. On October 17, 1989: M.G. Krein died in Odessa (USSR). There he is buried.

The passage of time brought about considerable changes. In the Institute of Mathematics of the Ukrainian Academy and the Institute of Mathematics of Odessa University some of M.G. Krein's followers and outstanding former students are now working. Preparation of an international conference on the occasion of the ninetieth anniversary of the birth of M.G. Krein is in progress. This conference is being organized by the two above mentioned institutes, which just half a century earlier had dismissed M.G. The conference is even partially supported by the city of Odessa. The conference will take place in Arcadia (in the suburbs of Odessa) on the shores of the Black Sea, not far from the Krein family dacha, where M.G. liked so much to work. There he spent many hours working with students and friends. And there he also spent the last days of his life. Unfortunately the dacha no longer belongs to the Krein family. It was taken away from them recently.

We hope that this conference will help everyone to understand in full measure who was Mark Grigorievich Krein, and also to understand the bitter and tragic mistakes and injustices of the past and the present.

10. References

The following sources were used in these recollections:

1. M.G. Krein. Autobiography, 1966. Private communication (Russian).
2. I. Gohberg and M. Kac. Biography of M.G. Krein. Topics in functional analysis. Essays dedicated to M.G. Krein on the occasion of his seventieth birthday, I. Gohberg and M. Kac editors, Academic Press, 1979.
3. I. Gohberg, Mathematical Tales. Gohberg Anniversary Collection, vol. 1, edited by H. Dym, S. Goldberg, M.A. Kaashoek and P. Lancaster. Birkhäuser Verlag, OT 40, 1989, pp. 17–56.
4. I. Gohberg. Mark Grigorievich Krein (1907–1989), Notices of the American Mathematical Society, vol. 37, no. 3, 1990, pp. 284–285.

Heinz Langer and his Work

Aad Dijksma and Israel Gohberg

1. Introduction

This volume is dedicated to Professor Heinz Langer to honor him for his outstanding contributions to mathematics. His results in spectral analysis and its applications, in particular in spaces with an indefinite metric, are fundamental. Five main themes emerge in Heinz Langer's work, some of them are closely connected or have much in common:

(1) Spectral theory of operators in spaces with indefinite inner product.
(2) Pencils of linear operators (nonlinear eigenvalue problems).
(3) Extension theory of operators in spaces with indefinite inner product.
(4) Block operator matrices.
(5) One-dimensional Markov processes.

Heinz has written more than 130 research papers with 45 coauthors from 11 countries. He advised about 25 Ph.D. students and always enjoyed cooperation with colleagues, students and friends. As a teacher, Heinz has the ability to clarify connections and to point out the important. His work has numerous followers and great influence in the world centers of operator theory.

The occasion marking the origin of this book is Heinz Langer's sixtieth birthday. Two of his collaborators Martin Blümlinger and Fritz Vogl, with the help of Gabi Schuster, organized a two day Colloquium on Thursday and Friday, 12 and 13 October 1995, at the Technical University of Vienna. Friends and colleagues from all over the world attended. At the end of the conference it was decided to prepare this anniversary volume.

2. Biography of Heinz Langer

Heinz Langer was born on August 8, 1935 in Dresden. He went to school and Gymnasium there and attended the Technical University of Dresden from 1953 to 1958. Originally he wanted to study physics, but the selection principles in the then eastern part of divided Germany were against him. Thanks to some personal connections of a friend to a professor in mathematics, he was enrolled in

Originally published in *Operator Theory: Advances and Applications* **106** (1998), pp. 1–5.

mathematics. In 1958, after the diploma (with a thesis on perturbation theory of linear operators), staying at the TU Dresden as an assistant, he began studying linear operators in indefinite inner product spaces.

It was his teacher, Professor P. H. Müller, who recommended this topic: At a conference in Hungary he had listened to a lecture of János Bognár, and had encountered such questions in his investigations of operator polynomials. Heinz studied the fundamental papers by I. S. Iohvidov and M. G. Kreĭn, and, in the fall of 1959, he showed his results to Professor Szőkefalvi-Nagy from Szeged, who visited Dresden on his first trip abroad after the 1956 revolution in Hungary. Sz.-Nagy reacted positively but, since he did not consider himself a specialist in this field, recommended to send it to Professor M. G. Kreĭn. Heinz sent a handwritten (Kreĭn mentioned this later, and not only once!) manuscript to him. Its main result was a generalization of the Theorem of L. S. Pontrjagin about the existence of a maximal nonnegative invariant subspace of a self-adjoint operator in a Pontrjagin space to a Kreĭn space. This result attracted Kreĭn's attention, and so he invited Heinz to stay for one year in Odessa.

There existed exchange programs for graduate students between the German Democratic Republic (GDR) and the Soviet Union (mainly used by students from the GDR), and Heinz applied. Before being admitted one had to undergo a preparatory course at a special faculty of the University of Halle, which lasted one month in 1960. Heinz attended this course in the summer of 1960, but after two weeks he was sent home: He was found ideologically unsuitable for a longer stay in the Soviet Union.

In the fall of 1960 Heinz received his Ph.D. at the TU Dresden. In September 1961, after someone from the ministry had encouraged him to apply again for a stay in Odessa and the preparatory course had been shortened from one month to two weeks, students and graduate students separated, he was finally admitted to go to the University of Odessa on a post doc fellowship for one year. However, until he arrived in Odessa, Heinz did not know that M. G. Kreĭn did not work at the University of Odessa, but held the chair of Theoretical Mechanics at the Odessa Civil Engineering Institute.

At that time in Odessa each week there was a special lecture of M. G. Kreĭn and about three seminars at different institutions (regularly at the Civil Engineering Institute, the Pedagogical Institute and monthly at the House of Scientists). In the second part of his Ph.D.Thesis, Heinz had applied the Livšic-Brodskiĭ model for operators with finite-dimensional imaginary part in order to obtain a model for self-adjoint operators in Pontryagin spaces. So he was quite well acquainted not only with indefinite inner product spaces, but also with some of the other main topics of interest of the Odessa school of functional analysis when he arrived, and he could actively take part in it.

The intense mathematical life in the circle around M.G. Kreĭn to which his former students belonged, among them M.S. Brodskiĭ, M.L. Brodskiĭ, I.S. Iohvidov,

I. S. Kac, Ju. P. Ginzburg, Ju. L. Smul'jan, but also V. P. Potapov and L. A. Sakhnovic, deeply impressed Heinz and had a great influence on his entire carrier and interests. In the Introduction to his Habilitationsschrift Heinz describes M. G. Kreĭn's influence as follows: 'Wer das Glück hatte, eine längere Zeit in der Umgebung von Professor M. G. Krein in Odessa arbeiten zu können, weiß, welche Fülle von Gedanken und Anregungen er ständig ausstrahlt. Diese habe ich in wesentlich größerem Maße ausgenutzt, als in der Einleitung zum Ausdruck gebracht werden konnte.' Heinz's high regard for M. G. Kreĭn was reciprocated: M. G. Kreĭn considered Heinz one of his most brilliant students and collaborators.

This fruitful collaboration lasted for almost twenty five years and ended with the death of M. G. Kreĭn. M. G. Kreĭn has worked with many mathematicians. Of his joint publications, most are written with I. Gohberg, next in number come those with Heinz. In Odessa Heinz also met I. Gohberg for the first time. The friendship with him and with the doctoral students of M. G. Kreĭn of that time (V. M. Adamjan, D. S. Arov, V. A. Javrjan and S. N. Saakjan) lasts until today!

During this year in Odessa Heinz completed the main result of his thesis by adding a statement about the location of the spectrum of the restriction of the selfadjoint operator in this invariant subspace which was now the full generalization of Pontryagin's theorem. Jointly with M. G. Kreĭn he proved the existence of a spectral function (with certain critical points) for a selfadjoint operator in a Pontryagin space. In the following years (until 1965) Heinz showed the existence of a spectral function for the more general situation of a definitizable operator in a Kreĭn space. Thus, besides the existence of a maximal nonnegative invariant subspace, a second cornerstone of the spectral theory of selfadjoint and other classes of operators in Kreĭn spaces was laid.

In 1955 R. J. Duffin proved a result in connection with network theory about strongly damped selfadjoint second order matrix pencils $\lambda^2 I + \lambda B + C$, which M. G. Kreĭn ingeneously interpreted as the existence of a solution Z of the quadratic matrix equation $Z^2 + BZ + C = 0$. Heinz realized that the main result of his Ph.D. thesis, which was proved just as an abstract generalization without any application in mind, could easily be applied in order to get a corresponding result for the infinite dimensional case. This was worked out by M. G. Kreĭn and Heinz in the summer of 1962, and thus finally the spectral theory of selfadjoint operators in Kreĭn spaces found an important application. It should be added that at about the same time M. G. Kreĭn was working with I. Gohberg on the two books on nonselfadjoint operators and quite a few results of this theory also turned out to be useful for second order pencils.

The main results of the years 1961–1965 were summed up in Heinz's Habilitationsschrift. This Habilitationsschrift became well-known among people interested in spaces with an indefinite inner product and had a big impact on the development of the spectral theory of operators in such spaces. The results about the spectral function for definitizable operators were published without proof only in

1971 in [22], and the original proofs were published only seventeen years later in the Lecture Notes [63].

At this time, Heinz received an offer to work at the Mathematical Institute of the Academy of Sciences of the GDR, but he preferred to remain at the Technical University of Dresden as this also involved teaching and working with Ph.D. and post doctoral students, which he always liked and still likes to do.

The next important period in Heinz's development was his one year stay in Canada during the academic year 1966/67. After having met him at an operator theory conference in Balatonföldvar in Hungary in 1964, Professor I. Halperin invited Heinz to spend the academic year 1965/66 with a fellowship of the National Research Council of Canada at the University of Toronto. As was to be expected in these years of the Berlin wall, the authorities of the GDR did not allow Heinz to accept this invitation. However, Halperin insistently renewed it for the following year, and so Heinz could spend the academic year 1966/67 at the University of Toronto. Shortly before, Heinz had married, and in May 1967 his only daughter Henriette was born. In Canada he also met Peter Lancaster, with whom he shared interests not only in operator pencils, but also personal ones like skiing and hiking in the mountains.

After returning in 1967, Heinz was appointed Professor at the Technical University of Dresden. Shortly afterwards in 1968, with the '3rd Hochschulreform' in the GDR, research at the universities was reorganized. It turned out that officially there should be no research group in analysis at the Technical University of Dresden, but only groups in 'Numerik', 'Mathematische Kybernetik und Rechentechnik' and 'Stochastik'. Heinz joined the last group and became interested in semigroup theory and Markov processes, in particular one-dimensional Markov processes. A fairly wide class of such processes, which contains diffusion processes and birth and death processes, can be described by a second order generalized or Kreĭn-Feller derivative, which Heinz had come across in Odessa. Together with his students he considered in particular processes with nonlocal boundary conditions and the time reversal of such processes. Nevertheless, he continued to be interested in operators in indefinite inner product spaces. The disadvantage of the situation was that he could not lecture about the topics he liked best. Instead, he lectured on 'Semigroups', 'Spectral theory of Kreĭn-Feller differential operators', 'Markov processes' etc and the topics for graduate students also had to have a probabilistic touch. At this point, it turned out to be useful that Heinz was known abroad: From 1970 on some mathematicians from other countries came to Dresden to do their Ph.D. work with him (Pekka Sorjonen, Björn Textorius, Karim Daho, and later Branko Ćurgus, Muhamed Borogovac), which was, of course, in operator theory, the topic which always was his favourite.

In 1969 Heinz again visited M. G. Kreĭn for one month. By then the famous papers of Adamjan, Arov and Kreĭn were finished. Because of these results, it seemed to be necessary and promising to generalize the extension theory for symmetric operators in Hilbert space to Pontryagin spaces. In fact, it was clear that this would give an operator theoretic approach to the Adamjan, Arov, Kreǐn results. This turned out to be an interesting and fruitful program: Already the generalization of the classical von Neumann-Kreǐn-Naimark extension theory showed new and interesting features in the indefinite case. In the following years also M. G. Kreǐn's theory of generalized resolvents, resolvent matrices and entire operators was extended to the indefinite situation.

In the seventies and early eighties Heinz visited Odessa quite regularly, sometimes officially, sometimes not quite officially, and then it was difficult to get the permission for the stay in Odessa. Sometimes the Rector of the Odessa Civil Engineering Institute could help. Another time, Heinz, without permission, just stayed at the apartment of Ju. L. Smul'jan, which was of course completely illegal and certainly a risk for the Smul'jan family.

During these years, besides the abstract lines of extension theory of symmetric operators, applications to indefinite moment problems, interpolation problems, to the continuation problem for hermitian functions with a finite number of negative squares, and to boundary eigenvalue problems were also studied. In the abstract results as well as in the applications the classical Hilbert space results were sometimes also completed, that is, the role of the Q-function in extension theory was worked out. These were the main topics of Heinz's work in these years, often done jointly with students and friends, mostly from outside Dresden or even outside the GDR. In addition, he also returned to work on operator pencils in the seventies. One of the main results he proved was the equivalence of a factorization of an n-th order pencil with the existence of a properly supported invariant subspace of the companion operator. At that time operator pencils were also studied by A. S. Markus and V. M. Matsaev. While Heinz used results from operator theory in indefinite inner product spaces as a tool, Markus and Matsaev applied results from factorization theory of analytic operator functions by I. Gohberg and J. Leiterer. It happened that on the same day in Kreǐn's seminar both methods were presented in a kind of friendly competition. However, only after Markus and Matsaev had moved to Israel and Heinz to Vienna, did the three of them start working together.

In the seventies Heinz could travel to the West almost every year, keeping the number of trips in balance with trips to the East. He lectured at the Universities of Jyväskylä, Uppsala, Linköping, Antwerp, and the KTH Stockholm during stays of a few weeks. Nevertheless, his applications for journeys to the West were not always successful, and it was impossible until the middle of the eighties to go to conferences there. In one case the permission to attend was granted only after the conference was over. While abroad he usually contacted Israel Gohberg and other friends (which was forbidden by the GDR authorities). Once Israel sent a letter for Heinz to Sweden, which arrived only after he had left. It was forwarded by the secretary to Dresden. Luckily someone gave him the open letter before it was read by some department officials (which was the rule for mail from abroad), and so it did not cause Heinz any problems.

In the eighties an intense collaboration developed with Aad Dijksma and Henk de Snoo from Groningen. They studied thoroughly the classes of analytic functions which arose in the extension theory of symmetric operators in spaces with indefinite inner product and applied this theory in order to get a unified treatment for selfadjoint boundary eigenvalue problems of ordinary differential operators containing the eigenvalue parameter in the boundary condition. On the invitation of Rien Kaashoek, he also regularly visited the Vrije Universiteit Amsterdam. All these contacts with colleagues from operator theory, which usually grew into friendship, were very important for Heinz's work. Also at that time, a result of R. Beals appeared about the half range completeness of Sturm-Liouville operators with an indefinite weight. Under the influence of Åke Pleijel, Heinz had considered such problems with Karim Daho already in the seventies in the context of the Kreĭn space generated by the weight function. So he understood that Beals' result could be interpreted as the regularity of the critical point infinity of the spectral function of the selfadjoint operator which can be associated with the problem in this Kreĭn space. This was further elaborated in cooperation with Branko Ćurgus.

In January 1988 Heinz was allowed for the first time to accept an invitation to West Germany: Reinhard Mennicken had invited him to spend one month at the University of Regensburg. They started joint work on the connections of operator pencils and special functions. Following an idea of F. W. Schäfke, certain systems of special functions were interpreted as eigenfunction systems of pencils of differential operators. They also began studying block operator matrices, which has been another main topic of Heinz's interests since then. The problem is to express the spectral properties of an operator, acting in the product of two spaces and given as a block operator matrix, by the properties of the entries of this matrix.

At the beginning of October 1989, shortly before the fall of the Berlin wall, Heinz fled from the GDR and went to West Germany. His first contact point was Regensburg. The decision to leave had ripened for a longer time and was certainly hard: a secure position at the university, pupils, friends and a part of his life had to be left behind. However, the pressure was stronger. Thanks to the assistance and intercession of Albert Schneider, Heinz obtained first a one year position as a professor at the University of Dortmund, and then, with the support of Reinhard Mennicken, a professorship at the University of Regensburg. Since August 1991 Heinz has held a chair in 'Anwendungsorientierte Analysis' at the Technical University of Vienna. Released from the psychological tension of life in the GDR, Heinz's life and work has come to a new blossoming. Within the last seven years he has organized three workshops on operator theory and its applications in Vienna, one of them in cooperation with the Schrödinger Institute, and he enjoys attending conferences all over the world. He has created a center of active research in operator theory in Vienna, attracting visitors from many countries, and still keeping a nice balance between those coming from the West and those from the East.

3. Some main results

In this section we explain some of Heinz's main results in detail. We focus on some theorems from the first three themes mentioned in the Introduction and relate them to the work of others, but first we recall some definitions.

An inner product space $(\mathcal{K}, [\cdot, \cdot])$ is called a Kreĭn space if $\mathcal{K}$ is a complex linear space which has a fundamental decomposition with respect to the inner product $[\cdot, \cdot]$, that is, a decomposition of the form

$$\mathcal{K} = \mathcal{K}_+ [\dot{+}] \mathcal{K}_-,$$

where $[\dot{+}]$ denotes the direct $[\cdot, \cdot]$-orthogonal sum and $(\mathcal{K}_\pm, \pm[\cdot, \cdot])$ are Hilbert spaces. The fundamental decomposition induces a Hilbert space inner product on $\mathcal{K}$, given by

$$(x, y) = [x_+, y_+] - [x_-, y_-], \quad x = x_+ + x_-, \ y = y_+ + y_-, \quad x_\pm, y_\pm \in \mathcal{K}_\pm.$$

The operator $J = P_+ - P_-$, where $P_\pm$ is the $(\cdot, \cdot)$-orthogonal projection onto $\mathcal{K}_+$, is called the fundamental symmetry corresponding to the fundamental decomposition. Note $(x, y) = [Jx, y], x, y \in \mathcal{K}$. Although the fundamental decomposition is not unique, different ones generate equivalent Hilbert space norms. Topological notions refer to this Hilbert space topology. For example, a subspace of $\mathcal{K}$ is a linear manifold in $\mathcal{K}$ which is closed and continuity of an operator means continuity with respect to this norm topology, etc. We denote by $\mathbf{L}(\mathcal{K})$ the set of bounded linear operators on $\mathcal{K}$.

The numbers $\dim \mathcal{K}_\pm$, each either a nonnegative integer or infinity, do not depend on the fundamental decomposition $\mathcal{K} = \mathcal{K}_+ [\dot{+}] \mathcal{K}_-$ of $\mathcal{K}$. If $\dim \mathcal{K}_+ = 0$ $\mathcal{K}$ is sometimes called an anti-Hilbert space. The Kreĭn space $(\mathcal{K}, [\cdot, \cdot])$ is called a π_κ-space or a Pontryagin space of index κ if $\kappa := \min(\dim \mathcal{K}_+, \dim \mathcal{K}_-) < \infty$. In the sequel we consider only Pontryagin spaces for which $\kappa = \dim \mathcal{K}_-$.

A linear subset is called nonnegative if its elements x have a nonnegative self inner product: $[x, x] \geq 0$; a nonpositive subset is defined in a similar way.

The linear operators, which we consider in the Kreĭn space $\mathcal{K}$, will in general be densely defined and closed or closable. If A is a densely defined operator then its adjoint A^+ is defined as follows: $\operatorname{dom} A^+$ is the set of all $u \in \mathcal{K}$ for which there exists a $v \in \mathcal{K}$ with

$$[Ax, u] = [x, v] \text{ for all } x \in \operatorname{dom} A,$$

and in this case $A^+ u = v$. We have $A^+ = JA^*J$, where A^* is the adjoint of A with respect to the Hilbert space inner product $[J\cdot, \cdot]$. The operator A in the Kreĭn space $\mathcal{K}$ is called selfadjoint if $A = A^+$, symmetric if $A \subseteq A^+$, unitary if

$A^+A = AA^+ = I$, isometric if $A^+A = I$, contractive if $[Ax, Ax] \leq [x, x]$ and expansive if $[Ax, Ax] \geq [x, x]$, for all $x \in \mathcal{K}$.

As mentioned before Heinz started his work in indefinite inner product spaces by studying the two articles by I. S. Iokhvidov and M. G. Kreĭn that had appeared in 1956 and 1959. Later these papers were incorporated in the joint work [57] with Heinz. It is a clear and comprehensive introduction to spectral theory in Pontryagin spaces and contains the basic results, not only on hermitian and isometric operators, but also on contractive and expansive operators in Pontryagin spaces. The material about the latter classes of operators has recently been reconsidered and completed in the joint publication [105] with Tomas Azizov.

From this early period also dates the Hilbert space result called "Langer's Lemma" (terminology from N.K. Nikol'skiĭ, Treatise on the shift operator, Grundlehren der mathematischen Wissenschaften 273, Springer-Verlag, Berlin, 1985) about the orthogonal decomposition of a Hilbert space contraction into a unitary part and a completely nonunitary part; see [2]. The same result was obtained independently by B. Sz.-Nagy and C. Foiaş.

(1) The maximal nonnegative invariant subspace theorem in Kreĭn spaces proved by Heinz reads as follows: If A is a selfadjoint operator in the Kreĭn space $\mathcal{K}$ and for some fundamental symmetry $J = P_+ - P_-$, ran $P_- \subset \operatorname{dom} A$ and $P_+ A P_-$ is compact, then A has a maximal nonnegative invariant subspace. In particular, a (bounded or unbounded) selfadjoint operator A in a Pontryagin space has a maximal nonnegative and a maximal nonpositive invariant subspace. Since the latter is finite dimensional, A has eigenvalues with corresponding nonpositive eigenvectors. This statement can be made more precise by considering multiplicities and the location of the eigenvalues with respect to the real axis, see [57].

The spectral theory of definitizable operators in Kreĭn spaces developed in the Habilitationsschrift "Spektraltheorie linearer Operatoren in J-Räumen und einige Anwendungen auf die Schar $L(\lambda) = \lambda^2 + \lambda B + C$", TU-Dresden, 1965, is a cornerstone in the operator theory in spaces with an indefinite metric. The spectral function of a definitizable operator and the description of its behavior in the critical points are powerful tools in the abstract theory as well as for the investigation of operators in function spaces, such as differential operators. We review the main definitions and results:

The selfadjoint operator A in the Kreĭn space $\mathcal{K}$ is called definitizable if the resolvent set $\rho(A)$ of A is nonempty and there exists a polynomial p such that

$$[p(A)x, x] \geq 0, \quad x \in \operatorname{dom} A^k,$$

where k is the degree of p; p is called a definitizing polynomial of A. The set $c(A)$ of critical points of A is the set of all $\lambda \in \mathbf{R}$ such that $p(\lambda) = 0$ for each definitizing polynomial p of A, and $\infty \in c(A)$ if one (and hence each) definitizing polynomial is of odd degree and $\sigma(A)$ contains arbitrarily large positive and negative numbers. It can be shown that $c(A) \subseteq \sigma(A)$, the spectrum of A. We denote by $\mathcal{R}_A$ the

Boolean algebra generated by all intervals of $\mathbf{R} \cup \{\infty\}$ whose endpoints are not in $c(A)$. Heinz proved that there exists a unique mapping $E : \mathcal{R}_A \to \mathbf{L}(\mathcal{K})$ with the following properties (Δ and Δ' are arbitrary elements of $\mathcal{R}_A$):

(a) $E(\Delta) = E(\Delta)^+$, $E(\emptyset) = 0$.
(b) $E(\Delta \cap \Delta') = E(\Delta)E(\Delta')$ (in particular, $E(\Delta)$ is a projection).
(c) $E(\Delta \cup \Delta') = E(\Delta) + E(\Delta') - E(\Delta)E(\Delta')$.
(d) $E(\mathbf{R}) = 1_{\mathcal{K}} - E_0$, where E_0 is the Riesz-Dunford projection associated with the nonreal spectrum $\sigma(A) \setminus \mathbf{R}$ of A.
(e) If $p|_\Delta > 0$ (or $p|_\Delta < 0$) for some definitizing polynomial p of A then $E(\Delta)\mathcal{K}$ is a positive (negative, respectively) subspace.
(f) $E(\Delta) \in \{(A - z)^{-1}\}''$, the double commutant of the resolvent $(A - z)^{-1}$ of A, $z \in \rho(A)$.
(g) If Δ is bounded then $E(\Delta)\mathcal{K} \subseteq \operatorname{dom} A$; if Δ is unbounded then $E(\Delta)\mathcal{K} \cap \operatorname{dom} A$ is dense in $E(\Delta)\mathcal{K}$. In both cases $\sigma(A|_{E(\Delta)\cap\mathcal{K}}) \subset \Delta^c$.
(h) If A is bounded then

$$p(A) = \int_{\mathbf{R}} p(\lambda)E(d\lambda) + N,$$

where N is a bounded nonnegative operator in $\mathcal{K}$ with $N^2 = 0$. (The integral here is improper with respect to the points $c(A)$ as at these points E is not defined.)

The mapping E is called the spectral function (with critical points) of the definitizable operator A. The statement (e) implies that $\mathcal{K}_\Delta = E(\Delta)\mathcal{K}$ is a Hilbert space or anti–Hilbert space if p is positive or negative on $\Delta \cap \sigma(A)$ for some definitizing polynomial p. This space $\mathcal{K}_\Delta$ reduces A and $A|_{\mathcal{K}_\Delta}$ is a bounded or unbounded and densely defined Hilbert space selfadjoint operator in $\mathcal{K}_\Delta$. Thus, with the exception of the (finitely many) points in $c(A)$, the definitizable operator A has locally the same spectral properties as a selfadjoint operator in a Hilbert space. The critical points $\lambda \in c(A)$ can be characterized as follows:

$$\Delta \in \mathcal{R}_A, \; \lambda \in \Delta \Rightarrow \text{ the inner product } [\cdot, \cdot] \text{ is indefinite on } \mathcal{K}_\Delta.$$

In [22] and [12] the results from the Habilitationsschrift in a completed form were published. It was proved in [22] among other more general results that a bounded definitizable selfadjoint operator in a Kreĭn space has a maximal dual pair of invariant subspaces. The question of uniqueness of such pairs was considered in the note [18].

The paper [12] concerns selfadjoint operators in Kreĭn spaces that arise from a fundamentally reducible selfadjoint operator by perturbations of Matsaev class. Under some additional assumptions it was proved that these operators possess a local spectral function. This paper is the starting point for the study of locally definitizable operators in papers by Peter Jonas and in the joint work [125] with Alexander Markus and Vladimir Matsaev. In the latter paper the sign classification of the spectrum of a selfadjoint operator in a Kreĭn space is obtained with the help of approximating eigensequences. This new approach is applied to the study of bounded and compact perturbations of selfadjoint operators in Kreĭn spaces. The main result in [125] on compact perturbations contains the result from [12] and has applications to the block operator matrices.

The papers [49] with Peter Jonas and [65] with Branko Najman deal with the perturbation theory of definitizable operators. In the first paper, for example, it is shown that within the class of selfadjoint operators, finite-rank perturbations of the resolvent preserve definitizability and that the emerging new critical points are of a special type. In the second paper, stability properties of the spectral function and its critical points are studied. In the paper [103] with Peter Jonas and Björn Textorius a model for an arbitrary selfadjoint operator in a Pontrjagin space is established. The model is closely related to selfadjoint differential operators with inner singularities arising in mathematical physics, presently under investigation with Aad Dijksma and Yuri Shondin.

(2) The spectral theory of selfadjoint operator pencils, of which Heinz is co-founder, is closely connected with indefinite metrics. The maximal nonnegative invariant subspace therorem has a large number of applications, for example, to the existence of an operator root and hence to the factorization of operator pencils. The joint papers [8] and [7] contain new ideas and methods which determined the development of this area for decades, and lead to new publications in spectral theory and in applications to mechanics and physics. To be more specific, with the pencil $L(\lambda) = \lambda^2 + \lambda B + C$ there is associated the quadratic operator equation

$$Z^2 + BZ + C = 0$$

and M.G. Kreĭn and Heinz looked for a root Z whose spectrum coincides with a specified part of the spectrum of the pencil. This problem is closely connected with the problem of factorizing the pencil, that is, the problem of representing it in the form

$$L(\lambda) = (\lambda I - Y)(\lambda I - Z).$$

This approach can be used even when the pencil is not selfadjoint. But in the selfadjoint case $B = B^*$, $C = C^*$, they proved that the quadratic equation has a root with the help of the invariant subspace theorem mentioned above applied to a certain companion matrix for the pencil. In the seventies Heinz proved general yet strong results on the factorization of operator polynomials of arbitrary degree; see [27], [30], [33], and [38]. For example, the operator polynomial

$$L(\lambda) = \lambda^n I + \lambda^{n-1} A_{n-1} + \cdots + \lambda A_1 + A_0$$

with operators A_j in a Hilbert space $(\mathcal{H}, (\cdot, \cdot))$ admits a factorization

$$L(\lambda) = N(\lambda)M(\lambda), \quad M(\lambda) = \lambda^k I + \sum_{j=0}^{k-1} \lambda^j B_j, \quad N(\lambda) = \lambda^{n-k} I + \sum_{j=0}^{n-k-1} \lambda^j C_j,$$

if and only if the companion operator

$$\tilde{A} = \begin{pmatrix} -A_{n-1} & \cdots & -A_1 & A_0 \\ 1 & & 0 & 0 \\ \vdots & & \vdots & \vdots \\ 0 & & 1 & 0 \end{pmatrix}$$

acting in $\tilde{\mathcal{K}} = \mathcal{H}^n$ has a specific invariant subspace. This result has been applied by many authors both for the selfadjoint and for the nonselfadjoint case. When the operators A_j are selfadjoint, the companion operator is selfadjoint with respect to the $\tilde{G}$–inner product $(\tilde{G}\cdot, \cdot)$ on $\tilde{\mathcal{K}}$, where

$$\tilde{G} = \begin{pmatrix} 0 & 0 & \cdots & 0 & 1 \\ 0 & 0 & & 1 & A_{n-1} \\ \vdots & \vdots & & & \vdots \\ 0 & 1 & & & A_2 \\ 1 & A_{n-1} & \cdots & A_2 & A_1 \end{pmatrix}.$$

In examples the operators $A_0, \ldots, A_{n-1}$ are often unbounded. Sometimes $\tilde{A}$ can also be considered in this situation, sometimes by a simple transformation the given pencil can be transformed into one with bounded operators.

Because of the formula

$$L(\lambda)^{-1} = Q(\tilde{A} - \lambda)^{-1} P, \quad P = \begin{pmatrix} I \\ 0 \\ \vdots \\ 0 \end{pmatrix}, \quad Q = \begin{pmatrix} 0 & \cdots & 0 & I \end{pmatrix},$$

where P is a mapping from $\mathcal{H}$ into $\tilde{\mathcal{K}}$ and Q maps $\tilde{\mathcal{K}}$ into $\mathcal{H}$, the companion matrix $\tilde{A}$ is sometimes called the linearization of the pencil $L(\lambda)$. In the lecture series [120] other eigenvalue problems whose linearization lead to selfadjoint operators in Kreĭn spaces are discussed.

We also mention the following natural and beautiful result, which has a simple formulation but a complicated proof. We use the same notation as above. If $L(\lambda)$ is a selfadjoint polynomial and for some segment $[a, b]$ on the real axis,

$$L(a) \ll 0, \quad L(b) \gg 0, \quad L'(\lambda) \gg 0 \ (a < \lambda < b),$$

then $L(\lambda)$ admits the above factorization with $k = 1$, $M(\lambda) = \lambda I - Z$ and the spectrum of the operator Z lies in (a, b). The operator Z not only has a real spectrum, but it is also similar to a selfadjoint operator.

Finally, in 1971–1973 Heinz studied the important class of weakly hyperbolic selfadjoint operator polynomials of arbitrary degree (or polynomials with real zeros) and proved theorems about their so-called spectral zones and factorizations. For the quadratic case this class is the class of "strongly damped pencils" and was considered earlier jointly with M.G. Kreĭn in [7] and [8].

(3) In the four "Fortsetzungsprobleme" papers [35], [40], [54], and [75] M.G. Kreĭn and Heinz formulate and study indefinite analogues of interpolation, moment and continuation problems. These papers contain a wealth of interesting results, which have subsequently been generalized by many authors. The indefiniteness comes in by requiring that certain kernels have κ negative squares, $\kappa \in \{0, 1, \ldots\}$. A kernel K on a nonempty set Ω is a function $K : \Omega \times \Omega \to \mathbf{C}$ which is hermitian: $K(z, w) = \overline{K(w, z)}$. It has κ negative squares on Ω if for every natural number n and arbitrary points $z_1, z_2, \ldots, z_n \in \Omega$, the hermitian matrix $(K(z_i, z_j))_{i,j=1}^n$ has at most and for at least one choice of $n, z_1, \ldots, z_n$ exactly κ negative eigenvalues counting multiplicities. Special kernels yield special classes of functions; we mention two examples from [35] and [75]:

(a) A function Q belongs to the class N_κ of generalized Nevanlinna functions if it is meromorphic on $\mathbf{C}^+$ and the kernel

$$N_Q(z, w) = \frac{Q(z) - \overline{Q(w)}}{z - \bar{w}}$$

has κ negative squares. For $\kappa = 0$, the class N_0 coincides with the class of Nevanlinna functions; by definition these functions are holomorphic on $\mathbf{C}^+$ and have a nonnegative imaginary part there. By N_κ^+ we denote the set of $Q \in N_\kappa$ for which $zQ(z) \in N_0$.

Like Nevanlinna functions, the functions in class N_κ have an operator and an integral representation, they are given in [35]. The latter is rather complicated because N_κ-functions have singularities which account for the negative squares; those at a nonreal point are just poles, but the ones on the real axis may be embedded.

(b) A function f belongs to the class $\mathcal{P}_\kappa$ if it is defined and continuous on $\mathbf{R}$, $f(t) = \overline{f(-t)}$, and the kernel $H_f(s, t) = f(s - t)$ has κ negative squares.

¿From the many interpolation, moment and continuation problems studied by M.G. Kreĭn and Heinz, we single out the following two. The Stieltjes moment problem:

Given a sequence $(s_j)_{j=0}^\infty$ of complex numbers such that of the Hankel forms

$$\sum_{j,k} s_{j+k} x_j \bar{x}_k, \quad \sum_{j,k} s_{j+k+1} x_j \bar{x}_k$$

the first has κ negative squares and the second is nonnegative, find all $Q \in N_\kappa^+$ such that

$$Q(z) \sim -\frac{s_0}{z} - \frac{s_1}{z^2} - \cdots, \quad z = iy, \ y \to \infty,$$

and the continuation problem:

Given the continuous function $f : [-2a, 2a] \to \mathbf{C}$ such that $f(t) = \overline{f(-t)}, t \in [-2a, 2a]$ and $H_f(s, t)$ has κ negative squares on $[-a, a]$, find all $\tilde{f} \in \mathcal{P}_\kappa$ such that

$$\tilde{f}(t) = f(t), \quad t \in [-2a, 2a].$$

For $\kappa = 0$, these problems where studied before by A.I. Akhiezer and M.G. Kreĭn, but even when restricted to this case some of the results in the Fortsetzungsprob-leme were new.

The conditions on the data are necessary and sufficient for the existence of a solution, and there is either one solution or there are infinitely many solutions.

If the moment problem has infinitely many solutions, a 2×2 matrix function $W(z) = (w_{ij}(z))^2_{i,j=1}$ exists such that the formula

$$Q(z) = \frac{w_{11}(z)N(z) + w_{12}(z)}{w_{21}(z)N(z) + w_{22}(z)}$$

gives a one-to-one correspondence between all solutions $Q(z)$ and all functions $N(z) \in N_0^+ \cup \{\infty\}$.

A similar result holds for the continuation problem, but extra assumptions on the function f are needed: Assume that (i) f has an accelerant, that is, there is a hermitian function $H \in L^2(-2a, 2a)$ such that

$$f(t) = f(0) - \frac{1}{2}|t| - \int_0^t (t-s)H(s)ds, \quad t \in [-2a, 2a],$$

and that (ii) -1 does not belong to the spectrum of the integral operator $\mathbf{H}$ on $L^2(0, 2a)$ defined by

$$\mathbf{H}\varphi(t) = \int_0^{2a} H(t-s)\varphi(s)ds, \quad t \in [0, 2a].$$

Then if the continuation problem has infinitely many solutions, a 2×2 matrix function $\tilde{W}(z) = (\tilde{w}_{ij}(z))^2_{i,j=1}$ exists such that the formula

$$-i \int_0^\infty e^{-izt} \tilde{f}(t)dt = \frac{w_{11}(z)N(z) + w_{12}(z)}{w_{21}(z)N(z) + w_{22}(z)}, \quad \text{Im } z \le -\gamma,$$

for some $\gamma \ge 0$ gives a one-to-one correspondence between all solutions $\tilde{f}$ and all functions $N(z) \in N_0 \cup \{\infty\}$.

Suitably normalized, the resolvent matrices $W(z)$ and $\tilde{W}(z)$ are unique; their entries are entire and have finite order. The matrix $W(z)$ coincides essentially with the transmission matrix of a string that can be associated with the moment problem. The string has a special structure: besides positive masses, it also has a finite number of negative masses and certain new elements called dipoles; see [54], part II. Under certain conditions on the accelerant, the matrix $\tilde{W}(z)$ is a solution of a Hamiltonian system of differential equations; see [75]. These results are closely related to a theorem of Louis de Branges in his theory of Hilbert spaces of entire functions.

The method M.G. Kreĭn and Heinz used to obtain the above fractional linear transformation representation of the solutions is based on the extension theory of a symmetric operator or isometric operator in a Pontryagin space, developed in for example [20],[21], [28], and [40]: The data of the problem at hand give rise to a symmetric operator S in a Pontrjagin space $\mathcal{P}$ of index κ and an element u from

$\mathcal{P}$. The solutions correspond 1–1 to the u-resolvents $[(\tilde{A} - z)^{-1}u, u]$ of S, where $\tilde{A}$ runs through the class of selfadjoint extensions of S with nonempty resolvent set acting in spaces of the form $\tilde{\mathcal{P}} = \mathcal{P} \oplus \mathcal{H}$, $\mathcal{H}$ a Hilbert space. These u-resolvents can be written as a fractional linear transformation over the functions from the class $N_0^+ \cup \{\infty\}$ or $N_0 \cup \{\infty\}$, depending on the problem.

Extension theory also entails the study of Straus extensions of a symmetric operator and the description of these involves unitary colligations and characteristic functions. For the indefinite case this has been worked out in, for example, [79], [80] and [82] with Aad Dijksma and Henk de Snoo and [91] also with Branko Ćurgus, and applied to the study of nonstandard boundary eigenvalue problems associated with Sturm-Liouville and Hamiltonian systems of differential operators; see [68], [85], [87], and [104]. Here nonstandard means that the boundary conditions contain the eigenvalue parameter. Earlier on eigenfunction expansions were obtained for the Hilbert space case in [62], [69], [74] with Björn Textorius using Kreĭn's method of directing functionals. Basis properties of the eigenfunctions for certain classes of boundary eigenvalue problems have been obtained recently with Reinhard Mennicken and Christiane Tretter in [121], [128] and [129].

We thank Paul A. Fuhrmann for the picture of Heinz, Peter Jonas, Alex Markus and Wilfried Schenk for their valuable contributions to this section, and Christa Binder who in part helped us with the bibliography below.

List of publications of Heinz Langer

[1] On J–Hermitian operators, Doklady Akad. Nauk SSSR 134, 2 (1960), 263–266 (Russian); English transl.: Soviet Math. Dokl. 1 (1960), 1052–1055.

[2] Ein Zerspaltungssatz für Operatoren im Hilbertraum, Acta Math. Acad. Sci. Hung. XII, 3/4 (1961), 441–445.

[3] Zur Spektraltheorie J–selbstadjungierter Operatoren, Math. Ann. 146 (1962), 60–85.

[4] Über die Wurzeln eines maximalen dissipativen Operators, Acta Math. Acad. Sci. Hung. XIII, 3/4 (1962), 415–424.

[5] Eine Verallgemeinerung eines Satzes von L.S. Pontrjagin, Math. Ann. 152 (1963), 434–436.

[6] The spectral function of a selfadjoint operator in a space with indefinite metric, Doklady Akad. Nauk SSSR 152, 1 (1963), 39–42 (Russian); English transl.: Soviet Math. Dokl. 4 (1963), 1236–1239 (with M.G. Kreĭn).

[7] A contribution to the theory of quadratic pencils of selfadjoint operators, Doklady Akad. Nauk SSSR 154, 6 (1964), 1258–1261 (Russian); English transl.: Soviet Math. Dokl. 5 (1964), 266–269 (with M.G. Kreĭn).

[8] On some mathematical principles in the linear theory of damped oscillations of continua, Proc. Int. Sympos. on Applications of the Theory of Functions in Continuum Mechanics, Tbilissi, 1963, Vol. II: Fluid and Gas Mechanics, Math. Methods, Moscow, 1965, 283–322 (Russian); English transl.: Integral Equations Operator Theory 1 (1978), 364–399 and 539–566 (with M.G. Kreĭn).

[9] Eine Erweiterung der Spurformel der Störungstheorie, Math. Nachr. 30, 1/2 (1965), 123–135.

[10] Invariant subspaces of linear operators on a space with indefinite metric, Doklady Akad. Nauk SSSR 169, 1 (1966), 12–15 (Russian); English transl.: Soviet Math. Dokl. 7 (1966), 849-852.

[11] Einige Bemerkungen über dissipative Operatoren im Hilbertraum, Wiss. Zeitschrift der Techn. Universität Dresden 15, 4 (1966), 669–673 (with V. Nollau).

[12] Spektralfunktionen einer Klasse J–selbstadjungierter Operatoren, Math. Nachr. 33, 1/2 (1967), 107–120.

[13] Über stark gedämpfte Scharen im Hilbertraum, J. Math. Mech. 17, 7 (1968), 685–706.

[14] Über Lancaster's Zerlegung von Matrizenscharen, Arch. Rat. Mech. Anal. 29, 1 (1968), 75–80.

[15] Über einen Satz von M.A. Neumark, Math. Ann. 175 (1968), 303–314.

[16] Über die schwache Stabilität linearer Differentialgleichungen mit periodischen Koeffizienten, Math. Scand. 22 (1968), 203–208.

[17] A remark on invariant subspaces of linear operators in Banach spaces with an indefinite metric, Matem. Issledovanija Kišinev 4, 1 (1969), 27–34 (Russian).

[18] On maximal dual pairs of invariant subspaces of J-selfadjoint operators, Matem. Zametki 7 (1970), 443–447 (Russian).

[19] Über die Methode der richtenden Funktionale von M.G. Kreĭn, Acta. Sci. Math. Hung. 21, 1/2 (1970), 207–224.

[20] Über die verallgemeinerten Resolventen und die charakteristische Funktion eines isometrischen Operators in Raume Π_κ, Colloquia Math. Soc. Janos Bolyai, Tihany (Hungary), 5. Hilbert Space Operators, 1970, 353–399 (with M.G. Kreĭn).

[21] Defect subspaces and generalized resolvents of an hermitian operator in the space Π_κ, Funkcional. Anal. i Priložen. 5,2 (1971), 59–71; 5,3 (1971), 54–69 (Russian); English transl.: Functional Analysis Appl. (1971), 136–146; (1972) 217–228 (with M.G. Kreĭn).

[22] Invariante Teilräume definisierbarer J-selbstadjungierter Operatoren, Ann. Acad. Sci. Fenn. A I, 475 (1971), 1–23.

[23] Generalized coresolvents of a π-isometric operator with unequal defect numbers, Funkcional. Anal. i Priložen. 5,4 (1971), 73–75 (Russian); English transl.: Functional Analysis Appl., 5 (1971), 329–331.

[24] Verallgemeinerte Resolventen eines J-nichtnegativen Operators mit endlichem Defekt, J. Functional Analysis 8,2 (1971), 287–320.

[25] Zur Spektraltheorie verallgemeinerter gewöhnlicher Differentialoperatoren zweiter Ordnung mit einer nichtmonotonen Gewichtsfunktion, Universität Jyväskylä (Finland), Mathematisches Institut, Bericht 14 (1972), 1–58.

[26] Über verallgemeinerte gewöhnliche Differentialoperatoren mit nichtlokalen Randbedingungen und die von ihnen erzeugten Markov–Prozesse, Publ. Res. Inst. Math. Sci. (Kyoto) 7,3 (1972), 655–702 (with L. Partzsch and D. Schütze).

[27] Über eine Klasse polynomialer Scharen selbstadjungierter Operatoren im Hilbertraum, J. Functional Analysis 12,1 (1973), 13–29.

[28] Über die Q-Funktion eines π-hermitschen Operators im Raume Π_κ, Acta Sci. Math. Szeged 34 (1973), 191–230 (with M.G. Kreĭn).

[29] Über eine Klasse nichtlinearer Eigenwertprobleme, Acta Sci. Math. Szeged 35 (1973), 73–86.

[30] Über eine Klasse polynomialer Scharen selbstadjungierter Operatoren im Hilbertraum, II, J. Functional Analysis 16,2 (1974), 221–234.

[31] Verallgemeinerte Resolventen hermitescher und isometrischer Operatoren im Pontrjaginraum, Ann. Acad. Sci. Fenn. A I, 561 (1974), 1–45 (with P. Sorjonen).

[32] Über indexerhaltende Erweiterungen eines hermiteschen Operators im Pontrjaginraum, Math. Nachr. 64 (1974), 289–317 (with M. Grossman).

[33] Zur Spektraltheorie polynomialer Scharen selbstadjungierter Operatoren im Hilbertraum, Math. Nachr. 65 (1975), 301–319.

[34] Invariant subspaces for a class of operators in spaces with indefinite metric, J. Functional Analysis 19, 2 (1975), 232–241.

[35] Über einige Fortsetzungsprobleme, die eng mit der Theorie hermitescher Operatoren im Raume Π_κ zusammenhängen, Teil I: Einige Funktionenklassen und ihre Darstellungen, Math. Nachr. 77 (1977), 187–236 (with M.G. Kreĭn).

[36] A class of infinitesimal generators of one–dimensional Markov processes, J. Math. Soc. Japan 28, 2 (1976), 242–249.

[37] Zu einem Satz über Verteilungen quadratischer Formen in Hilberträumen, Math. Nachr. 61 (1974), 175–179 (with G. Maibaum and P.H. Müller).

[38] Factorization of operator pencils, Acta Sci. Math. Szeged 38, 1/2 (1976), 83–96.

[39] On the indefinite power moment problem, Doklady Akad. Nauk SSSR 226, 2 (1976), 261–264; English transl.: Soviet Math. Doklady 17 (1976), 90–93 (with M.G. Kreĭn).

[40] Über einige Fortsetzungsprobleme, die eng mit der Theorie hermitescher Operatoren im Raume Π_κ zusammenhängen, Teil II: Verallgemeinerte Resolventen, u-Resolventen und ganze Operatoren, J. Functional Analysis 30, 3 (1978), 390–447 (with M.G. Kreĭn).

[41] Spektralfunktionen einer Klasse von Differentialoperatoren zweiter Ordnung mit nichtlinearem Eigenwertparameter, Ann. Acad. Sci. Fenn. A I, Vol. 2 (1976), 269–301.

[42] Absolutstetigkeit der Übergangsfunktion einer Klasse eindimensionaler Fellerprozesse, Math. Nachr. 75 (1976), 101–112.

[43] Generalized resolvents and Q-functions of closed linear relations (subspaces) in Hilbert space, Pacific J. Math. 72, 1 (1977), 135–165 (with B. Textorius).

[44] Sturm–Liouville operators with an indefinite weight function, Proc. Royal Soc. Edinburgh, A 78 (1977), 161–191 (with K. Daho).

[45] Some remarks on a paper by W.N. Everitt, Proc. Royal Soc. Edinburgh, A 78 (1977), 71–79 (with K. Daho).

[46] Sturm–Liouville problems with indefinite weight function and operators in spaces with indefinite metric, Differential Equations Proc. Uppsala 1977, Int. Conference, Symp. Univ. Uppsala 7, 1977, 114–124.

[47] A generalization of M.G. Kreĭn's method of directing functionals to linear relations, Proc. Royal Soc. Edinburgh, A 81 (1978), 237–246 (with B. Textorius).

[48] Singular generalized second order differential operators with accessible or entrance boundaries, Preprint TU Dresden 07–10–78.

[49] Compact perturbations of definitizable operators, J. Operator Theory 2 (1979), 63–77 (with P. Jonas).

[50] Random spectral functions of a random string, Preprint TU Dresden 07–18–79.

[51] A factorization theorem for operator pencils, Integral Equations Operator Theory 2 (1979), 344-363 (with K. Harbarth).

[52] Szökefalvi–Nagy, Bela 65 éves, Matematika Lapok 27, 1–2 (1976–79), 7–24.

[53] A class of infinitesimal generators of one–dimensional Markov processes II. Invariant measures, J. Math. Soc. Japan 31, 1 (1980), 1–18 (with W. Schenk).

[54] On some extension problems which are closely connected with the theory of hermitian operators in a space Π_κ, III. Indefinite analogues of the Hamburger and Stieltjes moment problems, Beiträge zur Analysis 14 (1979), 25–40; 15 (1980), 27–45 (with M.G. Kreĭn).

[55] A class of infinitesimal generators of one–dimensional Markov processes III, Math. Nachr. 102 (1981), 25 44 (with W. Schenk).

[56] Generalized resolvents of contractions, Acta Sci. Math. Szeged 44 (1982), 125–131 (with B. Textorius).

[57] Introduction to the Spectral Theory of Operators in Spaces with an Indefinite Metric, Mathematical Research, Vol. 9, Akademie Verlag, Berlin, 1982 (with I.S. Iohvidov and M.G. Kreĭn).

[58] Some propositions on analytic matrix functions related to the theory of operators in the space Π_κ, Acta Sci. Math. Szeged 43 (1981), 181–205 (with M.G. Kreĭn).

[59] Continuous analogues of orthogonal polynomials with respect to an indefinite weight on the unit circle, and extension problems associated with them, Doklady Akad. Nauk SSSR 258, 3 (1981), 537–540 (Russian); English transl.: Soviet Math. Dokl. 23 (1981), 553–557 (with M.G. Kreĭn).

[60] Generalized resolvents and spectral functions of a matrix generalization of the Kreĭn–Feller second order derivative, Math. Nachr. 100 (1981), 163–186 (with L.P. Klotz).

[61] Generalized resolvents of dual pairs of contractions, Operator Theory: Adv. Appl., Vol. 6, Birkhäuser Verlag, Basel, 1982, 103–118 (with B. Textorius).

[62] L–Resolvent matrices of symmetric linear relations with equal defect numbers; applications to canonical differential relations, Integral Equations Operator Theory 5 (1982), 208–243 (with B. Textorius).

[63] Spectral functions of definitizable operators in Kreĭn spaces, Proc. Graduate School "Functional Analysis", Dubrovnik 1981. Lecture Notes in Math. 948, Springer Verlag, Berlin, 1982, 1–46.

[64] Some questions in the perturbation theory of J–nonnegative operators in Kreĭn spaces, Math. Nachr. 114 (1983), 205–226 (with P. Jonas).

[65] Perturbation theory for definitizable operators in Kreĭn spaces, J. Operator Theory 9 (1983), 297–317 (with B. Najman).

[66] On measurable hermitian–indefinite functions with a finite number of negative squares, Acta Sci. Math. Szeged 45 (1983), 281–292.

[67] Knotting of one–dimensional Feller processes, Math. Nachr. 113 (1983), 151–161 (with W. Schenk).

[68] Selfadjoint π_κ–extensions of symmetric subspaces: An abstract approach to boundary problems with spectral parameters in the boundary conditions, Integral Equations Operator Theory 7 (1984), 459–515 (with A. Dijksma and H.S.V. de Snoo).

[69] Spectral functions of a symmetric linear relation with a directing mapping. I, Proc. Royal Soc. Edinburgh, 97 A (1984), 165–176 (with B. Textorius).

[70] Spectral properties of selfadjoint differential operators with an indefinite weight function, Proc. 1984 Workshop "Spectral Theory of Sturm–Liouville differential operators", ANL–84.47, Argonne National Laboratory, Argonne, Illinois (1984), 73–80 (with B. Ćurgus).

[71] Matrix functions of the class $N_\kappa^{n \times n}$, Math. Nachr. 120 (1985), 275–294 (with K. Daho).

[72] A characterization of generalized zeros of negative type of functions of the class N_κ, Operator Theory: Adv. Appl., Vol. 17, Birkhäuser Verlag, Basel, 1985, 201–212.

[73] Some interlacing results for hermitian indefinite matrices, Linear Algebra Appl. 69 (1985), 131–154 (with B. Najman).

[74] Spectral functions of a symmetric linear relation with a directing mapping. II, Proc. Royal Soc. Edinburgh, 101 A (1985), 111–124 (with B. Textorius).

[75] On some continuation problems which are closely related to the theory of Hermitian operators in spaces Π_κ. IV: Continuous analogues of orthogonal polynomials on the unit circle with respect to an indefinite weight and related continuation problems for some classes of functions, J. Operator Theory 13 (1985), 299–417 (with M.G. Kreĭn).

[76] A model for π–selfadjoint operators in π_1–spaces and a special linear pencil, Integral Equations Operator Theory 8 (1985), 13–35 (with P. Jonas).

[77] Duality of a class of one–dimensional Markov processes, Math. Nachr. 125 (1986), 69–81 (with W. Schenk).

[78] Some spectral properties of operators which are related to one–dimensional Markov processes, Math. Nachr. 127 (1986), 51–63 (with B. Zagany).

[79] Unitary colligations in π_κ–spaces, characteristic functions and Štraus extensions, Pacific J. Math. 125, 2 (1986), 347–362 (with A. Dijksma and H.S.V. de Snoo).

[80] Representations of holomorphic functions by means of resolvents of unitary or selfadjoint operators on Kreĭn spaces, Operator Theory: Adv. Appl., Vol. 24, Birkhäuser Verlag, Basel, 1987, 123–143 (with A. Dijksma and H.S.V. de Snoo).

[81] Characteristic functions of unitary operator colligations in π_κ–spaces, Operator Theory: Adv. Appl., Vol. 19, Birkhäuser Verlag, Basel, 1986, 125–194 (with A. Dijksma and H.S.V. de Snoo).

[82] Kreĭn space extensions of an isometric operator in Hilbert space, characteristic functions and related questions, Proc. Graduate School "Functional analysis II", Dubrovnik 1985. Lecture Notes in Math. 1242, Springer Verlag, Berlin, 1987, 1–42 (with A. Dijksma and H.S.V. de Snoo).

[83] Sturm–Liouville operators with an indefinite weight function: The periodic case, Radovi Matematicki 2 (1986), 165–188 (with K. Daho).

[84] Time reversal of quasidiffusions, Lecture Notes in Control and Information Sciences, Vol. 96, Springer Verlag, Berlin, 1987, 156–163 (with W. Schenk).

[85] Symmetric Sturm–Liouville operators with eigenvalue depending boundary conditions, in: Oscillation, Bifurcation and Chaos, Canad. Math. Soc.–Amer. Math. Soc. Conference Proc., Vol. 8 (1987), 87–116 (with A. Dijksma and H.S.V. de Snoo).

[86] Generalized zeros of negative type of matrix functions of the class N_κ, Operator Theory: Adv. Appl., Vol. 28, Birkhäuser Verlag, Basel, 1988, 17–26 (with M. Borogovac).

[87] Hamiltonian systems with eigenvalue depending boundary conditions, Operator Theory: Adv. Appl., Vol. 35, Birkhäuser Verlag, Basel, 1988, 37–84 (with A. Dijksma and H.S.V. de Snoo).

[88] Remarks on the perturbation of analytic matrix functions. II, Integral Equations Operator Theory 12 (1989), 392–407 (with B. Najman).

[89] A Kreĭn space approach to symmetric ordinary differential operators with an indefinite weight function, J. Differential Equations 79, 1 (1989), 31–61 (with B. Ćurgus).

[90] Time reversal of transient gap diffusions, Mathematical Research, Vol. 54, Akademie-Verlag, Berlin, 1989, 104–114 (with W. Schenk).

[91] Characteristic functions of unitary colligations and of bounded operators in Kreĭn spaces, Operator Theory: Adv. Appl., Vol. 41, Birkhäuser Verlag, Basel, 1989, 125–152 (with B. Ćurgus, A. Dijksma, and H.V.S. de Snoo).

[92] Generalized second–order differential operators, corresponding gap diffusions and superharmonic transformations, Math. Nachr. 148 (1990), 7–45 (with W. Schenk).

[93] Definitizing polynomials of unitary and hermitian operators in Pontrjagin spaces, Math. Annalen 288 (1990), 231–243 (with Z. Sasvari).

[94] A transformation of right-definite S-hermitian systems to canonical systems, Differential and Integral Equations 3,5 (1990), 901–908 (with R. Mennicken).

[95] Generalized coresolvents of standard isometric operators and generalized resolvents of standard symmetric relations in Kreĭn spaces, Operator Theory: Adv. Appl., Vol. 48, Birkhäuser Verlag, Basel, 1990, 261–274 (with A. Dijksma and H.S.V. de Snoo).

[96] A second order differential operator depending nonlinearly on the eigenvalue parameter, Operator Theory: Adv. Appl., Vol. 48, Birkhäuser Verlag, Basel, 1990, 319–332 (with R. Mennicken and M. Möller).

[97] Linearization of boundary eigenvalue problems, Integral Equations Operator Theory 14 (1991), 105–119 (with M. Möller).

[98] On spectral properties of regular quasidefinite pencils $F - \lambda G$, Results in Mathematics 19 (1991), 89–109 (with A. Schneider).

[99] Remarks on the perturbation of analytic matrix functions.III, Integral Equations Operator Theory 15 (1992), 796–806 (with B. Najman).

[100] Perturbation of the eigenvalues of quadratic matrix polynomials, SIAM J. Matrix Anal. Appl. 13, 2 (1992), 474-489 (with B. Najman and K. Veselic).

[101] Some remarks about polynomials which are orthogonal with respect to an indefinite weight, Results in Mathematics 21 (1992), 152–164 (with A. Schneider).

[102] On Floquet eigenvalue problems for first order differential systems in the complex domain, J. Reine Angew. Math. 425 (1992), 87–121 (with R. Mennicken and M. Möller).

[103] Models and unitary equivalence of cyclic selfadjoint operators in Pontrjagin spaces, Operator Theory: Adv. Appl., Vol. 59, Birkhäuser Verlag, Basel, 1992, 252–284 (with P. Jonas and B. Textorius).

[104] Eigenvalues and pole functions of Hamiltonian systems with eigenvalue depending boundary conditions, Math. Nachr. 161 (1993), 107–154 (with A. Dijksma and H.V.S. de Snoo).

[105] Some spectral properties of contractive and expansive operators in indefinite inner product spaces, Math. Nachr. 162 (1993), 247–259 (with T.Ja. Azizov).

[106] Leading coefficients of the eigenvalues of perturbed analytic matrix functions, Integral Equations Operator Theory 16 (1993), 600–604 (with B. Najman).

[107] Expansions of analytic functions in series of Floquet solutions of first-order differential systems, Math. Nachr. 162 (1993), 279–314 (with R. Mennicken and M. Möller).

[108] Expansions of analytic functions in products of Bessel functions, Results in Math. 24 (1993), 129–146 (with R. Mennicken, M. Möller, and A. Sattler).

[109] Sturm-Liouville problems with coefficients which depend analytically on the eigenvalue parameter, Acta Sci.Math. (Szeged) 57 (1993), 25–44 (with F.V. Atkinson and R. Mennicken).

[110] On an elliptic boundary value problem arising in magnetohydrodynamics, Quaestiones mathematicae 17 (1994), 141–159 (with M. Faierman, R. Mennicken, and M. Möller).

[111] On the papers by M.G. Kreĭn in the theory of spaces with indefinite metric, Ukrainskiĭ Matem. Žurnal 46, 1/2 (1994), 5–17 (Russian, with T.Ya. Azizov and Yu.P. Ginzburg).

[112] Eigenvalues of a Sturm-Liouville problem depending rationally on the eigenvalue parameter, Mathematical Research, Vol. 79: Systems and Networks, Vol. II (1994), 589–594 (with V. Adamyan and R. Mennicken).

[113] The essential spectrum of some matrix operators, Math. Nachr. 167 (1994), 5–20 (with F.V. Atkinson, R. Mennicken, and A. Shkalikov).

[114] Generalized matrix functions of the class $N_\kappa^{m \times m}$, in: Linear and Complex Analysis, Problem Book 3, part I (V.P. Havin and N.K. Nikolski (eds)), Lecture Notes in Mathematics 1573, Springer Verlag, Berlin, 1994, 201–204 (with A. Dijksma and H.S.V. de Snoo).

[115] Spectral properties of a class of rational operator valued functions, J. Operator Theory 33 (1995), 259–277 (with V.M. Adamjan).

[116] Selfadjoint extensions of a closed linear relation of defect one in a Kreĭn space, Operator Theory: Adv. Appl., Vol. 80, Birkhäuser Verlag, Basel, 1995, 176–203 (with P. Jonas).

[117] Selfadjoint extensions for a class of symmetric operators with defect numbers (1,1), Topics in Operator Theory, Operator Algebras and Applications, Vol. 1 (1995), 115–145 (with S. Hassi and H.S.V. de Snoo).

[118] Spectral components of selfadjoint block operator matrices with unbounded entries, Math. Nachr. 178 (1996), 43–80 (with V. Adamjan, R. Mennicken, and J. Saurer).

[119] The essential spectrum of a non-elliptic boundary value problem, Math. Nachr. 178 (1996), 233–248 (with M. Möller).

[120] Operator Theory and Ordinary Differential Operators, in: Lectures on Operator Theory and its Applications, Fields Institute Monographs, Vol. 3, Amer. Math. Soc, Providence, RI, 1996, 75–139 (with A. Dijksma).

[121] A selfadjoint linear pencil $Q - \lambda P$ of ordinary differential operators, Methods of Functional Analysis and Topology 2, 1 (1996), 38–54 (with R. Mennicken and C. Tretter).

[122] Spectral properties of a compactly perturbed span of projections, Integral Equations Operator Theory 26 (1996), 353–366 (with V. Pivovarcik and C. Tretter).

[123] Elliptic problems involving an indefinite weight, Operator Theory: Adv. Appl., Vol. 87, Birkhäuser Verlag, Basel, 1996, 105–124 (with M. Faierman).

[124] Instability of singular critical points of definitizable operators, Integral Equations Operator Theory 28 (1997), 60–71 (with B. Najman).

[125] Locally definite operators in indefinite inner product spaces, Math. Annalen 308 (1997), 405–424 (with A. Markus and V. Matsaev).

[126] Notes on a Nevanlinna-Pick interpolation problem for generalized Nevanlinna functions, Operator Theory: Adv. Appl., Vol. 95, Birkhäuser Verlag, Basel, 1997, 69–91 (with A. Dijksma).

[127] Nonnegative solutions of algebraic Riccati equations, Linear Alg. Appl. 261 (1997), 317–352 (with A.C.M. Ran and D. Temme).

[128] Spectral properties of the Orr-Sommerfeld problem, Proc. Royal Soc. Edinburgh 127A (1997), 1245–1261 (with C. Tretter).

[129] Spectral decomposition of some nonselfadjoint block operator matrices, J. Operator Theory 39, 2 (1998), 339–359 (with C. Tretter).

[130] Resolvents of symmetric operators and the degenerated Nevanlinna-Pick problem, Operator Theory: Adv. Appl., Vol. 103, Birkhäuser Verlag, Basel, 1998, 233–261 (with H. Woracek).

[131] Classical Nevanlinna-Pick interpolation with real interpolation points, Operator Theory: Adv. Appl. (with D. Alpay and A. Dijksma).

[132] Variational principles for real eigenvalues of selfadjoint operator pencils, Integral Equations Operator Theory (with D. Eschwé and P. Binding).

[133] The spectral shift function for certain block operator matrices, Math. Nachr. (with V. Adamjan).

[134] Direct and inverse spectral problems for generalized strings, Integral Equations Operator Theory (with H. Winkler).

Vladimir Maz'ya: Friend and Mathematician. Recollections

Israel Gohberg

It gives me great pleasure to write about Vladimir Maz'ya whom I have known for almost forty years. For me his name is closely connected with Leningrad, its mathematical school and, particularly, with the name of Solomon G. Mikhlin, and I would like to begin my recollections a decade before I met Maz'ya.

When I was in my fourth year at the Kishinev University (autumn 1949) a young D.Sc. from Leningrad came to work there. As he himself said, he taught us everything he knew. He was knowledgeable as well as a good lecturer and I listened to everything he had to say. That was I.A. Itscovitch, a very recent Ph.D. student of S.G. Mikhlin. I am grateful to him for introducing me to Mikhlin's school. This way I learned about singular integral equations, Fredholm operators, the theory of the operator index, and other topics in which I later started to work actively. In 1949, I obtained my first results in this area and the following year I met Mikhlin for the first time when visiting Leningrad. He made a very strong impression on me, a kind and wise man with a fatherly behavior. After that visit I made several trips to Leningrad. I lectured at the seminar of V.I. Smirnov and made many good friends in Leningrad. By the way, my wife and I got married in Leningrad and I also defended my Ph.D. thesis in Leningrad with Mikhlin as an external examiner (opponent). I am proud to consider Mikhlin as one of my teachers.

In one of my trips to Leningrad Mikhlin introduced to me Volodia Maz'ya (Volodia is the diminutive of Vladimir). I had heard from Mikhlin that the young man was very promising. He had already written several interesting papers discovering, in particular, the equivalence of Poincaré-Sobolev imbedding theorems and isoperimetric inequalities for measures and capacities. In the future these ideas of Maz'ya had a broad influence on the development of Sobolev spaces, potential theory, and partial differential equations on nonsmooth domains and manifolds. His monograph "Sobolev spaces" [1] played an important role in this, continuing to inspire numerous researchers.

Originally published in *Operator Theory: Advances and Applications* **109** (1999), pp. 1–5.

After the first encounter, I seldom met Maz'ya, since we lived in different cities and later on even in different countries. Although our mathematical directions did not show much similarity, I followed his work with great interest. Thus in 1968, I learned of his counterexamples for the nineteenth and twentieth Hilbert problems for higher order analytic elliptic equations and regular variational problems. The amazing fact I remember was that these equations and problems can have nonanalytic and even nonsmooth solutions. For experts this created a sensation.

Mikhlin started to pay attention to the 18 year old Maz'ya when the latter became a freshman at the Faculty of Mathematics and Mechanics in Leningrad. The student had solved all the problems for both first and second year while participating in the traditional olympiad of the Faculty. As Maz'ya did not make this a secret, his fellow students all decided not to submit their solutions. An unexpected result was that the jury deemed the contest a failure and no prizes were awarded to anyone, including the winner. Instead, Maz'ya got another, much greater prize, which came to shape his entire mathematical life. Professor Mikhlin invited him to his home, soon making this a habit. Vladimir's father died at the front in 1941, and as he once confessed to me, his deep longing for a father plagued him during his childhood and youth. Maz'ya never was a formal student of Mikhlin, but Mikhlin was for him more than a teacher. Maz'ya had found the topics of his dissertations by himself, while Mikhlin taught him mathematical ethics and rules of writing, referring and reviewing. He also expressed his opinions about the work of others and his views on relationships between people. Maz'ya's prevailing style of summarizing his work on a particular theme by writing a book is certainly inherited from Mikhlin. By the way, the professor had mixed feelings about Vladimir's early work on Sobolev spaces, being himself more application-oriented. He once expressed his opinion on Maz'ya's geometrical counterexamples in the following way: "Your domains are very interesting, but no mother would let her child play in such ravines."

Mikhlin considered the theory of singular integral equations as his favorite creation. Very soon his results led to pseudodifferential operators, and his notion of the symbol (1936) became a cornerstone of this new theory which revolutionized partial differential equations. I myself was attracted to Mikhlin's theory even in the 1950s and followed this theory with great interest.

Maz'ya was very enthusiastic about this theory and became interested in the case of the degenerate symbol, which emerged at that time thanks to Mikhlin. I would especially like to mention Maz'ya's later study of the nonelliptic boundary value problem with oblique derivative, which can be reduced to a singular integral equation with a degenerate symbol. He provided an ingenious proof of unique solvability for the so called generic case, when the behavior of the vector field in the boundary operator is quite complicated.

Another area of his research, that has lasted for many years, and that was close to my interests, is boundary integral equations on nonsmooth contours and surfaces. Certainly, Maz'ya is a leader in this rapidly developing field, which is highly important for applications.

In the new Springer monograph by Kozlov and Maz'ya [2], an asymptotic theory of ordinary differential equations originating from Poincaré and Birkhoff, is extended to general classes of differential equations with unbounded operator coefficients in a Banach space. These results build the foundation of a powerful theory of singularities of solutions to linear and nonlinear partial differential equations developed in recent years by the authors.

Having given a quick glance at the list of Maz'ya's publications, I am very far from giving a complete analysis of his entire work. In general this cannot be done by one expert. I know that other contributors to this volume fulfil the task of doing justice to the man, who as a result of four decades of incessant labour, has authoured and coauthoured over 300 papers and more than a dozen books.

Maz'ya's exceptional productivity is due to a rare combination of talent, working ability, and inner drive. The development of this drive was stimulated by the hardships of his childhood during and after World War II. The obstacles raised by Soviet antisemitism in his youth strengthened his will to survive. The stifling atmosphere did not make the mature years easier, hindering Maz'ya's international contacts, publications and trips abroad.

The enormous amount and variety of his work go hand in hand with excellent quality. Whatever he writes is beautiful, his love for art, music and literature seeming to feed his mathematical esthetic feeling. It appears as if Maz'ya intuitively aims to bring harmony to domains previously ruled by disorder. Degenerations and singularities form the sphere in which he feels at home. He either gives necessary and sufficient conditions or constructs striking counterexamples, as in the topics of his work mentioned above.

In fact, his work has many other facets. One example is concerned with the linear theory of time-harmonic water waves. In 1977, he proved a uniqueness theorem for the waves generated by a submerged body [3], thus solving an important problem posed by F. John in 1950.

Of quite another style are the Kresin and Maz'ya studies of the maximum modulus principle for elliptic and parabolic systems. They discovered necessary and sufficient conditions, giving an answer to a long-standing classical problem.

Coefficients of partial differential equations and even symbols of pseudodifferential operators can be considered as functions, which map one Sobolev space to another, by pointwise multiplication. Together with his wife Tatyana, Maz'ya described spaces of such multipliers and investigated their properties. Their Pitman monograph of 1985 [4] is a comprehensive account of this theory and its applications.

Maz'ya's long-standing interest in elasticity resulted in a number of excellent achievements. I mention only two of them here. The first is the Maz'ya and Plamenevskii method for calculating stress intensity factors in fracture mechanics [5, 1974]. The second, due to Maz'ya and Nazarov [6, 1986], is an asymptotic analysis of the well-known polygon-circle paradox: when a thin circular plate is approximated by regular polygons with freely supported edges, the limit solution does not satisfy the conditions of the free support on the circle.

Maz'ya's character leaves a stamp of perfectionism on his work. Definitive solutions are a must when he deals with problems of any kind. I know a couple of examples when he lost priority because of his reluctance to publish partial results. Some of the areas which he once explored either alone or with a few colleagues later became popular.

Finishing the short review of Maz'ya's mathematical results, I would like to name his other areas of work, following the list published in the article [7]:

- Theory of capacities and nonlinear potentials
- Boundary behaviour of solutions to elliptic equations
- Estimates for general diferential operators
- The Cauchy problem for the Laplace equation
- Characteristic Cauchy problem for hyperbolic equations
- Boundary value problems in domains with piecewise smooth boundaries
- Iterative procedures for solving ill-posed boundary value problems
- Asymptotic theory of singularly perturbed boundary value problems
- "Approximate approximations" and their applications

During the last 12 years, everything concerning the life and work of Jacques Hadamard has been a hobby of Vladimir and Tatyana. Their book on the subject, which was recently published by the American and London Mathematical Societies [8] is a masterpiece both from a biographical and a mathematical point of view and is definitely worth keeping on your bedside table.

Vladimir is a good friend and I would like to extend my most sincere congratulations to him on the occasion of his 60th birthday. May his creativity and energy never leave him.

References

[1] Maz'ya, V. G., *Sobolev spaces,* Springer, 1985.

[2] Kozlov, V., Maz'ya, V., *Differential equations with operator coefficients,* Springer, 1999.

[3] Maz'ya, V. G., *On the steady problem of small oscillations of a fluid in the presence of a submerged body,* Proc. Semin. Sobolev. No. 2, 57–79, Novosibirsk, Inst. of Math. 1977.

[4] Maz'ya, V. G., Shaposhnikova, T. O., *Theory of multipliers in spaces of differentiable functions,* Pitman, 1985.

[5] Maz'ya, V. G., Plamenevskii, B. A. *On the coefficients in the asymptotics of solutions of elliptic boundary value problems in domains with conical points,* In: Elliptic boundary value problems, American Mathematical Society Translations, Ser. 2, vol. 123, 1984.

[6] Maz'ya, V. G., Nazarov, S. A., *Paradoxes of limit passage in solutions of boundary value problems involving the approximation of smooth domains by polygonal domains,* Math. USSR Izvestia **29**, No. 3 (1987).

[7] Eidus, D., Khvoles, A., Kresin, G., Merzbach, E., Prössdorf, S., Shaposhnikova, T., Sobolevskii, P., Solomiak, M., *Mathematical work of Vladimir Maz'ya,* Functional Differential Equations **4**, No. 1–2 (1997), 3–11.

[8] Maz'ya, V. G., Shaposhnikova, T. O., *Jacques Hadamard, a universal mathematician,* American Mathematical Society and London Mathematical Society, 1998.

Reminiscences of Béla Szőkefalvi-Nagy

Israel Gohberg

Béla Szőkefalvi-Nagy was one of my important teachers. I never was a formal student of his, but I studied systematically his book as well as the books he co-authored with F. Riesz and with C. Foias; they were always on my table. His papers and his results impressed me very much; they played a crucial role in my mathematical education. The papers of B. Sz.-Nagy also served as examples for me of how to write mathematics. M.G. Krein told me about the work and personality of B. Sz.-Nagy, so even before our first meeting I was very well informed.

I met B. Sz.-Nagy for the first time in the Moscow University at the Conference of Functional Analysis in January 1956. These were the first years of my career and I knew about the importance of his contributions from M.G. Krein, and I was very happy to be introduced to him. We also had common interests, and I tried to have a talk with him. Unfortunately, I did not speak English and my German was not very good, but we soon discovered that we could converse in Rumanian, and this is how we communicated until I learned English. I very much liked B. Sz.-Nagy's papers, and on a few occasions I found in his work answers to questions which were bothering me. The meeting with B. Sz.-Nagy impressed me, and soon after the conference I received from him a wonderful present - a large set of his reprints accompanied by a warm presentation. Later I followed his joint papers with C. Foias with great interest.

I met B. Sz.-Nagy a few times in Moscow. Once he related to me the following story. During one of his first visits to Moscow he tried to buy a map of the City in order to be more independent. He quickly discovered that such maps are not available for sale. He then asked his guides (who were officially provided by the Soviet Academy) to make a plan for him of that part of Moscow which included the Moscow University, the administrative offices of the Academy, the metro and other sites. Even such a map he could not obtain, and only many years later did he receive a map of Moscow. He was not always able to use this map because the distances were wrongly shown.

B. Sz.-Nagy and M.G. Krein had a high regard for each other. They met frequently at conferences and congresses in the Soviet Union, and at least once

Originally published in *Operator Theory: Advances and Applications* **127** (2001), xiii–xvi.

B. Sz.-Nagy visited M.G. Krein in Odessa. In 1968 B. Sz.-Nagy visited Moscow together with A. Rényi. The aim of their visit was to improve cooperation between the Academy of Sciences of Hungary and the Soviet Academy of Sciences in Mathematics. At the request of B. Sz.-Nagy, M.G. Krein was invited by the Soviet Academy to Moscow. By chance I was in Moscow during these days to receive an exit visa for a visit to Hungary at the invitation of B. Sz.-Nagy. From our conversations I understood that B. Sz.-Nagy was considering cooperation with M.G. Krein and his school as an important part of his plans. Very soon after the formal meetings B. Sz.-Nagy understood that the administration of Steklov Institut was against such plans and the cooperation was restricted to a formal cooperation with the Steklov Institute. At the same time B. Sz.-Nagy continued to discuss the matter of cooperation with M.G. Krein and he continued to get advice from him.

B. Sz.-Nagy was planning that after Moscow he would visit Chisinau (Kishinev), where he was invited by the Academy. He understood that in this case I would probably have to return my exit visa for my visit to Hungary, and it was questionable whether I would be able to receive it again. He therefore sacrificed his visit and said that he would visit Chishinau another time. Unfortunately there never was an opportunity for another visit. B. Sz.-Nagy visited me in Tel Aviv where he presented a very nice Toeplitz Lecture. We also met in Amsterdam at the invitation of our friend Rien Kaashoek. B. Sz.-Nagy was very interested in hearing about how I succeeded in emigrating from the Soviet Union, and about life in Israel.

B. Sz.-Nagy and his family were very religious Catholics. During my first visit to Szeged in 1968 he took me on a walking tour through the town, and the first place he showed me was the great synagogue in Szeged. I was very impressed by it, but I was also afraid that my visit to the synagogue could lead to problems in regard to other trips abroad. Many years later I learned that the synagogue was built by the grandfather of my friend Terry Horvath (the wife of my colleague and friend, John Horvath from College Park, Maryland).

In 1970 B. Sz.-Nagy organized a conference on functional analysis in Tihany on the Balaton Lake in Hungary. As with everything organized by him, it was perfect from every point of view. The timing and the selection of participants were very good; it came after the International Mathematical Congress in Nice and many Westerners came, as well as a large group of mathematicians from the East. I participated in this conference and enjoyed it very much. Moreover this conference proved to be very important for me later after emigration because here I met for the first time the majority of my Western colleagues.

M.G. Krein never traveled abroad. The reason for this was that he had never been granted an exit visa. Only once he was given per mission to travel abroad. That was in 1970 to attend the conference in Tihany. It probably worked out this time because he used a private invitation which did not have to go through the high official channels. But this time he could not use the visa because precisely at that time there was an epidemic of cholera in Odessa and no one was allowed to leave Odessa. At the request of M.G. I gave Professor B. Sz.-Nagy regards from

M.G. and told him the reason why M.G. could not come. Professor B. Sz.-Nagy smiled and answered, "So it's now called cholera, is it?" In the West people were already used to the various reasons that were invented to justify the absence of M.G. Krein. This was the only time that the reason given was the true reason, but no one believed it already.

The following event took place during the conference in Tihany. Lewis Coburn presented a nice talk; this talk was held before a long break. During the talk he used up all the available chalk. Béla - the organizer of the conference - immediately passed to him a piece of soft yellow chalk. Lewis finished his talk with this piece of chalk which colored his hands yellow. During the break that followed the talk Lewis looked for somewhere to wash his hands. On the way he showed everyone his yellow hands and explained what was the cause. After a while he dropped his left hand and showed only his right hand. In this way he approached also Béla who was involved in a discussion with a participant. Being busy Béla did not really take in what Lewis was saying to him, and he also did not notice the color of the hand which was extended to him. He had the impression that Lewis just wanted to shake hands with him, so he grasped Lewis' yellow hand in a handshake. At this point Lewis started to jump around and yell. Béla immediately understood what had happened – he was under the impression that Lewis wanted to pay him back for giving him the yellow chalk. With a smile he started to stroke Lewis on his head and face, saying, "You are a good boy, a good boy", thereby turning Lewis' head and face yellow. All those who witnessed this incident started laughing. This laugh resounded throughout the courtyard.

The last time Béla Sz.-Nagy and I met was in the summer of 1993 during the conference in Szeged dedicated to his eightieth birthday. His wife was already very sick and this affected him strongly. Nevertheless he made an effort to attend all the talks given at the conference, and to be active during them. Together with C. Foias we discussed the organization of the next edition of the Sz.-Nagy-Foias book. I understood that he had detailed plans for it and he very much wanted to see this new edition. There was an atmosphere of sadness at this meeting, and I felt that this was probably the last time we would meet. Unfortunately, this sad prediction turned out to be correct.

Béla died in 1998, but I and my colleagues always feel his presence among us. We see him smiling at us through his books and papers, he is with us at our desks, in our Seminar rooms and classrooms. His influence will be felt for a very long time.

Peter Lancaster, my Friend and Co-author

Israel Gohberg

The first time I heard about the mathematician Peter Lancaster was from Heinz Langer. In 1967–68 Heinz spent a year in Toronto University at the invitation of Professor Israel Halperin. During his visit to Canada he was also invited by Peter Lancaster to spend a short period in Calgary. After his return Heinz told us about the work of Peter, about his book on vibrations of systems and about his personality. He also brought the book to Odessa and Peter's results were often quoted in the seminars and discussions, and very soon Peter became popular in Odessa.

In July 1974 I immigrated to Israel. At the end of that year I visited the USA at the invitation of my colleagues from Stony Brook: R.G. Douglas and J. Pincus. In the beginning of 1975 I was a guest of Seymour Goldberg in the University of Maryland, College Park. During my latter visit I was invited by Chandler Davis to visit the University of Toronto and by Peter Lancaster to visit Calgary.

From the start I was impressed by Peter. We discussed mathematics and started to look for an area for joint work. After my return to Israel in March I obtained my first Ph.D. student in Tel Aviv University. He was Leiba Rodman. I started to work with him on matrix polynomials. The problem which we were interested in was the problem of reconstruction of a matrix polynomial from its spectral data. The spectral data consists of the eigenvalues, eigenvectors and chains of generalized eigenvectors. We were looking for something which would generalize for matrix polynomials the Jordan form for matrices. I thought that this problem would fit Leiba because he had a much better education in algebra than in analysis. In 1975–76 we obtained our first results. The results were difficult to formulate and to present. On my second visit to Calgary Peter and I continued to look for an area of common interest and Peter showed me a number of his recent papers. Reading those papers I understood that some of Peter's results could form the piece of information which we were missing in the work with Leiba. We invited Peter to join our team. Peter accepted and soon visited Tel Aviv University and we started to work actively together.

Originally published in *Operator Theory: Advances and Applications* **130** (2001), 23–28.

In 1978 our first two papers were published in Linear Algebra and its Applications. These papers contained the solution of the above mentioned problem, and much more. They contained important formulas for the polynomials via the spectral data results for factorization matrix polynomials including spectral factorization and applications to differential and integral equations. We continued to work on matrix polynomials, but we already understood that our papers could serve as a basis for a book on matrix polynomials. This book was written in a relatively short period of time when Leiba spent his post doctoral period at the University of Calgary as a guest of Peter. During this time our other friends also joined us in our studies. I have in mind H. Bart, M.A. Kaashoek and L. Lerer. Some of their results were also included in the book.

During the writing of this book the authors became interested in matrix polynomials which have important applications. I have in mind the theory of matrix polynomials with selfadjoint coefficients. Our first results in this area were also included in the book.

In particular, whilst developing the theory of matrix polynomials with selfadjoint coefficients we discovered new invariants for such polynomials. We studied these invariants and in the process of this work we got the feeling that we had a good basis for a second book "Matrices and indefinite scalar products".

An important part of this second book was dedicated to problems of perturbations in the presence of an indefinite scalar product. It also contained important applications to differential equations. Included are deep results of M.G. Krein, I.M. Gelfand and V.B. Lidsky, which were obtained within our framework.

The work on our books was pleasant and all authors enjoyed it. Leiba spent long periods in Calgary and I visited Calgary for one to two months each year. Peter also visited us in Tel Aviv many times. Each of us worked with full dedication and responsibility. Many times during our work we had differences of opinion. I do not remember a single case where we were not able to reach a consensus after a friendly discussion. Except once, during a hike when we could not reach a consensus as to whether we should have lunch immediately or wait till the end of the hike. Peter's taste, his knowledge, talent and command of English combined with the working capability, breakthrough power and talent of Leiba, were very important for our team. Peter's wife Edna, and their three daughters (Jane, Jill, and Joy) were an integral part of our team. This warm family displayed outstanding hospitality and understanding of my and Leiba's Russian-Jewish mentality.

I would like to tell separately about the wonderful weekends that we spent in the Lancasters' cottage in the Rocky Mountains, between Calgary and Banff. It is a nice small cottage built of wood in the middle of the wilderness. All around is wild forest and a nearby river visited by beavers and bears. How many people enjoyed the wonderful hospitality of the Lancasters in the modern comfort of this cottage. How many picnics, barbecues and dinners. The Lancasters have a book containing remarks of their guests. In this book there is a very long list in many languages of mathematicians and other guests. Often we stayed overnight at the cottage. The nights spent there were beautiful. Around the brightly burning fireplace the company gathered, continuing discussions in mathematics, projects and plans; rarely did we allow ourselves a discussion on general topics or politics. Very often during those visits we made long and difficult hikes. I am proud that I introduced the Lancasters to mushroom hunting. We often enjoyed the combination of hiking and mushroom hunting, followed by mushroom dinners on our return. During our walks and hikes we also discussed many issues, but mathematics of our books and our research was one of the main topics. My wife, Bella, and I will never forget the wedding of Jane (Peter's eldest daughter) to Ken held in a small hotel in the middle of the Rocky Mountain wilderness. It was one of the nicest events that we have ever attended. The atmosphere was extremely friendly, the background was beautiful and the food was excellent. The folk dancing was very enjoyable and fitting for the event.

Peter and Edna highly appreciated their English heritage. This did not exclude criticism of some aspects of it. The family considers itselve Canadians but every year Peter and Edna, sometimes together with their children, would visit their parents and other relatives in England. I had the privilege of meeting Peter's entire family: his parents, brothers and sister. Once after a conference in Great Britain organized by N. Young, Peter took me to Appleby to meet the family. Such a warm family; I very much enjoyed meeting them, as well as an English brunch with the family.

During our work on the second book the plan of the third book was born. The plan was to present matrix theory, geometrical, algebraical, analytical and topological aspects from one point of view, namely from the point of view of invariant subspaces. This book is "Invariant subspaces of matrices with applications". In linear algebra the set of tools is so large that often the invariant subspaces are lost. Our plans were to put invariant subspaces as the basis of the entire theory. I would like to note that all our books contain applications to system theory.

In the late '80s my former doctoral student, Israel Koltracht from the Weizmann Institute of Science, spent a successful two year post doctoral period in Calgary with Peter. This led to the cooperation of the three of us in a new area in numerical analysis for structured matrices and integral operators. We published a number of papers in this area, and again the expertise and experience of Peter was important and the work enjoyable.

In 1988 Peter, Rien Kaashoek and myself published a large paper. It is in fact a small monograph (more than one hundred pages), which contains the complete spectral theory of regular matrix polynomials and applications to band block Toeplitz matrices. In fact it contains also the most complete theory for inversion of Toeplitz band block matrices.

Peter Lancaster organized many conferences in Calgary and its environs. These conferences were always well prepared and interesting and were very important in view of the far distance Calgary is from the main mathematical centers. Peter understood this and made every effort to overcome the problem. I would like to mention especially the conference held in Calgary in 1988. This was a comparatively large conference with a very good representation from all over the world including a large group of Soviet and Eastern European participants. I am proud and grateful that this conference was dedicated to my sixtieth birthday. It was certainly a milestone in the development of operator theory and its applications.

Though now emeritus and retired from his academic duties, Peter still continues as usual his research, organization of conferences and other work that he likes. Our joint work continues even today and there are new plans and new projects. Recently Peter went through a tragically difficult period; he lost Edna, his wife and best friend. He and the entire family took it very hard. Peter is very strong and on a recent visit to him I saw how he was struggling with those difficulties; there were the first signs that he is recovering and starting a new life.

On Joint Work with Harry Dym

Israel Gohberg

1. How it started

I immigrated with my family to Israel at the end of July 1974. In the beginning we studied Hebrew very intensively. I also started to look for work in Israeli institutions of higher education; very soon I received an offer from the Tel-Aviv University which I decided to accept. In March 1975 I was invited by the Dean of the Faculty of Mathematics, Professor S. Karlin, to take a part time position at the Weizmann Institute of Science. I accepted this position and started working there two days a week. The Pure Mathematics Department in the Weizmann Institute of Science was very small; apart from S. Karlin the Dean, Harry Dym and Yakar Kannai were the only senior researchers. There was also a group of doctoral students. I started to lecture different courses in advanced operator theory and applications.

I met Harry and from our conversations I understood that he was very well informed in operator theory in general and in the work of the school of M.G. Krein especially. Already then he was the author of two books with H.P. McKean and was active in research.

I accepted a Ph.D. student. This was Sofia Levin and I started to work with her. During one of my visits to the Institute Harry expressed interest in joint work with me. I was also interested in this offer and we started to look for an appropriate problem.

I soon found such a problem during my visit to Amsterdam. The problem was proposed by a colleague from the Free University, Professor G.Y. Nieuwland. He in his turn obtained the problem from a colleague who was working in theoretical chemistry.

2. Band extension problems

The first problem consisted of the following: a function $k(t)$ $(-T < t < T)$ has to be extended to the full line to $f(t)$ in such a way that the function $1 - \hat{f}(\lambda)$, where $\hat{f}$ is the Fourier transform of $f(t)$, is positive (or more generally different from

Originally published in *Operator Theory: Advances and Applications* **134** (2002), 25–30.

zero on the line) and the function $1/(1 - \hat{f}(\lambda))$ has the form $1 - \hat{g}(\lambda)$, where $\hat{g}$ is the Fourier transform of a function $g(t)$ that vanishes outside the interval $(-T, T)$. Both functions f and g belong to $L_1(-\infty, \infty)$.

In a short time we had a solution to this problem and we started to write it down. We did not succeed in finishing this work before the summer. One of the reasons was that Harry decided to take a four month sabbatical. The last half was spent in Stanford with Tom Kailath.

By a coincidence, in Stanford Harry discovered that the thesis of Tom Kailath's doctoral student, A.C.G.Vieira, was relevant to our problem. In fact he was dealing with a matrix discrete analogue of the above mentioned problem for the positive definite case. More than that, in applications this problem is important and the solution is called the maximum entropy solution, or the autoregressive extension of statistical estimation theory. The discrete case of the scalar solution was solved and analyzed before by J.P. Burg in 1975. He came to it within the framework of spectral analysis in geophysics problems. After Harry returned to Israel we wrote our first joint paper [1] where we solved the generalized problem of extension of matrix valued functions, including the positive definite case with the maximum entropy solution. Explicit formulas for the solution based on Szegő orthogonal polynomials were also presented.

The following year, 1980, we published the paper [2] which contained the complete solution of the continuous analogue in the matrix valued case. This is a large paper (more than 70 pages) and it contains probably the first solution of the maximum entropy extension problem in this setting, together with a new definition of entropy under some natural technical conditions.

As a byproduct of the two papers described, we obtained new results in the theory of completion of finite matrices. The results were published in 1981 in [3]. The problem of extension in this case is the following: Let a symmetric band of width $2m + 1$ in an $n \times n$ matrix with complex entries be given and let the rest of the entries of this matrix be unspecified. The problem is to complete the matrix in such a way that the inverse of the completed matrix is a symmetric band matrix of $\leq 2m + 1$. Of special interest is the case where the completion is additionally required to be positive definite. In this case under natural conditions the solution exists, it is unique and can be characterized to have the maximum determinant between the determinants of all other positive definite completions. An explicit algorithm for this solution is also presented. This result contains Burg's maximal entropy inequality in the theory of covariance extensions. This is a result that follows from the case that the band is Toeplitz and in this case the solution is also Toeplitz. The described results are also generalized for block matrices. This paper became much more popular than the first two. A number of interesting results for the more general non-band case were obtained by other colleagues. Till today the non-band case in general has not been solved to the end.

Our next paper [4] can be considered as the solution of a continuous analogue of the previous problem. It is about extensions of kernels of Fredholm integral operators given in a band. The positive case generalized Burg's maximal entropy inequality. This result can be considered for the time dependent noncovariance case. In this paper is developed the beginning of the general theory of extensions and completions in an abstract algebra with multiplication subject to some special features that generalize the features in the concrete examples. This abstract approach served to clarify the band extension and completion problem, and to unify the results of the latter paper with the previous ones. The abstract approach became popular. It was used as a basis for a far-reaching development. This led to the band method presented in a number of papers of I. Gohberg, M.A. Kaashoek and H. Woerdemann, and of J. Ball, I. Gohberg and M.A. Kaashoek, in which new extensions and interpolation problems were solved. The results of the beginning of this section intersect with some results of D.Z. Arov and M.G. Krein.

3. Working together

In the first years of cooperation both of us made serious efforts to progress in the extension and completion problems mentioned above. The problems were new in an area which we had not considered before and we worked with interest and enthusiasm. We presented these results at different conferences and our results were nicely received by our colleagues.

I came to the Weizmann Institute twice a week and most of this time was used for joint work with Harry. A small part of the time I spent with my graduate students. Soon they were three, Sofia Levin, Israel Koltracht and Nir Cohen. The joint work with Harry was very pleasant. The work was continuing also during the lunch break and during the tea break in the afternoon. Sometimes we worked in unexpected places. I remember a few hours work in the foyer of the Van Gogh Museum in Amsterdam (while Harry's wife Irene was enjoying the exhibition). Harry is a very fine coauthor; he is talented, has good taste and a wide knowledge in theoretical mathematics as well as in applications. He is hardworking and has a wonderful command of English, and he very easily puts mathematics on paper.

I learned many things from Harry in mathematics and also in everyday life. I was used to the Soviet mentality and rules of behavior. Harry helped me to understand the new situation and to become used to it. In view of our friendly relations I could ask his advice on any question without hesitation. For instance, he was the first to notice and explain to me the difference between the practice regarding very good Ph.D. students in USSR and in the West after graduation. In the USSR the best Ph.D. students were kept for permanent work in the university (chair) where they studied. In the West on the other hand they would have to leave and spend at least a short time in other universities. There is a big difference between the USSR and the West in the evaluation of various areas of mathematics and mathematicians. Harry explained these things which looked like contradictions to me. On my part I told Harry a lot about M.G. Krein, his work and his school, about the difficulties of Jewish life in the USSR. All of this interested him. He especially enjoyed hearing jokes from the USSR.

4. Triangular extensions

Let $f(z)$ $(|z| = 1)$ be a function with specified Fourier coefficients f_j for $j = 0, 1, 2, \ldots$ and $|f_0| + |f_1| + |f_2| + \cdots < \infty$, and let $\psi_{-1}, \psi_{-2}, \ldots$ also be given complex numbers with $|\psi_{-1}| + |\psi_{-2}| + \cdots < \infty$. The problem consists of specifying the Fourier coefficients of $f(z)$ with negative indices in such a way that $f(z) \neq 0$ $(|z| = 1)$ and the Fourier coefficients of $1/f(z)$ with negative indices to be equal to $\psi_{-1}, \psi_{-2}, \ldots$. Of special interest is this problem with the additional condition that $|f(z)| = 1$ $(|z| = 1)$.

Our next two papers [5, 6] were dedicated to different generalizations of this problem. We solved it in the block discrete case as well as for the matrix continuous analogue. In the latter case with the additional condition this result was stated by M.G. Krein and F.E. Melik-Adamyan without proof in their study of scattering theory. This is probably the first published proof of this theorem. We also solved the finite matrix block analogue of the triangular extension problem. As far as we know this was a new result for matrices. The triangular completion problem for scalar matrices is stated in the following way: Let the entries of the upper triangular part (including the diagonal) of an $n \times n$ matrix be specified. Complete the matrix in such a way that it is invertible and the inverse has a priori given entries in the lower triangular part (without the diagonal). We also solved the problem of completing a matrix to be unitary if the entries of the upper triangular part is given. In the triangular extension problems some technical conditions were required. In particular the canonical factorization or the partial indices equal to zero were required for the solution.

5. Unitary interpolants and factorization indices

Three papers [7, 8, 9] deal with the problem of extending a matrix function $f(z)$ $(|z| = 1)$ with specified Fourier coefficients $f_0, f_1, \ldots$; $|f_0| + |f_1| + \cdots < \infty$ to a unitary matrix function without assumptions of canonical factorization as in the previous section. The solution if it exists certainly admits a factorization in general with nonzero partial indices. In paper [7] are described all unitary interpolants. One of the central results is the expression of the number of nonnegative factorization indices of the interpolants and their individual size via the given data $f_0, f_1, f_2, \ldots$. The set of the negative indices, when not empty, can be chosen arbitrarily, and hence in this case there exist an infinite number of unitary interpolants. Paper [8] contains the matrix continuous analogues of the previous results. In paper [9] is considered a more general problem when the condition $|f_0| + |f_1| + \cdots < \infty$ is eliminated and the factorization is replaced by generalized factorization. The results of these papers intersect with results of F.E. Melik-Adamyan and M.G. Krein and are related to a paper of J. Ball.

In 1983 Harry and I organized a workshop on applications of linear operator theory to systems and networks in the Weizmann Institute; as we now call it, an IWOTA workshop. It was the second in this series and it was a satellite workshop just before the MTNS conference in Beersheva. The workshop attracted mathematicians and engineers. A volume of the proceedings was published in the OT series – OT12 [13].

6. Contractive interpolants and a maximum entropy principle

This section is based on two papers [10, 11]. In paper [10] are studied all $n \times n$ matrix contractive interpolants on the unit circle when the Fourier coefficients with positive indices are given. It turns out that for this problem a maximum entropy solution can be found with an appropriate entropy formula and inequality. The solution is obtained by a reduction to a generalized band problem. Paper [11] contains further generalizations of these results.

7. Nevanlinna-Pick problem and maximum entropy

Our last paper was written after a long break. Starting with 1984, I did not work regularly in the Weizmann Institute. The Institute was going through a financial crisis and all part time positions were disbanded. For a while, by inertia, I continued to visit the Institute and by the way continued to work with Harry. During these visits we wrote papers [10, 11]. Then the breaks became longer, but we again started to work systematically for a period in 1995. My dentist's office was located in Rehovot and for some part of 1995 I had to visit him at least once a week. Sometimes I would visit Harry in his office before the dental appointment, sometimes after. This time we worked on the Nevanlinna-Pick problem for matrix valued functions in the disc and we wrote paper [12]. In this paper we studied maximum entropy solutions and an extremal problem for the Pick matrix. A generalization for the half plane was also obtained.

8. This is not the end

Harry Dym is a very good friend and an excellent coauthor. We worked together for almost twenty years. Some of the periods were more intensive, some less. In parallel with this research, each of us was involved in many other research activities, so the joint work was never a burden. Our joint work influenced and enriched our individual research, as well as research with other colleagues, and led to cross-fertilization and influence.

Now, after going over all our papers as a reader, I look back with satisfaction and gratitude. This was a fruitful and enjoyable period which I hope will continue.

References

[1] H. Dym, I. Gohberg, Extensions of matrix valued functions with rational polynomial inverses. *Integral Equations Operator Theory* **2** (1979), 503–528.

[2] H. Dym, I. Gohberg, On an extension problem, generalized Fourier analysis and an entropy formula. *Integral Equations Operator Theory* **3** (1980), 143–215.

[3] H. Dym, I. Gohberg, Extensions of band matrices with band inverses. *Linear Algebra and its Applications* **36** (1981), 1–14.

[4] H. Dym, I. Gohberg, Extensions of kernels of Fredholm operators. *Journal d'Analyse Mathématique* **42** (1982/83), 51–97.

[5] H. Dym, I. Gohberg, Extensions of triangular operators and matrix functions. *Indiana University Mathematics Journal* **31** (1982), 579–606.

[6] H. Dym, I. Gohberg, Extensions of matrix valued functions and block matrices. *Indiana University Mathematics Journal* **31** (1982), 733–765.

[7] H. Dym, I. Gohberg, Unitary interpolants, factorization indices and infinite Hankel block matrices. *Journal of Functional Analysis* **54** (1983), 229–289.

[8] H. Dym, I. Gohberg, Hankel integral operators and isometric interpolants on the line. *Journal of Functional Analysis* **54** (1983), 290–307.

[9] H. Dym, I. Gohberg, On unitary interpolants and Fredholm infinite block Toeplitz matrices. *Integral Equations Operator Theory* **6** (1983), 863–878.

[10] H. Dym, I. Gohberg, A maximum entropy principle for contractive interpolants. *Journal of Functional Analysis* **65** (1986), 83–125.

[11] H. Dym, I. Gohberg, A new class of contractive interpolants and maximum entropy principles. *Operator Theory: Advances and Applications* **29**, Birkhäuser, Basel (1988), pp. 117–150.

[12] H. Dym, I. Gohberg, On maximum entropy interpolants and maximum determinant completions of associated Pick matrices. *Integral Equations Operator Theory* **23** (1995), 61–88.

[13] H. Dym, I. Gohberg (editors), *Topics in Operator Theory, Systems and Networks. Workshop on Applications of Linear Operator Theory to Systems and Networks, Rehovot (Israel), June 13–16 (1983)*. Operator Theory: Advances and Applications **12**, Birkhäuser Verlag (1984).

In Memoriam Seymour Goldberg (1928–2004)

Robert L. Ellis and Israel Gohberg

The editors and editorial board of *Integral Equations and Operator Theory* announce with great sorrow the death of Professor Seymour Goldberg. Professor Goldberg passed away on December 11, 2004 at Suburban Hospital in Bethesda, Maryland after a long and valiant fight against a difficult illness. He is one of the founding members of the editorial board of this journal and the series of books *Operator Theory: Advances and Applications* (Birkhäuser). His influence, talent and dedication played an important role in the development of the journal and the series. We will always remember him as an outstanding mathematician, a very

Originally published in *Integral Equations and Operator Theory* **51** no. 1 (2005), 1–3.

kind personality and a dedicated friend. Our heartfelt condolences go out to his wife and family.

Seymour Goldberg was born March 24, 1928 in Brooklyn, New York. After serving in the army, he studied mathematics at Hunter College in New York and graduated with an AB degree in 1950. Seymour was one of the first males admitted to this traditionally all-female college. There he met his future wife, Lillian Slominsky, with whom he enjoyed a loving and dedicated marriage until death separated them. They have two lovely children, daughter Florence, a lawyer in Ohio, and son Benjamin, a professor of computer science at NYU. There are four grandchildren.

Seymour continued his mathematical education, receiving a master's degree in 1952 from Ohio State University, and a Ph.D. in 1958 from UCLA under the direction of Professor Angus Taylor. He interrupted his education from 1952 to 1954 to serve as a mathematical analyst for Lockheed Aircraft. Later he spent a year at Hebrew University as a fellow and three years at New Mexico State University as an assistant professor. From 1962 until his retirement in 1999 he worked at the University of Maryland in College Park, until 1967 as an associate professor and then as a full professor. Seymour visited for long periods many important universities, among them Cambridge University, the University of Beijing, the Vrije Universiteit in Amsterdam, Tel-Aviv University and the Weizman Institute. He was promoted to emeritus status when he retired.

Seymour's main research interests were in operator theory and its applications. In 1964 he was one of only three mathematicians to receive an Air Force grant for the purpose of writing a book. With an invitation from Professor Stampacchia, he spent most of that year working on the manuscript at the University of Pisa in Italy. This culminated in 1966 in the publication of his well known book *Unbounded Linear Operators*. It was one of the first books on operator theory and received very favorable reviews. It stayed in print at McGraw Hill more than 20 years and is still available from Dover.

The Air Force grant included funds for travel to the Soviet Union. Accordingly, Seymour sent his manuscript to I.M. Gelfand in Moscow, O.A. Ladyzhenskaya in Leningrad, M.G. Krein in Odessa, and I.Ts. Gohberg in Kishinev. This resulted in a visit to discuss the manuscript and to speak in their seminars. The visit to Kishinev was the start of a long collaboration and friendship between Seymour and Israel Gohberg, which continued in Tel-Aviv and College Park until Seymour passed away.

Of all his professional accomplishments, Seymour was proudest of the books he wrote. *Unbounded Linear Operators* was only the first of six books by Seymour. The next book was *Basic Operator Theory* (1981, Birkhäuser), coauthored by Israel Gohberg. In this book the traditional syllabus in functional analysis was revised. It has been used as a graduate text at many universities, has been reprinted several times, and is still available and in use. Recently Seymour published with I. Gohberg and M.A. Kaashoek *Basic Classes of Linear Operators* (2003, Birkhäuser), an expanded and enriched version of *Basic Operator Theory*.

This book also serves as an introduction to the books *Classes of Linear Operators* (Vol. 1, 1990; Vol. 2, 1993, Birkhäuser), written by the same three authors. The sixth of Seymour's book was *Traces and Determinants of Linear Operators*, coauthored by I. Gohberg and N. Krupnik (2000, Birkhäuser), which presents a general abstract approach to the trace and determinant with many important applications.

In addition to his books, Seymour published many interesting papers and directed several thesis students. One of his Ph.D. students, Richard Herman, later became a dean at the University of Maryland. Seymour taught a large variety of courses over the years, including many at the graduate level . He had a reputation for being an excellent lecturer, particularly in the graduate real analysis courses.

Seymour appreciated beauty in mathematics, especially very short and elegant proofs. He never tired of searching for such proofs and enjoyed presenting them to students and colleagues. He also appreciated beauty in classical music and opera, travelling the globe in the pursuit of this passion.

Seymour was a man of great wit, humor, and kindness. His presence at a gathering or a party invariably brought laughter. Based on his humanitarian principles, he spent the last 52 years of his life as a strict vegetarian.

Seymour has made significant contributions to his department and to mathematics in general. The mathematical community will always remember this outstanding and dedicated member.

In Memoriam Georg Heinig (1947–2005)

Albrecht Böttcher, Israel Gohberg and Bernd Silbermann

On May 10, 2005, Georg Heinig died unexpectedly of a heart attack in his apartment in Kuwait. We have lost one of the top experts in the field of structured matrices, an irreplaceable colleague, and a good friend. He was an active member of the editorial boards of the journal *Integral Equations and Operator Theory* and the book series *Operator Theory: Advances and Applications* since 1993. Our heartfelt condolences go out to his wife and his family.

Georg Heinig was born on November 24, 1947 in the small town of Zschopau in the Ore Mountains (Erzgebirge) in East Germany. From 1954 to 1964 he attended

Originally published in *Integral Equations and Operator Theory* **53** no. 2 (2005), 297–300.

the school in Zschopau and from 1964 to 1966 the elite class for mathematics at Chemnitz University of Technology. Such elite classes were established to provide especially gifted pupils with an extraordinary education in mathematics (but also in the natural sciences and in languages) under the guidance of experienced university teachers. The careers of many successful East German scientists started at elite classes. None of these classes has survived the German reunification.

He studied mathematics at Chemnitz University of Technology from 1966 to 1970 and graduated with the diploma degree in 1970. His diploma thesis was written under the supervision of Siegfried Prössdorf and was devoted to certain properties of normally solvable operators in Banach spaces.

After defending his diploma thesis with the best possible grade, Georg Heinig was given the opportunity of entering a PhD program abroad. He decided to continue his studies at Kishinev (now Chisinau) University under the supervision of the second of us. His wife Gerti accompanied him in Kishinev and also completed a dissertation during that period. Georg Heinig was a very talented and dedicated researcher. In Kishinev he embarked on research into the theory of Toeplitz, Wiener-Hopf, and singular integral operators with scalar and matrix-valued symbols, and it was during those wonderful years that he has fallen in love with all the exciting mathematics of structured matrices. His deep results in this area formed the basis of his excellent PhD thesis, which he defended in Spring of 1974. Many other mathematical insights gained by Georg during the years in Kishinev went into his habilitation thesis, which he completed in Chemnitz. The early paper Gohberg/Heinig, Inversion of finite Toeplitz matrices consisting of elements of a non-commutative algebra (Russian), Rev. Roumaine Math. Phys. Appl. 19, 623–663 (1974) became one of his most frequently cited works.

Georg Heinig returned to Chemnitz in 1974. In the following five years the first of us had the pleasure of attending his classes as a student, the third of us received an outstanding member of his research group, and the second of us was proud of Georg's outstanding mathematical achievements. Georg Heinig integrated several young people into his research, Karla Rost being the most prominent figure of them. In 1979 he defended his habilitation thesis, which was on the spectral theory of operator bundles and the algebraic theory of finite Toeplitz matrices. His two children Peter and Susanne were born in 1974 and 1977.

The scientific outcome of the research directed by Georg Heinig in the 1970s and early 1980s is summarized in his and Karla Rost's book *Algebraic Methods for Toeplitz-like Matrices and Operators,* which was originally published by Akademie-Verlag, Berlin in 1984 and was republished by Birkhäuser Verlag, Basel in the same year. This book has found a warm reception and perpetual interest by a large community for now about twenty years. Some of its basic ideas, such as the so-called UV reduction (which later received more popularity under the name displacement operation), have become important tools for workers in the field of structured matrices. Moreover, the scientific collaboration of Georg Heinig with Karla Rost lasted three decades until the day of Georg's death. Their joint research resulted in more than 30 papers. The results and methods of these papers are an essential ingredient to the present-day mathematical high-technology one is encountering in connection with structured matrices.

In 1982, Georg Heinig was a guest professor at Aleppo University in Syria, and from 1987 to 1989, he held a guest professorship at Addis Ababa University in Ethiopia. In the late 1980s he was appointed full professor at Leipzig University.

After the political events in Germany at the turn to the 1990s the life for Georg changed dramatically. All people working at East German universities were formally dismissed and had to apply for a position anew. Those who had shown a certain extent of political proximity to the former socialist system had no chance of receiving a new position at a German university, neither in East Germany nor in the subsequently reunified Germany. The situation was extremely difficult, and the efforts of Georg's friends to help him did not bring any positive results. Certainly Georg was very disappointed and despaired. Some time he planned to take over his father's store for vegetables, but eventually he looked for a job at a foreign university.

In 1993, Georg Heing went to Kuwait University, where he worked as a professor until his tragic death. The scientific conditions at Kuwait University were excellent and Georg has always thankfully acknowledged the recognition and friendship he received from his Kuwaiti colleagues. In 2003, he was awarded as the Scientist of the Year by the Amir of Kuwait. Despite all these successes, his and his wife's dream was to endure the university job only until the age of 60 years and then simply to relish life together, including travelling around the world. His unexpected death at the age of 57 abruptly dispersed this dream.

Georg Heinig's scientific legacy is immense. In more than 100 publications he made outstanding contributions to a variety of fields, including

- theory and fast algorithms for several classes of structured matrices,
- periodic Jacobi, Toeplitz, and Wiener-Hopf operators,
- classes of singular integral operators,
- resultants and Bezoutians for operator-valued polynomials,
- continual analogs of resultants and Bezoutians,
- numerical methods for convolution equations,
- applications in systems and control theory and signal processing.

Discoveries by Georg and his co-workers, such as the structure of the kernel and of the pseudoinverse for certain classes of structured matrices, significantly shaped the development of numerical algorithms. He also remarkably enriched various areas of operator theory, for example by deep results on the spectral theory of Jacobi matrices and of Toeplitz and Wiener-Hopf operators. He supervised 6 dissertations.

Georg Heinig was a very pleasant person and an inspiring colleague. His sense of humor and his characteristic bright laughing will be missed by everyone who was lucky enough to meet him. His permanent endeavor for disclosing the absolute essence of a matter and his untiring aspiration for clearness and brevity were challenges for his co-workers on the one hand and have resulted in grateful appreciation by his students and the readers of his publications on the other.

Another dream of Georg Heinig was a joint textbook with Karla Rost on structured matrices, ranging from the basics for beginners up to recent developments. About one year ago they started writing this book and three chapters are already more or less complete. It is unimaginable that he will never have this book in his hands some day. This tragedy bitterly reveals the gap that Georg has left and painfully reminds us of the projects and ideas that passed away with him. However, his work will endure and we will always remember this outstanding mathematician, excellent colleague, and wonderful friend.

Recollections: Vasile Grigorievich Ceban, on his 90th Birthday

Israel Gohberg

In a short while Vasile Grigorievich Ceban will celebrate his ninetieth birthday and it is my pleasure for this date to recall in a short note our joint work and our warm friendship during many years. I hope it will also be pleasant for the celebrator of the Jubilee.

Our friendship started in 1951 after I graduated from the Kishinev University and came to work in the Soroca Teachers Institute by an assignment of the Minister of Education: Agripina Nikitichna Crachun. I understand that it was not simple for her in the anti-Semitic atmosphere of that time to cancel my assignment to a school in Ungheni (to replace a drunken teacher) and to send me to the Teachers Institute in Soroca. Until today I am thankful to her. I am also thankful to my docents V.A. Andrunakievich and I.F. Volkov which recommended me for this assignment. Already then A.N. told me that the Director in Soroca is V.G. Ceban, a very pleasant man and a very good scientist. V.G. Ceban also held the chair of Physics and Mathematics there. So from the beginning he was in a sense my double boss.

V.G received me nicely and asked me to teach in the Moldavian language. It was very difficult for me to do that but I appreciated it later when the knowledge of Moldavien helped me a lot in my career. V.G. Ceban received the final part of his mathematical education in Moscow State University (MGU) which certainly was the best university for mathematics in the USSR. Very soon I started with him to discuss different mathematical questions, mostly in his area of mathematical applications to mechanics which was far from mine. I remember well the interesting stories he told me about his studies in Moscow and about his distinguished teachers at MGU. A big impression made on me the story how in the beginning of the World War Two, he helped to evacuate from Moscow the great mathematician Professor N.N. Luzin.

The birthday was on February 3, 2007. This note was published in a booklet about V. Ceban. Unfortunately V.G. Ceban died January 30, 2008.

I was also introduced to the lovely family of V.G.. I was lonely in Soroca and his wife Tamara Dmitrievna and daughter Ninochka were very nice to me. Many times Tamara Dmitrievna helped me in daily life. I appreciate highly the help which I received in organizing my visits to Odessa, to my teacher and master Professor M.G. Krein.

In the institute also happened a few unpleasant events. For instance some Jewish colleagues which were good teachers were fired without any serious reasons. Usually the reasons were based on denunciations by students which were so numerous at that time. This was executed and organized by the Vice director and party secretary, and the time was always chosen by him when V.G. was in the two months vacation.

After two years of our friendship in Soroca, there followed six years in Belts. I remember well these years. V.G. Ceban was assigned as the Director of the newly created Belts Pedagogical Institute. I was moved also to this Institute. As I found out later it was a result of the new Director's strong insistence and help of the minister A.G. Krachun. I am thankful to Vasile Grigorievich for helping me to find the time for preparing my candidate thesis (external) and later to set up my three months visit to Leningrad to defend it. Later I was appointed to be the holder of the chair of mathematics and V.G. helped me a lot in getting my first apartment in Belts which was very non-trivial at that time. In these years I married Bella and we became friendly with the entire families. Once in this period I had a serious accident and got on fire and was burning. It happened during a trip with him from Odessa where we discussed mathematics with the famous mathematicians M.G. Krein and S.G. Michlin. This was in in the summer of 1956 when he was a member of the Supreme Soviet of MSSR and his assistance with my medical treatment was of great help.

In 1959 I moved to Kishinev and worked in the Institut of Mathematics of the Academy of Sciences of MSSR. Very soon V.G. came also to work in the same institution. Now we exchanged roles and for a short time he worked in my department and I became his boss. During these years we wrote a joint paper on numerical methods for Wiener-Hopf equations, an area which was of interest to both of us. It was in the intersection of Numerical Analysis and Wiener-Hopf equations. This paper became quite popular and was used and quoted many times by other authors. Our friendship continued even after I emigrated to Israel. Here in Israel we often recall Vasile Grigorievich with my colleagues from Soroca, Belts and Kishinev. We remember his dedication and his kindness. It was a pleasure to meet him in 2003 when we visited Kishinev for my Dr. h.c. degrees in Kishinev and Belts.

Vasile Grigorievich was always for me an admirable person: a kind man, a dedicated scientist and a wonderful colleague. He made important contributions to higher education of Moldova. It was a pleasure to work with him both when he was my boss or I was his.

Today my family and myself congratulate Vasile Grigorievich with his ninetieth birthday and thank him and his family for a long and wonderful friendship and permanent support. According to the old Jewish tradition we rise our glasses with wine to drink a "lehaim" for your health and wish you 120 happy years of life. Our warm regards and congratulations to your family from all of us.

Heinz Langer, my Friend and Colleague Recollections. After Dinner Speech

Israel Gohberg

Dear Heinz, Dear Colleagues, Dear Guests,

Today we celebrate the retirement of Heinz. All his friends and colleagues are happy and in a cheerful mood. Is there a good reason for this? It is not clear. In general what is retirement? We are used to a scientific approach, so I looked for the answer in the Oxford Advanced Dictionary of Current English (New Edition). There I found RETIRE and listed what it means:

1. retire from to withdraw; go away;
2. retire to bed;
3. retire, of an army, go back; retire to prepared positions;
4. retire give up ones work, position, business, etc;
5. to retire the head clerk;
6. retire from the world enter a monastery or become a hermit, become a recluse;
7. retire into oneself become unsociable because one is wrapped up in ones thoughts.

Looking for RETIREMENT did not bring much more:

1. retirement retiring or being retired;
2. go into retirement retire;
3. retirement pension old-age pension, or pension.

After spending some time with the dictionary I became absolutely confused and decided that Heinz deserves a special definition.

I think that according to a widely accepted interpretation, retirement is the period when one receives ones pension and does not have any work obligations. But we were used to Heinzs advice, experience, knowledge, and talent, which he so generously shared with us and which was so important for our community. What will happen to it? Will it just disappear?

Does this make sense? Is this not a complete absurdity? Especially if we take into account that he is certainly the leading expert in the world in Functional

Analysis with Indefinite Metric or Indefinite Scalar Product. I would even suggest introducing a nickname for him: Heinz the Indefinite, the Second. The first was our teacher M.G. Krein; and Heinz Langer is the second. We cannot accept the fact of the retirement of Heinz and we cannot accept this celebration (we are lucky that the dinner is almost over. A pity about the dessert).

But I feel we are moving too fast with the conclusions.

In ideology very often things depend on interpretation. A small change in interpretation may lead to converse conclusions. This happened in our case also. As far as I am informed Heinz understands retirement differently.

He is certainly expecting to receive his pension and not to have any work obligations. Now comes the interpretation. What does it mean to receive his pension? This has only one trivial interpretation. What does it mean without obligations? This has many interpretations and one of them means without any obligation to do what he does not like to do and certainly to continue, to a much greater extent, to do what he loves to do. I do not want to enter into details as to what he loves to do and what he doesnt. I would like to assure you that Heinz will continue to do the work in which our community is so interested. More than that, for these activities he will have much more free time and interest. So after this analysis is clear we have good reason to celebrate and to be happy for Heinz and for our community. We now have every reason to accept the dessert. This is the end of the introduction, which is the first part of my speech.

The second part consists of recollections. I met Heinz for the first time in 1961 when he was spending an academic year in Odessa. Not all institutions of higher education in Odessa had the right to accept foreign students. Only the University had. It has the appropriate facilities including dormitories, and trained staff to watch and listen to follow the foreigners even when they do not want it. The members of this staff were busy reporting their findings not only to the academic authorities. The authorities of the University at that time were very anti-Semitic. This was demonstrated during the end of the 40s when M.G. Krein with other Jewish professors was fired without reason. The relations between the University and M.G. were far from friendly. Heinz Langer was sent to Odessa University and was caught in the middle. Mathematics in this University was very weak and Heinz was not interested in working and studying there. He had to stay in the University, use the mensa, receive there his financial support from an exchange program, and study Russian. All mathematical activities of Heinz were connected with M.G. Krein and his informal school. This included lectures, seminars, research, discussions and other activities. Such a co-operation was very strange and even impossible to imagine, but in this special case it worked, and as far as I can witness Heinz did not feel any animosity. To describe the situation I would like to tell you a short story about that time.

Mark Grigoreivich Krein was very careful in his behavior towards the authorities. In this respect he was strongly under the influence of his family and they knew well what type of problems one can expect from them. Also M.G. had his own reasons too. M.G. knew that it was forbidden for him to have Jewish graduate students at that time. He told me this. The selection was made via entrance exams and everything was organized in such a way that the Jews would not pass the compulsory entrance exam on Marxism. It is clear that M.G. was unhappy with this situation. During these years he instructed a number of excellent students on a voluntary basis. The students where working full time some time out of town and studding with M.G.in his apartment when it was possible. This way received the doctoral degree Yu.L. Shmulyan, I.S.Kac, A.A. and P.P. Nudelman and others. In fact I was also working with M.G. on such a private basis.

Once when Mark Grigoryevich worked in the Civil Engineering Institute he asked for an audience with the Rector. After making him wait some length of time the Rector received M.G. Krein and M.G. asked the following question: *Is it not dangerous politically for me and for the Institute that I have four graduate students who are Armenians? Can the Party authorities not accuse me in Armenian nationalism?*

I would like to explain that indeed the Armenian scientific authorities were sending students to study with M.G. Krein. The students were financially supported in Armenia and studying with M.G.

Certainly not, answered the Rector and he started to explain in detail the meaning of the friendship between Republics and Nationalities. He did not catch the mockery and sneering tone of M.G.

M.G. Krein highly appreciated Heinzs work and his research plans and in general very quickly Heinz became an outstanding member of M.G. Kreins school. Very soon he spoke fine Russian and was called in the Russian style with the patronymic Heinz Kurtovich.

Heinz had a very good relationship with M.G. and his family and with his entire neighborhood. Much later in August 1988 M.G. wrote to me on my 60th birthday. He started this letter with the following phrase: *I have had a number of students I feel proud of. The list includes M.A. Krasnoselsky, H.K. Langer, M.S. Livsic, M.A. Naimark and V.P. Popov. In this brilliant company a conspicuous place is occupied by Israel Gohberg.*

Heinzs first visit was very successful. He obtained with M, G. very interesting deep results. He started to work in other areas but the main focus remained the Indefinite Metric. He made new friends. I would like to mention I.S. Iohvidov and his family. He continued to work in Dresden and to visit Odessa regularly. He and his family visited us in Kishinev, and me and my family visited the Langers in Dresden. In Dresden the situation was not simple. Heinz was forced to spend time on intensive study of Marxist ideology and some communist party officers made serious efforts to force Heinz to join the party. His correspondence was controlled and some conversations also. He was able to fight back and he never joined the party. (By the way, M.G. was never a member of the communist party; some people from his neighborhood were.)

Heinz avoided it by becoming the chairman of the faculty of the committee of German-Soviet friendship. This clearly was in his own best interest. I remember meetings of this committee, with tea and cookies.

Heinzs visit to Canada and his stay there for an academic year at the invitation of Professor Israel Halperin was also very important. He was introduced to North America. He visited many universities there. Let me also mention Calgary and Professor Peter Lancaster, for example; after his return our information about Canada and mathematics there was much more complete. This was the beginning of our friendship with Peter. During his stay in Canada somebody in Odessa spread a rumour that Heinz would probably not return to DDR. Should this have happened it would have meant a lot of problems for M.G. Krein with the Soviet authorities. I remember how this topic was discussed in Kreins family. They were afraid of it. It is worth mentioning that this did not happen and in fact I was the first of this group to leave the Soviet Union for the East. It happened in 1974 when M.G. was already retired and at that time, as far as I know, it did not make any big problems for him.

The next period was the period of preparation to leave for the West. This period had its excitements. Finally it ended successfully and Heinz is with us in the TU Vienna already many years. We hope that this pleasure and privilege will continue for many more years . We thank Heinz for his active participation and initiative in the life of the Operator Theory community.

Heinz I hug you brotherly and wish you all the best in the years to come.

Our Meetings with Erhard Meister

I. Gohberg and M.A. Kaashoek

Professor Erhard Meister was an outstanding expert in many areas of mathematics and a wonderful person. He was a pleasant colleague and a good friend of us both. We shall miss him.

We met Erhard Meister regularly at colloquia, conferences, and seminars, in Germany, in Israel, in the Netherlands, and in other countries. Our meetings with him were important for us. During these meetings we discussed progress and problems in the area of singular integral equations and Wiener-Hopf operators, with an emphasis on recent developments in factorization. But not only that, we also discussed organizational matters, including conferences, journals, books, and invitations.

The first author recalls with gratitude his meeting with Meister in the Spring of 1975 when he visited West-Germany for the first time after leaving the former Soviet Union. During this first meeting, which took place in Darmstadt, Meister put the first author in touch with C. Einsele, a member of the Birkhäuser family, who was then the head of the editorial branch of the publishing company Birkhäuser Verlag Basel. At that time Birkhäuser Verlag was preparing the translation of the Gohberg-Krupnik book[1] into German, under Meister's supervision and editorship. The connection with Einsele was very important and led to the foundation of the journal *Integral Equations and Operator Theory* in 1978. From the very beginning, Meister was a member of the editorial board of the journal, as well as of the later book series *Operator Theory: Advances and Applications*.

Meister was knowledgable in pure and applied mathematics. His scientific achievements in both areas are many and essential. He attracted students and co-authors, and was influential both in the German mathematical community and abroad. His wide knowledge made him a valuable contributor to discussions in conferences and seminars where he often raised unexpected questions and made interesting remarks.

Originally published in *Operator Theory: Advances and Applications* **147**, Birkhäuser Verlag, Basel, 2004, pp. 73–75.
[1]I. Gohberg, N. Krupnik. *Einführung in die Theorie der Eindimensionalen Singulären Integraloperatoren*. Birkhäuser Verlag, Basel, (translated from Russian E. Meister (Ed.)) 1979.

Together with Meister, the two of us organized two one-week Oberwolfach meetings on "Wiener-Hopf Problems", one in 1986 and the other in 1989. These were very special meetings which attracted participants from the West (including the USA) and from East-Europe. The second meeting was one of the first where experts from East-Germany met their colleagues from the West in person. These were exciting events which had a considerable influence on further development of Wiener-Hopf theory and its applications.

Meister was also an active participant in other Oberwolfach meetings on operator theory. At all these meetings his positive attitude and cheerful personality helped to create a stimulating atmosphere and an environment of friendship.

At the Oberwolfach meetings he had a double role: first as a scientific leader, and secondly as the chief organizer and guide of the hikes on Wednesday afternoons. In his second role he usually appeared in knickerbockers and colorful long socks, the traditional German outfit for hikers, and an impressive hat. He knew the geography of the Black Forest well, including the right cafes with good coffee and delicious tortes. His guiding was precise, and we always were back on time at the institute.

An unexpected meeting, which Meister organized, was a small one day conference in Darmstadt in 1997 when the first author received a honorary doctoral degree from the Technische Hochschule at Darmstadt, and the second author presented the laudatio lecture. Later we learned that Erhard had been preparing this event for many years, skilfully guiding his proposal through the university channels.

We remember gratefully his support for and participation in the International Workshops on Operator Theory and Applications (IWOTA). At the IWOTA meeting at Faro (Portugal) he could not give his plenary talk, and his lecture was presented by his wife Ludmilla. But even then no one expected that this IWOTA meeting would be his last public appearance at an international conference. His death is a great loss to the mathematical community and to his many friends.

Part VI
Honorary Doctorates: Laudatios and Replies

This part concerns the six honorary doctorates Israel Gohberg has received. Corresponding documents such as laudatios, acceptance speeches and other related material are presented here.

VI.1. The Technische Hochschule at Darmstadt, Germany

On June 27, 1997 professor Israel Gohberg received his first honorary doctorate. It was awarded by the Technische Hochschule at Darmstadt. The laudatio was presented by the late Prof. Dr. Erhard Meister, under the title Fredholm-Noether-Gohberg. Unfortunately, the text of the laudatio is missing. The program (in German) of the corresponding "Festkolloquium" is reproduced here below.

Waiting for the "Festkolloquium" to begin, from the left to the right: the late Leonid Frank, Bella and Israel Gohberg, Freek van Schagen, Wies Kaashoek, Harm and Greetje Bart, Albrecht Boettcher, Ernst Albrecht, and Rien Kaashoek.

Invitation

EINLADUNG ZUM FESTKOLLOQUIUM ANLÄSSLICH DER VERLEIHUNG DER WÜRDE EINES EHRENDOKTORS DER TECHNISCHEN HOCHSCHULE DARMSTADT

AN HERRN PROFESSOR DR. ISRAEL GOHBERG

AM 27. JUNI, 16 UHR, 64289 DARMSTADT, HOCHSCHULSTR. 1 (ALTES HAUPT-GEBÄUDE)

Program

BEGRÜSSUNG
16.00 Uhr: Prof.Dr. Johann-Dietrich Wörner, Präsident der TH Darmstadt

16.10 Uhr: Prof.Dr. Peter Rentrop, Dekan des Fb Mathematik

LAUDATIO
16.20 Uhr: Prof.Dr. Erhard Meister
Fredholm-Noether-Gohberg

VERLEIHUNG DES EHRENDOKTORATS
16.20 Uhr: Prof.Dr. Peter Hagedoorn, Vizepräsident der THD

PAUSE
17.00 Uhr: Sekt und Kaffee zum Hörsaal

FESTVORTRAG
17.30 Uhr: Prof.Dr. Marinus A. Kaashoek, Vrije Universiteit Amsterdam:
Linear Operators and the State Space Approach

18.00 Uhr: Ende

19.30 Uhr: Möglichkeit zur Teilnahme am gemeinsamen Abendessen im Restaurant

Note from Rien Kaashoek: When I told professor Gohberg that I was going to give the "FESTVORTRAG" at Darmstadt, he answered me in his usual direct way: *Well, then probably you will be the only one who has to work hard that day.* I believe he was right but I did it with great pleasure and satisfaction.

Israel Gohberg receives honory doctoral degree in Darmstadt

What hope is there to contain - in the few slim paragraphs here - the phenomenal energy of a man who has more than 365 journal articles and 19 books to his credit, who has founded and edits a major journal (see page 8) and book series of international standing, who has guided countless students, 40 doctoral candidates among them, through the modern realm of operator and system theory? Very little hope, indeed.

Israel Gohberg has been described as one of the exceptional mathematicians of our time, comparable in his manner perhaps only to Paul Erdös. Three qualities would appear to account for his unique influence: he is a charismatic and inspirational man who quickly pulls others into his orbit, keeping acolytes motivated for long periods of time; he possesses an uncanny instinct for what is mathematically possible and productive; and he has the talent and the stamina to see that his ideas and those of his co-authors are realized, and then communicated through publications to the larger interested audience.

A man almost larger than life. Next year, Professor Gohberg celebrates his seventieth birthday. Here, a thumbnail sketch of his professional biography to date:

Israel Gohberg, Russian born (23 August 1928), attended primary and middle schools in the former Soviet Socialist Republic of Kirgiz where he later spent two years at the College of Education in Frunze, the republic's capital. So impressed were the young man's teachers by his mathematical acumen, they urged him to transfer to university. He did so, beginning his university education in Kishinyov (Moldavian S.S.R.), and moving on to complete his doctorate in Leningrad in 1954 - three years after the publication of his first mathematical results. At Moscow State University he earned his qualification as university lecturer.

Thus academically girded, Gohberg was ready to embark on a career as researcher and educator that has come to span a vast amount of territory - both geographically and intellectually. In 1974, having risen to the rank of senior researcher and department head of functional analysis at the Academy of Sciences in Kishinyov and also professor at the University of Kishinyov, Gohberg emigrated to Israel, accepting a professorship at Tel Aviv University and working also at the Weizman Institute in Rehovot. In 1983 he was called upon by Amsterdam's Vrije Universiteit to accept a part-time professorship and he has repeatedly held guest professorships in Calgary and College Park (Maryland). He has also been professionally active in Germany, partly in connection with the Alexander von Humboldt Prize awarded him in 1992, but also by virtue of the long ties held with faculty members of the Institute of Technology in Darmstadt. On 27 June 1997 this school paid tribute to Israel Gohberg by conferring on him an honorary doctoral degree.

Israel Gohberg with his friend Erhard Meister and Marinus A. Kaashoek

A much more detailed account of Israel Gohberg's life and achievements until 1989 can be found in volume I of the following works:

OT 40 + 41
Dym, H. / Goldberg, S. / Kaashoek, M.A. / Lancaster, P. (Eds)

The Gohberg Anniversary Collection
Vol. I: The Calgary Conference and Matrix Theory Papers   Vol. II: Topics in Analysis and Operator Theory
1989. 504 pages. Hardcover • ISBN 3-7643-2307-8   *1989. 560 pages. Hardcover • ISBN 3-7643-2308-6*
Vols I + II (set price)
ISBN 3-7643-2283-7

VI.2. The Vienna University of Technology at Vienna, Austria

On January 18, 2001, the Vienna University of Technology awarded the degree of honorary doctor (Dr. rer. nat. h. c.) to Prof. Dr. Dr. h. c. Israel Gohberg. The laudatio was presented by Prof. Heinz Langer from the Institute of Analysis and Technical Mathematics of the Vienna University of Technology. In this chapter we reproduce the laudatio as well as the response of the honorary doctor on this high award.

Laudatio by Heinz Langer

Magnifizenz, dear Israel, ladies and gentlemen,

Israel Gohberg is one of the exceptional mathematicians of our time. By now he has written more than 400 research papers and about 20 monographs, he has supervised about 40 PhD students, and he has worked with more than 60 coauthors. In a Newsletter editorial of the Birkhäuser Verlag it is said that 3 qualities of Gohberg appear to account for his unique influence:

- He is a charismatic and inspirational man who quickly pulls others into his orbit, keeping them motivated for a long time.
- He has an uncanny instinct for what is mathematically possible and productive.
- He has the talent and the stamina to see that his ideas are realized and then communicated through publications to the larger interested audience.

In the following, I would like to elaborate a little on these qualities.

First, his charisma as a mathematician and a man. Gohberg has been working permanently or regularly at a number of different places all over the world: for 15 years he worked in Kisinev, after 1974 he held a chair in Tel Aviv, at the same time he worked in Stony Brook, at the Weizman Institute in Rehovot, and in Calgary. He also held a professorship in Amsterdam for 15 years and has regularly been a guest professor in Maryland. He is an excellent teacher and at each of these universities or institutes he inspired a group of mathematicians to follow his ideas. Between his emigration from the SU in 1974 and the end of the 80ies, when almost no official contact between him and his former students and collaborators was possible, his inspiration even overcame the Iron Curtain.

As soon as the Soviet Union had collapsed, he helped many of them to start a new life and he still tries to keep them close to him, geographically as well as mathematically. This also shows that Israel's relations with his numerous students and collaborators do not only concern mathematics, they often grew into close friendships. That Israel is highly appreciated as a mathematician and as a friend can also be seen from the fact that scientists from 7 countries spontaneously agreed to participate in the Colloquium 'Operator Theory and its Applications', which will follow today's ceremony to honour Israel Gohberg.

One of the reasons why mathematics and friendship are so important for Israel is that his life was not always easy. He was born on August 23rd, 1928, in Tarutino, a town in the southern part of Bessarabia, which then belonged to Rumania. His father owned a small one-man printshop, his mother worked as a nurse midwife. In June 1940 the area was taken over by the Soviet Union, and two months later, exactly on Israel's 12th birthday, his father was arrested by the KGB and sent to a Gulag in Siberia. The reason for the arrestment was not made clear and the father was never seen again by the family. 20 years later, his mother received a brief note that her husband was posthumously declared 'free of guilt'. After the arrestment of his father, his mother, whom Israel considers his first and most important teacher, was the sole support of her 2 children, Israel and Fanya, his 5 years younger sister. Only one year later, in 1941, the German Army invaded the Soviet Union forcing the Jewish family to flee for their lifes thousands of kilometers to the East, to Frunze, the capital of Kirgizia, not far from the Chinese border. There Israel completed elementary and high school and started his university education at the Pedagogical Institute. To his mother's disappointment he did not study medicine. Fortunately, his sister Fanya became a prominent pediatric surgeon (she even received a 'Woman of the year' award of an Israeli magazine), and Israel got a wonderful medical doctor for his life, his beloved wife Bella, and it is my pleasure to welcome both of you here, Bella and Fanya.

During the two years in Frunze, Israel's excellent mathematical abilities became already apparent. A professor helped him to transfer to Kisinev where he got his master's degree in Mathematics in 1951. Some time before, in 1950, he had already met Mark Krein, the mathematician who influenced Israel most, although he could never become his student officially. Krein had created a world famous school in functional analysis in Odessa and Israel wanted to do his PhD work with him. But this turned out to be impossible: Krein would not have been allowed to have more Jewish students since he had already had two or three after 1945, who were war veterans and therefore privileged. So Israel worked on his PhD thesis externally while teaching at a high school. Stalin's death in 1953 lead to an improvement of the situation. In 1954, Israel could defend his PhD thesis in Leningrad. Afterwards he worked at a teacher's college, then at the Pedagogical Institute in Beltsii (all in Moldavia), and finally, in 1959, he was invited to work in the Moldavian branch of the Soviet Academy, which later became the Academy of Sciences of the Moldavian SSR. There, from 1964 to 1974, he was the head of the Department of Functional Analysis. Although Krein could not accept Israel as a PhD student, they started to work together in the 50ies and this cooperation continued for many years. As Ralph Philipps formulated it, it was 'one of the most successful collaborations in Mathematics, the resulting books and papers are outstanding'.

And it was in Mark Krein's seminar in Odessa where Israel and I have met for the first time in December 1961, almost 40 years ago. Israel and I have never worked together, but we were both strongly influenced by Krein, and since the end of the 60ies we have met rather regularly, first in Dresden and Kisinev, later in Amsterdam and after 1990 also in Tel Aviv, and at many conferences. Over all these years, although Israel lived under very different conditions, the human being always remained unchanged. He has a deep optimism and a true sense of humor, which have helped him to overcome all the difficulties in his private life and all obstacles in his career.

Towards the end of the 60ies, the situation for Jewish scientists in the Soviet Union worsened again. An example for this is what happened to one of the first and best known students of Israel, Alexander Markus. In the summer of 1970, I visited Kisinev right after Alik Markus had defended his second doctorate with great success in Voronesh. There was a large banquet to celebrate this, which I attended. However, a successful defense was not the last step in the SU to obtain a higher academic degree. There existed a (widely anonymous) committee in Moscow which had to give a final approvement. Alik Markus never got this approvement and the second doctorate was never transferred to him. Also Israel suffered from the reviving antisemitism in the SU in various ways. For instance, it became impossible to travel even to neighbour socialist countries like Poland and Bulgaria or to become elected as a Full Member of the Academy of Sciences of the Moldavian SSR. After all, Israel finally decided to emigrate to the country of Israel with his family. He got the permission in 1974, and with an unbelievable amount of energy and ideas he started a new life in the West.

Much had to be left behind, but what Israel could take with him was *his mathematical instinct for what is important and potentially fruitful*. Israel Gohberg is the leading authority in

- integral equations,
- nonselfadjoint operators,
- spectral theory and factorization of matrix and operator functions,
- inversion problems for structured matrices.

His first research topic in the 50ies were Fredholm operators and their different generalizations, analytic operator valued functions, Fredholm properties of Toeplitz matrices and of Wiener-Hopf operators. Often he used ideas and techniques which at first glance did not seem very much related with the problem. E.g., he applied Banach algebra techniques in the theory of singular integral operators. The collaboration with Mark Krein led to fundamental results on the perturbation of Fredholm operators, systems of Wiener-Hopf equations, factorization of matrix functions in the Wiener algebra, the foundation of the theory of operator ideals, completeness properties of the systems of root vectors, and the theory of characteristic operator functions. From 1959 on, Gohberg started to build his own group in Kisinev. He and his students and collaborators worked on factorization of operator functions, decomposing algebras, Wiener-Hopf operators, singular integral operators, and on other topics. The famous Gohberg-Semencul and Gohberg-Heinig formulas for the inversion of finite Toeplitz matrices and their continuous analogues, which are often used in electrical engineering literature, were proved in this period.

After his emigration from the Soviet Union, he also started a fruitful interaction with control and electrical engineers. Inspired by these, he developed new directions in operator theory, which in turn led to new constructive operator theory methods in control theory. As an example, I would like to mention the so-called state space method, which has its roots in the mathematical systems theory of the 60ies. Due to the results of Gohberg and his co-workers, it became an effective tool for the solution of a wide range of problems like Wiener-Hopf equations, matrix versions of interpolation problems, e.g., of Nevanlinna-Pick and of Caratheodory-Toeplitz type, and, recently, for inverse problems for canonical differential systems with rational spectral density. As a by-product, modern control problems like the sensitivity minimization problem, model reduction and robust stabilization could be solved explicitly with the solution being described in the state space form.

But Israel Gohberg is not only an outstanding mathematician, he also has the rare *talent and the stamina that his ideas get realized*. All this makes him play a leading role in the international mathematical community. Still in Kisinev, he was the driving force behind the journal 'Matematiceskije Issledovanja'. After having moved to Israel, he contributed a lot to make operator theory and its applications more and more popular: He founded the journal 'Integral Equations and Operator Theory', which became the leading journal of this field in the world and which really can be considered as *his* journal. In the series 'Operator Theory: Advances and Applications', which was also founded and is still being edited by him, more than 120 volumes have appeared since 1980. About 20 years ago, Gohberg initiated the biannual 'International Workshops on Operator Theory and its Applications', IWOTA, of which so far 12 workshops took place, one also at the Vienna University of Technology in 1993. Israel is also heavily involved in the organization of the 'International Symposia on the mathematical theory of networks and systems', MTNS, where he successfully tries to bring together engineers and mathematicians. Finally, together with Bernhard Gramsch from Mainz, he also organized a series of meetings in Oberwolfach, a famous centre for mathematical research in Germany.

Maybe this is a good opportunity to thank you, Israel, on behalf of my colleagues, in Vienna and elsewhere, for giving us all these possibilities for publishing papers in your journals and book series, for meeting at these high level conferences, and, last but not least, for being a real good friend on whom one can always rely.

Of course, all these achievements have not been unnoticed so far. Israel Gohberg has got a number of prizes and academic awards. In 1970 he was elected Corresponding Member of the Academy of Sciences of the Moldavian SSR, reinstated again in 1996 after having been removed due to his emigration. In 1985 he was elected Foreign Member of the Royal Netherlands Academy of Arts and Sciences. He was awarded the Landau Prize and the Rothschild Prize in Mathematics, the Hans Schneider Prize by the International Linear Algebra Society, an Alexander von Humboldt Prize by the German Humboldt Foundation, and an honorary doctoral degree by the Technische Hochschule Darmstadt in 1997.

I would like to thank the Rector Magnificus and the Senate of the Vienna University of Technology that they have decided to award the honorary degree *Doctor rerum naturalium honoris causa* to Israel Gohberg and I would like to congratulate you, Israel, very much on this occasion.

Response of the honorary doctor Israel Gohberg

RECTOR MAGNIFICUS
MEMBERS OF THE SENATE,
DEAR GUESTS, COLLEAGUES AND FRIENDS

I am very happy to receive today the Honorary Doctoral Degree from the Vienna Technical University. I would like to thank with all my heart the Senate and the authorities of this university for this high appreciation of my research during 50 years of my career. Special thanks are due to my colleague and friend Professor Heinz Langer.

I am very sorry that my parents cannot enjoy this ceremony. My father died in 1942 in the Gulag and the place of his grave is unknown; my mother, who was a midwife living in difficult conditions dedicated all her life to the education of my sister and myself. I am very sorry that my teacher and friend, M.G. Krein, one of the great mathematicians of the twentieth century also cannot enjoy this ceremony.

Professor M.G. Krein created in Odessa a school of mathematics which was known worldwide for its achievements. After World War II, during the years of severe antisemitism in the Soviet Union (especially in Odessa), he was thrown out of the Odossa University. His famous mathematical school at this university came to an end. In fact it sort of moved underground to his apartment and became like a private institution without offices, secretaries or bureaucrats, but with outstanding achievements. Seminars were held in the Odessa Scientists' club and in the institutions where he worked. During my vacations I spent at least three months each year with M.G. Krein working in his apartment or dacha. I was never a Ph.D. student and never a formal student of M.G. Krein; to me he was much more than a teacher and friend. A large part of my achievements are due to his influence.

In Odessa in the school of M.G. Krein I met Professor Heinz Langer who came from Dresden after his promotion to work with M.G. for a year. At that time the status of M.G. Krein was not simple and as a result Heinz Langer was officially associated with the the Odessa University but for mathematics he was with M.G. Krein at his private school. There was a wonderful atmosphere in this school; goodwill, dedication and friendship, as well as efforts to be appreciated M.G. Krein and colleagues dominated.

From the Odessa period on Heinz and I had common interests and followed each other's work. Our friendship went through different stages, but it was always enjoyable and fruitful, and continues till this day. However we do not have joint work. I was also in that the second half of my career which was spent at the Tel Aviv University in Israel, my home university, where I had so many opportunities. I enjoy living in my own country and the free world and I have wonderful friends and colleagues; I have how second university homes in the Free University in the Netherlands and in the University of Maryland at College Park, USA.

To close I would like to thank my family, my wife, my daughters, my sister and her family for their continued support and understanding. To have a mathematician in the family is not easy.

Again I would like to thank the Vienna Technical University for bestowing on me such a high honour. I am especially happy that it is Vienna because I have special feelings towards this town; Vienna was the first town in the free world to receive us on our immigration to Israel.

VI.3. West University of Timişoara in Rumania

On January 28, 2002, the West University of Timişoara awarded the degree of honorary doctor to Prof. Dr. Dr. h. c. Israel Gohberg. The laudatio was presented by Prof. Gaspar from the Mathematics Department. In this chapter we reproduce the laudatio as well as the response of the honorary doctor on this high award.

Laudatio by Professor Dumitru Gaspar

Dear Rector Magnificus,
Dear Colleagues of the Senate of the West University of Timişoara,
Dear Colleagues, participants at the XIXth International Conference on Operator Theory,
Beloved Colleagues, mathematicians of Timişoara,
Dear Students,

It is a rare joy to welcome in our midst a distinguished master of world mathematics, a never-failing companion of the West University of Timişoara and a great and steady friend of Romanian mathematics and education, professor Israel Gohberg.

He is one of the outstanding mathematicians of our time. He is known, appreciated and respected by the whole international mathematical community. How else could one appreciates this mathematician: a great teacher, who published over 410 articles in journals of the highest standard, 20 monographs from several different publishers, who supervised over 40 Ph.D.' s, and who worked in more than 10 research teams with more than 60 collaborators? Of course, something contributing to this appreciation is also his most impressive career in various Research Institutes and Universities worldwide, starting with the University of Chişinău. He led a department of research of the Mathematical Institute of the Academy of the Moldavian (at that time Soviet) Republic. He had a professorship at the University of Tel Aviv, where now he is Professor Emeritus and he worked at the famous Weizman Research Institute in Rehovot. He held for 15 years a professorship at the Free University at Amsterdam, Holland, and was visiting professor for long periods of time at several American Universities, as Stony Brook in New York, Georgia in Athens, College Park in Maryland, Blacksburg in Virginia, Storrs in Connecticut Calgary in Canada. Also he was active in Germany at the Universities of Regensburg and Mainz, and at the Institute of Applied Analysis and Stochastics in Berlin.

After this brief exposition of the top position in world mathematics held by our distinguished guest, I take great pleasure in proceeding with his connections with us, Romanian mathematicians, and especially those from Timişoara. I had the opportunity to meet our distinguished guest via another great friend of Romanian mathematicians, the late Béla Szőkefalvi - Nagy. It was in November 1968 during my first visit at a foreign University, the Bolyái Institute in Szeged. Finding out that I was one of the students of the well known Romanian professor Ciprian Foiaş, professor Israel Gohberg accepted to share with me the wonderful invitation so elegantly extended to us by the one and only Béla Szőkefalvi - Nagy. Just imagine a young, barely known mathematician at an evening walk along the banks of the Tisza river with two of the greatest mathematicians, Béla Szőkefalvi - Nagy and his guest, Israel Gohberg, who with their well-known multilinguism and generosity agreed to have Romanian as our conversation language. I remember perfectly well that week spent in Szeged, which instilled in me for the rest of my life the calling to be a researcher, the attachment to science.

I learned from professor Israel Gohberg how to distinguish between my own mathematical ideas and truly new ones, how to present a mathematical result, what means priority in research, how to work with my younger colleagues. Professor Gohberg's papers, joint with Mark Kreĭn, on the theory of nonselfadjoint operators (see items 1 and 2 from the list of books in Gohberg's List of Publications[1]) and on the study of classes of compact operators with the aid of characteristic numbers, inspired me later on in my research. More than this, 4 of my younger colleagues and coauthors : Traian Ceauşu, Dobrinca Mihailov, Ilie Stan and Nicolae Cofan amplified on these ideas both in their Ph. D. Theses as in later papers. Also the interventions of Israel Gohberg in control theory had positive influences on another team of matematicians around prof. Mihail Megan, now Dean of our Faculty. The journal "Integral Equations and Operator Theory", founded and run by professor Gohberg, opened it's pages also for our colleagues, from whom I mention the youngest : Adina and Bogdan Sasu and Păstorel Gaşpar.

Obiously, the scarce space of this laudation does not allow us to present the fullness of the mathematical relations between the Romanian mathematicians and our distinguished guest. I only recall that two of the essential developments in the theory of nonselfadjoint operators, which were in the forefront of the whole literature on operator theory after 1970, are represented by our guest today and Ciprian Foiaş, professor at the University of Bucharest and at the University of Bloomington, Indiana (USA) and Dr. H. C. of the University of the West, Timişoara. Their starting points are the famous monographs I. Gohberg & M. G. Krein "Introduction á la Theorie des Opérateurs lineaires non autoadjoints dans un espace hilbertien", Dunod, Paris, 1971 and B. Sz.-Nagy & C. Foiaş "Analyse Harmonique des Opérateurs de l'espace de Hilbert", Paris, 1968. It is not necessary to expand further on the mathematical friendship and the deep scientific bonds of our guest with Ciprian Foiaş, bonds that include by the way the whole group of operator theory researchers from the Mathematical Institute of the Romanian Academy and the Mathematical Faculties of the Universities of Bucharest and Timişoara.

Although professor Gohberg was not able to attend our Conferences on Operator Theory, he appreciated them, offering for some of their proceedings publishing space in the monograph series "Operator Theory : Advances and Aplications", which he runs for Birkhäuser Verlag. These all are arguments and occurrences, which fully confirm what an editorial from Birkhäuser wrote about the qualities of Israel Gohberg: "He is a charismatic and inspirational man who quickly pulls others into his orbit, keeping them motivated for a long time. He has an uncanny

---

[1]See Part 2 of the present book.

instinct for what is mathematically possible and productive. He has the talent and the stamina to see that his ideas are realized and then communicated through publications to the larger interested audience." As an excellent teacher and as a competent editor, he inspired a large group of mathematicians to follow his ideas. As soon as the Soviet Union had collapsed, he helped many of his former students and collaborators to start a new life and he still tries to keep them close to him, geographically as well as mathematically. This also shows that Israel Gohberg's relations with his numerous students and collaborators do not only concern mathematics, but also in many instances grew into close friendships.

One of the reasons why mathematics and friendship are so important for Israel is that his life was not always easy. He was born on August 23rd, 1928, in Tarutino, a town in the southern part of Bessarabia. His father owned a small one-man printing-shop, his mother worked as a nurse and midwife. In June 1940 Bessarabia was taken over by the Soviet Union, and two months later, exactly on Israel's 12th birthday, his father was arrested by the KGB and sent to a Gulag in Siberia. The reason for the arrest was not made clear and the father was never seen again by the family. Twenty years later, his mother received a brief note that her husband was posthumously declared 'free of guilt'. After the arrest of his father, his mother, whom Israel considers his first and most important teacher, was the sole support of her 2 children, Israel and his 5 years younger sister. Only one year later, in 1941, the German Army invaded the Soviet Union forcing the Jewish family to flee for their lives thousands of kilometers to the East, to Frunze, the capital of Kirgizia, not far from the Chinese border. There Israel completed elementary and high school and started his university education at the Pedagogical Institute. During the two years in Frunze, Israel's excellent mathematical abilities became already apparent. A professor helped him to transfer to Chişinău where he got his master's degree in Mathematics in 1951. Some time before, in 1950, he had already met Mark Krein, the mathematician who influenced Israel most, although he could never become his student officially. Krein had created a world famous school in functional analysis in Odessa and Israel wanted to do his Ph.D. work with him. But this turned out to be impossible: Krein was not allowed to have more Jewish students since he already had two or three after 1945, who were war veterans and therefore privileged. So Israel worked on his Ph.D. thesis externally while teaching at a high school. Stalin's death in 1953 led to an improvement of the situation. In 1954, Israel could defend his Ph.D. thesis in Leningrad. Afterwards he worked at a teacher's college, then at the Pedagogical Institute in Balti and, finally, in 1959, he was invited to work in the Moldavian branch of the Soviet Academy, which later became the Academy of Sciences of the Moldavian SSR. There, from 1964 to 1974, he was the head of the Department of Functional Analysis. Although M. G. Krein could not accept Israel as a Ph.D. student, they started to work together in the fifties and this cooperation continued for many years. As Ralph Phillips formulated it[2], it was "one of the most successful collaborations in Mathematics,

---

[2]See the introduction to Gohberg's "Mathematical Tales" in Part 1 of the present book.

the resulting books and papers are outstanding". Towards the end of the sixties, the situation for Jewish scientists in the Soviet Union worsened again. Also Israel suffered from the reviving antisemitism in various ways. For instance, it became impossible to travel even to neighbor socialist countries or to become elected as a Full Member of the Academy of Sciences of the Moldavian SSR. Israel finally decided to emigrate to the country of Israel with his family. He got the permission in 1974, and with an unbelievable amount of energy and ideas he started a new life in the West. He became a leading authority in many mathematical areas: integral equations, nonselfadjoint operators, spectral theory and factorization of matrix and operator functions, inversion problems for structured matrices.

His first research topic in the fifties were Fredholm operators and their different generalizations, analytic operator valued functions, Fredholm properties of Toeplitz matrices and of Wiener-Hopf operators. Often he used ideas and techniques which at first glance did not seem very much related with the problem. E.g., he applied Banach algebra techniques in the theory of singular integral operators. The collaboration with Mark Krein led to fundamental results on the perturbation of Fredholm operators, systems of Wiener-Hopf equations, factorization of matrix functions in the Wiener algebra, the foundation of the theory of operator ideals, completeness properties of the systems of root vectors, and the theory of characteristic operator functions. From 1959 on, Israel Gohberg started to build his own group in Chişinău. He and his students and collaborators worked on factorization of operator functions, decomposing algebras, Wiener-Hopf operators, singular integral operators, and on other topics. The famous Gohberg-Semencul and Gohberg-Heinig formulas for the inversion of finite Toeplitz matrices and their continuous analogues, which are often used in the electrical engineering literature, were proved in this period. After his emigration from the Soviet Union, he also started a fruitful interaction with control and electrical engineers. Inspired by this, he developed new directions in operator theory, which in turn led to new constructive operator theory methods in control theory. As an example is to be mentioned the so-called state space method, which has its roots in the mathematical systems theory of the sixties. Due to the results of Gohberg and his co-workers, it became an effective tool for the solution of a wide range of problems like Wiener-Hopf equations, matrix versions of interpolation problems, e.g., of Nevanlinna-Pick and of Caratheodory-Toeplitz type and, recently, for inverse problems for canonical differential systems with rational spectral density. As a by-product, modern control problems like the sensitivity minimization problem, model reduction and robust stabilization could be solved explicitly with the solution being described in the state space form.

But Israel Gohberg is not only an outstanding mathematician, he also has the rare "talent and the stamina that his ideas get realized". All this makes him play a leading role in the international mathematical community. Still in Chişinău, he was the driving force behind the journal 'Matematiceskije Issledovanja'. After having moved to Israel, he contributed a lot to make operator theory and its applications more and more popular: He founded the journal 'Integral Equations and Operator Theory', which became the leading journal of this field in the world and which really can be considered as "his" journal. In the series 'Operator Theory: Advances and Applications', which was also founded and is still being edited by him, more than 120 volumes have appeared since 1980. About 20 years ago, Israel Gohberg initiated the biannual 'International Workshops on Operator Theory and its Applications', IWOTA. Israel Gohberg is also heavily involved in the organization of the 'International Symposia on the mathematical theory of networks and systems', MTNS, where he successfully tries to bring together engineers and mathematicians. Finally, together with Bernhard Gramsch from Mainz, he also organized a series of meetings in Oberwolfach, a famous centre for mathematical research in Germany.

Maybe this is a good opportunity to thank you, dear Professor Gohberg, on behalf of my colleagues, in Timişoara and elsewhere, for giving us all these possibilities for publishing papers and proceedings of the OT Conferences from Timisoara, in your journals and book series respectively, and for meeting each other at these high level conferences.

Of course, all these achievements have not gone unnoticed. Israel Gohberg has got a number of prizes and academic awards. In 1970 he was elected Corresponding Member of the Academy of Sciences of the Moldavian SSR, reinstated again in 1996 after having been removed due to his emigration. In 1985 he was elected Foreign Member of the Royal Netherlands Academy of Arts and Sciences. He was awarded the Landau Prize and the Rothschild Prize in Mathematics, the Hans Schneider Prize by the International Linear Algebra Society, an Alexander von Humboldt Prize by the German Humboldt Foundation, and honorary doctoral degrees by the Technische Hochschule Darmstadt in 1997 and by the Technische Universität Vienna in 2001.

I would like to thank the Rector Magnificus and the Senate of the University of the West in Timişoara that they have decided to award the honorary degree of Doctor Honoris Causa to Israel Gohberg and I would like to congratulate you, Professor Gohberg, very much on this occasion.

The above laudatio was written by
Prof. Dr. Dumitru Gaspar, Director of the Department of Mathematics, University of the West Timişoara,
with the cooperation of the Laudatio Committee consisting of
Prof. Dr. Mihail Megan, Dean of the Faculty of Mathematics, University of the West Timişoara,
Prof. Dr. Heinz Langer, Head of the Institute of Analysis and Technical Mathematics, Vienna University of Technology,
Prof. Dr. Dr. H.C. Ciprian Foias, Indiana University at Bloomington, Indiana,
Prof. Dr. Ion Colojoara, Faculty of Mathematics, University of Bucharest.

Response by the honorary doctor Israel Gohberg

RECTOR MAGNIFICUS
MEMBERS OF THE SENATE
DEAR GUESTS, COLLEAGUES AND FRIENDS

I am very happy to receive today the Honorary Doctoral Degree from the West of Timisoara University. I would like to thank with all my heart the Senate and the authorities of this university for this high appreciation of my research during 50 years of my career. Special thanks are due to my colleague Professor Dumitru Gaspar for his efforts and hospitality.

I am very, sorry that my parents cannot enjoy this ceremony. My father died in 1942 in the Gulag and the place of his grave is unknown; my mother, who was a midwife living in difficult conditions, dedicated all her life to the education of my sister and myself. I am very sorry that my teacher and friend, M.G. Krein, one of the greatest mathematicians of the twentieth century, also cannot enjoy this ceremony.

I was born in Roumania and a large part of my education was under the influence of the Roumanian culture. In fact four grades of primary school (Scoala primara) and the first grade of gymnasia ("Liceu Vasile Lupu din Orhei") I finished in Roumania. I completed my studies and graduated from the University of Kishinev in Soviet Moldavia. Many of my teachers and professors received their degrees in Roumanian universities. During my studies I heard many stories about the nice traditions in different Roumanian universities. In the first half of my career I taught in schools of higher education in the Roumanian language (Moldavian).

An important part of my advanced mathematical education I owe to M.G. Krein. Professor M.G. Krein created in Odessa a school of mathematics which was known worldwide for its achievements. After World War II, during the years of severe antisemitism in the Soviet Union (especially in Odessa), he was thrown out of the Odessa University. His famous mathematical school at this university came to an end. In fact it sort of moved underground to his apartment and became a sort of private institution without offices, secretaries or bureaucrats, but with outstanding achievements. Seminars were held in the Odessa Scientists' Club and in the institutions where he worked. During my vacations I spent at least three months each year with M.G.Krein working in his apartments or dacha. I was never a graduate student, and student, and never a formal student of M.G. Ki-eiii; to me he was much more than a teacher and friend. A large part of i-iiv achievements ai-(@ due to his influence.

I was lucky also in the second half of my career which was spent at the Tel Aviv University in Israel. my home university where I had so many opportunities; I was the incumbent of the Nathan and Lily Silver Chair in Mathematical Analysis and Operator Theory. I enjoy living in my own country and the free world and I have wonderful friends and colleagues. I have second university homes in the Free University in the Netherlands and in the University of Maryland at College Park,

USA. I would like to thank all the above mentioned universities and the Silver family foundation for many years of help and support.

To close I would like to thank my family, my wife, my daughters, my sister and her family for their continued support and understanding. Again I would like to thank the West of Timisoara University for bestowing on me such a high honour. It has a special meaning for me. This university is an important international center in operator theory, and this in a sense is closing the circle of my studies and education started in Rumania 67 years ago in my childhood.

VI.4. The Moldova State University at Chisinau, Moldova

In 2003 the Moldova State University at Chisinau awarded the degree of doctor hon-oris causa to Israel Gohberg. Two laudatios were presented. The first by Professor Petru Soltan, Member of the Academy of Sciences of the Republic of Moldova and Honorary Member of the Romanian Academy, and the second by Professor Con-stantin Gaindrik, Director of the Institute of Mathematics and Informatics of the Academy of Sciences of Moldova.

Laudatio by Petru Soltan

Dear Mr. Rector,
Distinguished members of the Senate,
Dear colleagues,
Dear students,

Moldova State University joins many other highly esteemed universities in the honorable tradition of distinguishing great personalities of our time by awarding them the title of DOCTOR HONORIS CAUSA. The present meeting of our Senate represents a new stage of this time-honored tradition; it is the first such award given by our university's Faculty of Mathematics and Informatics.

On this occasion we recognize a remarkable personality, who has over many years built strong ties with our faculty. Israel Gohberg is a professor of Tel-Aviv Univer-sity, a world renowned master in modern mathematics, a Member of the Academy of Science of the Republic of Moldova (1970) and of the Royal Netherlands Acad-emy of Arts and Sciences (1985). Over the years he has been awarded the Landau Prize (1976) the Rothschild Prize (1986), the Humboldt Prize, (1992), and the Hans Schneider Prize in Linear Algebra (1994), and he holds the title of Doctor Honoris Causa from Darmstadt (1997, Germany), Vienna (2001, Austria), and the West University of Timisoara (2002, Romania).

Allow me to recall for you a number of significant details from the curriculum vitae of this most valuable professor and colleague.

Israel Gohberg was born on the 23rd of August 1928 in the village of Tarutino (which at that time was in Bessarabia and became part of the Kingdom of Ro-mania, now the Ismail region of Ukraine). He went to the elementary school and completed one year of middle school in Romania (Orhei, Secondary school "Vasile Lupu"). The Second World War brought great trouble and strife to the world of the Gohberg family. Israel's father, Tsudick Gohberg, was arrested by the KGB on Israel's twelfth birthday in 1940, was exiled to Siberia, and died after two years of the crushing life of the GULAG. He was not rehabilitated until 20 years after his death. Israel's mother, Clara, remained with two children, but was eventually evacuated, being sent as far away as Kirgizia. Mrs. Gohberg was a midwife by profession who all by herself brought up her children, the young Israel and his five years younger sister. We can certainly conjecture that those difficult years contributed to the inner strength that has sustained Israel through his admirable career.

After some relative stability had returned to his life, Israel continued his education by completing two years of the Pedagogical Institute in the town of Frunze, now Bishkek. One of his professors, observing Gohberg's mathematical abilities, helped him to transfer and become a student at the State Kishinev University. In 1950 he met the great mathematician Professor Mark Krein, who had established his school of Functional Analysis in the town of Odessa, and to whom he became a valuable colleague. Gohberg graduated from KGU in 1951 with a first-class honors degree. Despite his excellent education and some original publications even at this early stage, he was obliged to teach mathematics in a rural school (1951). But he was soon transferred as a professor to Sorocca Pedagogical College (1951 - 1953), after which he was placed in Balti as a professor of mathematics at that state university's Pedagogical Institute (1953 - 1959). In 1954 he defended his doctoral dissertation in Mathematics at Saint-Petersburg (then Leningrad). In 1959 he was invited to join the Moldova Branch of the Academy of Sciences of the USSR. In 1961 the Academy of Sciences of the Moldovian SSR was formed, and his work became focussed on research in Functional Analysis. His activities in this field began at the Institute of Physics and Mathematics (1961), continued at the Institute of Mathematics and Computer Science (1964), and next at the Institute of Mathematics and Informatics. From 1964 to 1974, Gohberg served as Head of the Department of Functional Analysis, creating a prestigious and influential school, known in the international world of mathematics as the "Mighty Team".

In 1964 Gohberg successfully defended his Habilitation thesis at Lomonosov Moscow State University and in 1966 received the title of University Professor. From that time to 1973 he was an acting professor at the Department of Applied Mathematics of our University. After two myocardial infarctions, and with perhaps other personal motives as well, our highly appreciated mathematical colleague decided to leave Chisinau with his whole family (wife, two daughters, aged mother) for Israel.

For this great man a new and promising way opened up, a manifestation of his talent and the school he created. Moreover, the Mathematical World at large obtained a genuine pillar of stability and development.

Professor Israel Gohberg is the founder of the journal *Integral Equations and Operator Theory*, which has become recognized as the world's most important journal in its respective domain, a consideration which has led to the simple title *Gohberg's Journal*.

In 1980 he founded, and still serves as editor for, the monograph series *Operator Theory: Advances and Applications*, at present having published more than 120 volumes of the work of many hundreds of mathematicians from around the world. Two years later he initiated a bi-yearly "International Workshop on Operator Theory and its Applications" and then became profoundly involved in organization of the "International Symposium on the Mathematical Theory of Networks and

Systems" which successfully drew together like-minded mathematicians, engineers and other applied scientists.

Since his departure from Moldova, Gohberg has quadrupled both the number of his monographs, reaching now 21 titles, and his number of published papers, the list of which consists of more than 450 titles. In addition he serves on the editorial boards of five international journals. The number of students he has supervised in defending their doctoral theses is not exactly known; we can safely say however that it exceeds the number 40.

What are the concrete mathematical results of Israel Gohberg's excellent work? The answer to this question may only, in view of its vastness and importance, be summarized - and I wish to do that now.

Between 1950 and 1959 Professor Gohberg was dealing with Fredholm operators of Toeplitz matrices and Wiener-Hopf operators, and applying methods of Banach algebras. A significant aspect of this period is his collaboration with Mark Krein that led to fundamental results on perturbed operators.

In 1959, in his new position, Gohberg created a group of followers here in Kishinev, conducting research in linear spaces, factorization of operators, decomposition of algebras, Wiener-Hopf operators, integral and singular operators, etc. Among the results obtained were the excellent Gohberg-Sementzul and Gohberg-Heinig formulae that led to inverse finite Toeplitz matrices and continuous analogues of them that are applied to electro-engineering theory.

From 1974 on, Gohberg established himself as a first rate authority on integral equations, non-selfadjoint operators, spectral theory, factorization of matrix and operator valued functions, and problems of inversion for structured matrices. On all these topics he personally obtained excellent results.

Today Israel Gohberg is recognized as a pillar of international modern mathematics. He was the incumbent of a chair for Analysis and Operator Theory at Tel-Aviv University, where he is now Professor Emeritus. For 15 years he was professor at the Free University of Amsterdam, at the same time being Visiting Professor at Universities of Stony Brook, Maryland, and Blacksburg (USA), Calgary (Canada), Regensburg and Mainz (Germany) and he had a number of other similar appointments.

Professor Gohberg made an important contribution to the Republic of Moldova by nurturing a generation of productive mathematicians. Among the many we would like to mention M.S. Budianu, M.A. Barkari, I. Chebotaru, who all passed away, and V.A. Zolotarevschi, A.A. Sementsul, M.K. Zambitskki.

And we must not forget that this brilliant mathematician of world-wide caliber is also a philosopher of Natural Sciences.

Finally, Professor Israel Gohberg, by his enormous energy and penetrating insight, perfectly corresponds to what the great novelist Andre Maurois (1885-1967) said that in order to reach unexpected heights it is necessary to have a bit of a gift from God, to toil day and night, and to get a drop of luck.

So we believe that awarding the title DOCTOR HONORIS CAUSA of Moldova State University to the prominent mathematician and remarkable person Professor Israel Gohberg will represent an abundantly appropriate appreciation of the contribution of His Excellency to universal culture and an honor to our beloved University.

Laudatio by Constantin Gaindrik

Mr. Rector,

Distinguished members of the USM Senate,

Dear audience,

I am both proud and happy for this opportunity to salute one of the most distinguished mathematicians of our times, a man whose life and works attests to Moldova's role in the development of significant mathematics both in Europe and the world at large – Professor Israel Gohberg.

Professor Gohberg graduated with honors from the Kishinev State University in Moldova in 1951. Fifteen years later he returned to Moldova as a professional mathematician with a rich background of pedagogical and scientific experience at all levels from secondary and higher education to research. In the intervening years, he had, among other activities, taught at the Balti State University "Alecu Russo" and worked as a researcher and chief of the department of mathematical analysis of the Moldavian Academy of Sciences' Institute of Mathematics during a robust stage of that institution's growth and development. He also, during that time, successfully directed the dissertations of six PhD students. In addition, he published over 50 scientific works, including *Introduction to the Theory of Linear Non-Selfadjoint Operators in Hilbert Space* (1965 in Russian with the Moscow publishing house "Nauka", translated later into both English and French).

Gohberg's following years with the University were very productive for him, as well as for hundreds of students who had the opportunity to listen to his lectures, which were substantial from the scientific point of view and magnificent from the didactic side. I can attest to this from personal experience: during my student years, I had the good fortune to attend his lectures on mathematical analysis in Balti.

In his educational work, Gohberg has always searched for roads less traveled, trying to awake in his students an intrinsic interest in mathematics, showing them the logic of mathematics and the beauty that can be found even in its elementary problems. As an example, he organized at Balti University a competition in solving mathematical problems which, surprisingly, was won by a group of first year students. I was a member of that group and still treasure the special award, a book received as a gift with Gohberg's generous inscription. Moreover, he and Professor V.D. Belousov signed a recommendation for my employment at the Institute of Mathematics. So, I have ample reason to consider both of them as people who opened to me the way to mathematical research.

Israel Gohberg's teaching has always been inseparable from his scientific endeavors. Teachers' College in Soroki, the Balti Pedagogical Institute, the State University of Moldova, Tel Aviv University, Free University in Amsterdam, New York State University, University of Calgary in Canada, Universities of Georgia, Maryland, Connecticut (USA), the Polytechnic Institute of Virginia, Blacksburg (USA), Regensburg and Mainz Universities (Germany) – these are only some of the universities where he taught courses and lectured in seminars. More than 40 PhD's in mathematics are proud to be his scientific disciples; of this number, 19 were PhD students of the Institute of Mathematics of the Moldavian Academy of Sciences. I believe that, despite many hardships and injustices that he met in his life, these achievements both at home and abroad have made him a happy person.

Always very active in the publication of mathematics, Gohberg was one of the initiators of the Series *Mathematical Researches*, a publication that appears parallel to the *Bulletin of the Academy of Sciences of Moldova*. He is the founder of the journal *Integral Equations and Operator Theory* which has become the most important scientific publication in the world in this area of research.

He also founded the series *Operator Theory: Advances and Applications* in which, as of now, 120 monographs have been published and have become recognized as works of the highest scientific level. Many of these volumes have resulted from seminars and symposiums organized by Gohberg and contain the work of both world renowned mathematicians and beginners. A happy circumstance is that many of these beginners became recognized specialists in a short time due to the free and creative atmosphere of these meetings and the opportunities to talk with and learn from mathematicians of the first rank. The circumstances of these meetings and the resulting publications are largely due to the charismatic personality, erudition and good will, as well as the exacting writing standards of, Israel Gohberg.

Schopenhauer said that scientific honors for those who devote themselves to research come too late or never. In this sense Gohberg is an exception. The significance of his scientific activities, confirmed by more than 20 monographs (which include translations and re-editions, altogether 34 publications) published by prestigious publishing houses, and more than 400 articles published in highly ranked journals, is appreciated by experts in the field all over the world.

We are proud that those seven of his monographs which had been written by him during his activity in Moldova, have enjoyed a total of 16 translations by such organizations as the American Mathematical Society, the publishing house Birkhäuser and others in the English, German, French, Spanish and Magyar languages.

Professor Israel Gohberg is the recipient of many high academic awards. He is a member of the Academy of Sciences of Moldova and the Royal Netherlands Academy of Arts and Sciences. He has been awarded the Rothschild prize for mathematics, the Humboldt prize for research, and the Hans Schneider prize for achievements in the field of linear algebra. He holds the title of Doctor Honoris Causa from the Darmstadt Technical University, the Vienna Technological University and the West University of Timisoara.

Mr. Rector, I congratulate you for an excellent choice of Professor Gohberg to receive the title of Doctor Honoris Causa of the Moldova State University. This appreciation, although belated for generally known reasons, nevertheless does honor both to Professor Gohberg and to our University where in the past he held a position.

Dear Professor Israel Gohberg, I congratulate you on a well-deserved reward. I wish you much good health, new achievements, and new scientific disciples. Sincere congratulations for you and all the best for your family.

VI.5. Balti State University "Alecu Russo" at Balti, Moldova

In 2003 the Balti State University "Alecu Russo" at Balti awarded the degree of doctor honoris causa to Israel Gohberg. Here we only reproduce the certificate.

DIPLOMA

The SENATE
of Balti Alecu Russo State University,
Republic of Moldova
in accordance with the decision from 09.25,2003

CONFERS

the title of

HONORIS CAUSA DOCTOR
of BALTI ALECU RUSSO
STATE UNIVERSITY

on Mr. Israel GOHBERG

Chairman of tht Senate *Rector*	*Scientific Secretary* *of the Senate*
Nicolae FILIP Doctor, Professor, Academician	**Valeriu CABAC** Doctor, Professor

Picture of the hononary doctorate certificate in English

VI.6. The Israel Institute of Technology, Haifa, Israel

In June 2008 the Israel Institute of Technology at Haifa in Israel awarded an honorary doctorate to Israel Gohberg. The official citation appearing in the honorary doctorate certificate is reproduced here below together with the letter of nomination.

Citation appearing in the honorary doctorate certificate

Israel Gohberg:
In recognition of your monumental and pioneering contributions to mathematics, which span an exceptionally wide spectrum; in tribute to the originality, breadth, and depth of your numerous fundamental books, monographs, and papers that have vastly extended the frontiers of mathematics and created new, far-reaching applications in astrophysics, transport theory, control theory, signal processing, and numerical linear algebra; in acknowledgement of your role in making Israel one of the world centers of research in operator theory and its applications; and in appreciation of your role as advisor and mentor to innumerable students and colleagues, and your active cooperation with faculty at the Technion.

Letter of Nomination for Israel Gohberg

We are deeply privileged to nominate Prof. Israel C. Gohberg as a candidate for the Technion Honorary Doctor of Science degree, in recognition of his monumental contributions to Mathematics and its Applications. His outstanding work spans an exceptionally wide spectrum of areas from Analysis to Linear Algebra, from Operator Theory to Applied and Combinatorial Geometry.

Dr. Gohberg is the world's leading authority in the following central research areas:

- integral equations and discrete analogs;
- non-selfadjoint operators;
- factorization and spectral analysis of operator and matrix functions;
- inversion problems for structured matrices.

Gohberg has a phenomenal talent to define and uncover problems which eventually turn out to have important applications in diverse research areas, and in particular, in various engineering problems. As a result, his seminal contributions in the above areas include diverse applications of great importance.

The hallmarks of all his work include a penetrating vision, deep insight, and connections - often unexpected - to other questions. This is characteristic of only a select few of the world's greatest mathematical researchers. His seminal work serves as a powerful interface between mathematics and engineering sciences. In more than 470 research papers, monographs, and books, all based on his insight, he and his school of collaborators have vastly extended the frontiers of operator theory and other areas and created new applications to fields such as astrophysics, transport theory, control theory, numerical linear algebra, and signal processing. His outstanding results, papers and books are well known to and often quoted by experts in all the above fields. Some of his achievements are described in the

Appendices 1 - 2[1]. Here we list a number of major results of Prof. Israel Gohberg that made him world famous and will remain connected to his name forever:

1. His pioneering results on index, perturbations and representation of Fredholm operators.
2. Gohberg's outstanding contributions to the theory of nonselfadjoint operators, and in particular, the foundation of the theory of operator ideals, completeness of systems of generalized eigenfunctions (including applications to boundary value problems for differential equations), the theory of characteristic operator functions, the factorization theory for operators (which is now used in system theory, probability theory, integral equations and other areas).
3. His classical results on (systems of) Wiener-Hopf equations, singular integral equations, and other classes of convolution type equations, based on introducing and developing methods and ideas of functional analysis into this area.
4. Gohberg's fundamental results on factorization of operator functions which have important applications to astrophysics and linear transport theory.
5. Gohberg's Fredholm index theorem for Toeplitz operators with piecewise continuous symbols and the symbol calculus for Banach algebras generated by these operators.
6. His inversion formulas for finite sections of (block) Toeplitz and Wiener - Hopf operators and their numerous generalizations for other types of structured matrices, which turned out to be of great importance in signal processing.
7. His seminal contributions to the use of the projection method in solving systems of Wiener - Hopf integral equations, singular integral equations and infinite (block) Toeplitz equations.
8. Gohberg's creation of the spectral theory of matrix polynomials, rational and analytic matrix functions, extending the classical results of complex analysis for scalar functions and putting these into a new perspective.
9. The approach to employ state space methods from mathematical system theory to solve effectively problems in operator theory, mathematical physics, non - linear PDE's and other branches of analysis. As a significant by-product, modern control problems, like the sensitivity-minimization, model reduction, and robust stabilization, were solved explicitly in state space form.
10. Gohberg's invention of the band method which allows one to deal with all classical (e.g. Caratheodory, Nevanlinna-Pick, Nehari) and modern extension and interpolation problems, including the four block problem arising in engineering, from one point of view. This method reduces these problems to linear equations in appropriate C*-algebras that can be solved by factorization. It yields explicit formulae that are easy to implement for the suboptimal case.

---

[1]The first appendix is the short review of the mathematical work of Israel Gohberg, written by M.A. Kaashoek, on the occasion of Gohberg's seventieth birthday. This review is reproduced in Part II of the present book. The material of the second appendix appears in a somewhat more elaborate form in the paper *Gohberg's mathematical work in the period 1997-2007*, which can also be found in Part II of the present book.

The discovery of the connection between the central completion and maximum entropy solutions is one of Gohberg's outstanding achievements in this area.

In addition to developing his own ideas he has taken a leading role internationally in stimulating and disseminating the work of others. Already at the beginning of his career in Kishinev (Moldavia) he formed a group of young researchers and became the leading force behind the journal "Mathematicheskiye Issledovaniya". After making aliyah to Israel in 1974 he founded and edited the journal "Integral Equations and Operator Theory" and the series of monographs "Operator Theory: Advances and Applications," of which 170 volumes have appeared. He started the bi-annual "International Workshops in Operator Theory and Applications," the "Toeplitz Conferences" that attracted some of the world's outstanding mathematical analysts to Israel, and was co-organizer of the "International Symposia on the Mathematical Theory of Networks and Systems" with the explicit goal of bringing together engineers and mathematicians. These symposia aim at cross fertilization, importing mathematical ideas and results to engineering application fields and at the same time finding inspiration for novel mathematics discussions relevant to these applications. A hardworking, tireless and stimulating collaborator, he was invited to work and lecture at practically all important centers of scholarships around the world and has held long term visiting professorships at the Weizmann Institute, and the Universities of Stony Brook, Calgary, Maryland and Amsterdam.

Dr. Gohberg is not only an outstanding mathematician, but he also has the rare talent and the stamina that allows his ideas to get realized. As a gifted and inspiring teacher, Professor Gohberg has trained and advised more than 40 outstanding PhD students, who have in turn made a distinctive and important contribution to the world of theoretical and applied mathematics and are to be found as professors at leading institutions around the world, including the Technion. He had numerous scientific contacts with researchers at the Technion and took an active part in workshops and conferences held here. An outstanding lecturer and expositor he has written with his collaborators several popular textbooks. In particular, for about 25 years the book "Basic Operator Theory" served as the main textbook on functional analysis in both the faculties of mathematics and electrical engineering at the Technion.

Stellar in mathematics, he is equally outstanding as a person. Storm-tossed by WWII in the Soviet Union, where on his 13th birthday his father disappeared forever into the gulags, his talent nevertheless ultimately brought him to Odessa, where he collaborated with M.G. Krein, one of the greatest analysts ever, and, despite official anti-Semitism, was elected to membership in the National Academy of Moldavia. A proud Jew, he felt that his place for life and work is Israel and he resolved to settle there. As an early refusenik he lost all his positions and suffered extreme hardship before succeeding in his struggle to leave; he reached Israel in 1974 and started his career anew. As devoted to his friends and to mathematics, he has attracted more than 60 collaborators from all parts of the world, and he maintains his energy and humor up to this day.

Gohberg's work and influence are a strong component of Israel's renown in mathematics. He is one of the very few Israeli mathematicians whose biography is included in MacTutor History of Mathematics. Awarding the Honorary Doctorate of the Technion to Professor Gohberg would add a world wide famous and very prominent scholar to the list of distinguished researchers which have already been recipients of this major honor.

The above Letter of Nomination was signed by a group of professors from the Department of Mathematics, the Department of Electrical Engineering and the Department of Computer Sciences of the Technion. The group included Abraham Berman, Freddy Bruckstein, Daniel Hershkowitz, Leonid Lerer, Allan Pinkus, Baruch Solel, Allen Tannenbaum, Avraham Sidi, Moshe Zakai.

A response from professor Gohberg

On January 24, 2008 Professor Yitchak Apeloig, President of the Israel Institute of Technology (the Technion) wrote to professor Israel Gohberg that the Council of the Technion, on behalf of the Board of Governors and with the approval of the Senate, had decided to confer upon him the most honored title of this institution, namely the degree of Honorary Doctor of Science of the Technion.

Since this volume had to be submitted to the publisher a month before the doctoral award ceremony was held at the Technion, we cannot present here professor Gohberg's response at the ceremony. However we do have his answer to the letter of the President.

TO PROFESSOR YITCHAK APELOIG   2008-02-06
PRESIDENT OF THE TECHNION

DEAR PROFESSOR APELOIG:

THIS IS A LETTER OF GRATITUDE TO YOU, TO THE COUNCIL, TO THE SENATE, TO THE NOMINATORS, TO MY COLLEAGUES WHICH WERE INVOLVED, AND TO ALL AUTHORITIES OF THE TECHNION FOR AWARDING ME THE DEGREE OF HONORARY DOCTOR OF SCIENCES OF THE TECHNION.

IT IS CERTAINLY A PLEASURE AND HONOR FOR ME, ESPECIALLY TO RECEIVE IT ON THE NAME OF SUCH A PRESTIGIOUS INSTITUTION AS THE TECHNION. I APPRECIATE HIGHLY THE FACT THAT THE AWARD IS RELATED TO THE 60TH ANNIVERSARY OF THE STATE OF ISRAEL.

I WILL DO MY BEST TO DESERVE THIS AWARD. WITH BEST REGARDS,

ISRAEL GOHBERG

P.S. WITH PLEASURE I WILL TAKE PART IN THE CEREMONY OF JUNE 2, 2008.

Part VII
Festschrift 2008

This part consists of material comparable to that of Parts III and IV, but then from a younger date and written especially for the present occasion. In short articles seventeen friends and colleagues reflect on their experiences with Israel Gohberg. All of them have felt his influence. In some cases it changed their life.

Collaborating with Israel Gohberg: Some Personal Thoughts

Daniel Alpay* (Beer-Sheva)

> *"What are the important problems of your field?",* ... *"What important problems are you working on?",* ... *"If what you are doing is not important, and if you don't think it is going to lead to something important, why are you at Bell Labs working on it?"*
> *Richard W. Hamming*

Introduction. The above questions are taken from the transcription of the Bell Communications Research Colloquium Seminar - 'You and Your Research' given by Hamming, the creator of error correcting codes, in 1986. These are fundamental questions every mathematician should ask himself. Working with Israel Gohberg is being at the center where the main problems in operator theory and linear system theory are created and developed. I am fortunate to have been working with Israel since 1985 and to this very day. Thanks to the opportunity Rien Kaashoek is giving me, I would like to discuss in these pages some aspects of this collaboration. Each collaboration between two mathematicians is certainly unique in its kind, and, like a living organism, evolves in time. My work and interactions with Israel had a definite impact at all stages of my career, and these pages are a modest tribute to this fact.

The beginning. I first met Israel Gohberg via the Gohberg-Krein book on Volterra operators, while studying in Engineering School in Paris in 1976-1978. My then teacher, Prof. Hayri Korezlioglu, was studying this book, and in particular the theory of factorizations along chains, having in mind applications to prediction theory. My second meeting with Israel was in graduate school at the Weizmann Institute, attending classes he gave on topics in operator theory and on the theory of matrix polynomials. Although I could appreciate the lectures Israel gave in Rehovot, there was no interaction between us. The topics dealt with in my thesis included indefinite inner product spaces and J-unitary rational functions; still, I

---

*Earl Katz Chair in Algebraic System Theory

did not dream neither suspect then that there would be any interaction. But, toward the end of my thesis, I was fortunate enough to be given a research problem by Israel. This was the beginning of a long series of weekly meetings (then mostly on Friday mornings). These meetings lead to a draft on the theory of realization and factorization of unitary matrix-valued functions, and a number of notes on a new class of potentials for canonical differential equations with rational spectral data. These projects were left unfinished when I went to a postdoctoral stay in Groningen. I was very surprised (I should not have been), when in the winter 1987/88, in cold Groningen, I got a phone call at my apartment: Israel was on the line, calling from Amsterdam, and asking when we meet. The meetings in Tel-Aviv thus continued in Amsterdam, and lead to our first two joint publications in 1988. There were many more to come (publications and meetings). Although it is maybe too personal a note, I would like to mention that at that time I was in the middle of a divorce, and that Israel was also very helpful in his moral support during these meetings. This fact says a lot on the warmth and care of Israel toward his collaborators. Following Groningen, I went to a postdoc in Blacksburg. Israel suggested a continuation of our work on unitary rational functions together with Joe Ball and Leiba Rodman. This is one of the characteristic of Israel's way of working: making people meet and work together.

Choosing problems. In his book *Panorama des mathématiques pures, Le choix bourbachique*, Jean Dieudonné speaks of some mathematical theories which, due to excessive and non motivated level of abstraction, are in the process of dilution. The road to void generalization and dilution is very easy to take, and we all know colleagues (sometimes ourselves), which flirt with this road. One needs a good compass to avoid it. An important idea does not fear such a danger; on the opposite, it suggests the right generalizations. For instance, the Hamming code can be seen as a particular case of a cyclic code, and also as a particular case of a (rational) Goppa code, the latter being quite far away from the original Hamming code. Working with Israel, I can feel this constant concern of choosing the right abstract concepts from concrete examples, which, far from dilution, on the opposite give rise to new fruitful avenues of research. The best example is the *band method*, which Israel developed with a number of collaborators over the years, and which we used in some of our joint works. Other examples abound, and are discussed elsewhere in this book. It requires a vision and an outstanding intuition to define appropriate research areas. These are characteristic properties of Israel.

CAOT. When I invited Israel to be on the editorial board of *Complex Analysis and Operator Theory*, I was very anxious of his reaction and answer; after all, CAOT was to be published by Birkhauser, where Israel's journal, *Integral Equations and Operator Theory* appears. He was positive, very encouraging, and gave me a number of advices and made a number of critics. His main critic was that the proposed name of the journal (which was not CAOT, but another name, best left forgotten here), was not at all appropriate. The name *Complex Analysis and*

Operator Theory is his suggestion, and it is clear *a posteriori*, that the original name would have lead to a failure. Now, CAOT is in its third year, and hopefully will survive and flourish in these troubled times for journals.

Collaboration: One can say that all collaborations with Israel share a number of common points:
(*i*) The problems are part of a global vision, and being aware that a paper written now, or an argument found now, will be part of a larger construction.
(*ii*) Israel gives his full trust to his collaborators, and this leads, together with a definite boost in self-confidence, to a deep sense of responsability for the work to be done.

The paper I am currently working on with Israel is a perfect illustration of these lines. It solves an important and difficult question related to Krein's systems, it uses tools Israel developed in the 1980's with Israel Koltracht, and it is a five authors collaboration, each of us bringing its own expertise.

Future. Israel, you turned eighty now. I wish to thank you for this collaboration which last already for more than 23 years. This is just a beginning, since we have research projects for many years to come, and in particular this project of a book on first order discrete systems. I already mentioned your warmth and care for your collaborators. I would like to mention also your very high standards, and your critics, which are always sharp and to the point. There would be much more to say, but allow me to conclude with a quotation from the book *Ethics of the Fathers (Mishnah Pirkei Avot 2:15)* (translation taken from `http://www.shechem.org/torah/avot.html`):

Rabbi Tarfon said: The day is short, the task is great, the laborers are lazy, the wage is abundant and the master is urgent.

My Collaboration with Israel Gohberg

Joseph A. Ball (Blacksburg)

My first meeting with Israel Gohberg was in the early 1970s when Israel gave a colloquium lecture at the University of Virginia and I drove up from Blacksburg to hear his talk. As a student at U.Va. only a few years earlier, we had a student seminar working through parts of the first volume of Gohberg-Kreĭn and here was my chance to meet the master in person. As I was a somewhat shy beginner in the profession, this was more of a case that I met Israel than Israel met me; my former U.Va. professors did all the hosting while I followed along in the background.

A watershed event in my own personal odyssey of gaining a firmer footing on the international mathematical stage was the International Symposium on the Mathematical Theory of Networks and Systems (i.e., MTNS 1979 for short) hosted by Patrick Dewilde in Delft in July 1979. There I came upon a series of talks by Gohberg and friends (including Leiba Rodman, Leonid Lerer, Rien Kaashoek, Harm Bart, Peter Lancaster among others) on the state-space method applied to matrix-valued polynomials and, more generally, rational functions. Part of the plot here also, as I soon realized, was an orchestrated introduction to the book OT1 *Minimal Factorization of Matrix- and Operator-Valued Functions* by Bart-Gohberg-Kaashoek (we were warned to respect the ordering BGK rather than the reverse which meant something else unpleasant to refugees from the former Soviet Union) which was about to come out in the new Birkhäuser series *Operator Theory: Advances and Applications*. I thought that these guys must be good to have such a large number of individuals working as a team in such a well coordinated series of talks. I was also intrigued by the ideas as my student work was closely connected with the Sz.-Nagy-Foias/de Branges-Rovnyak model theory and here was a natural analogous version which had something to do with engineering (i.e., mathematical system theory). At this meeting, besides becoming reacquainted with Israel, I met a number of other prominent people in the Gohberg orbit for the first time.

My first sustained interaction with Israel came in 1983 when I (with my recent bride Amelia) used a sabbatical leave to visit the Weizmann Institute for a whole semester. There with Nir Cohen and Israel Koltrecht we made weekly treks from Rehovot to Tel Aviv for the Gohberg operator theory seminar at the Tel-Aviv University. At the same time Israel was coming to Rehovot to give a course on topics in operator theory at Weizmann. Having seen my lectures on my work with Bill Helton on the Grassmannian approach to Nevanlinna-Pick interpolation, Israel immediately saw that something similar should hold for finite triangular matrices. There was hatched our first series of papers on interpolation and eventually commutant lifting for finite triangular matrices. These ideas were eventually completed to an infinite-dimensional nest algebra setting by others.

It was not long after this (I believe at the CBMS conference in Lincoln, Nebraska with Bill Helton as principal lecturer) that the idea emerged that Israel, Leiba Rodman and I should put together a book on interpolation for rational matrix functions, with OT45 in the Birkhäuser series being the eventual result. It is described in the preface to OT45 how this enterprise evolved: *there followed a period of polishing and of 25 chapters and the appendix commuting at various times somewhere between Williamsburg, Blacksburg, Tel Aviv, College Park and Amsterdam (sometimes with one or two of the authors).* Upon completion of OT45 the project went into the next phase, *time-varying interpolation* with myself, Rien Kaashoek and Israel as the main participants (with independent efforts by Alpay-Dewilde-Dym and Kailath and associates); this project involved regular meetings among the three of us in Amsterdam for extended summer periods in the early 1990s. Since then Israel and Rien took up collaboration with Ciprian Foias and Art Frazho while I moved on to other projects.

Let me say that all this collaboration with Israel was quite a ride. Our relationship truly evolved from an aloof first meeting to an intensive collaboration sustained over a number of years. In my experience Israel was always demanding and precise, but at the same time warm, sensitive and understanding. I think that one can find the payoff for his persistence for precision and mathematical elegance in his work. It is my pleasure to congratulate Israel on the occasion of his eightieth birthday.

Israel Gohberg: a Teacher and a Friend

Harm Bart (Rotterdam)

My first encounter with Israel Gohberg left me confused.

It was late spring or early summer 1975, the precise date I don't remember. My former Ph.D. adviser Rien Kaashoek had written me from the University of Maryland (where he was at that moment) that Israel Gohberg, who a year before had emigrated from the Soviet Union, would come to Kaiserslautern to speak in the Functional Analysis seminar of Bernhard Gramsch. Rien told me that I should go there. I did – to meet the man whose work had been so important for us.

With "us" I mean: Rien Kaashoek, David Lay and myself. After having obtained my Ph.D. in the summer of 1973, David Lay from Maryland visited the Free University in Amsterdam for a couple of months and the three of us worked on meromorhic operator functions. The research was in line with what I had done for my thesis, but the new developments drew strongly on papers by Israel Gohberg and some of his students, among them Efim Sigal. These articles were in Russian, so not accessible to us in their original form. However, via Bernhard Gramsch we were able to get German translations. These were made by a person familiar with ordinary Russian but not with mathematics, so sometimes we were surprised (if not amused) by very strange formulations. But this did not prevent us to appreciate the quality of the material that served as an enormous source of inspiration for us.

In the first half of 1975, Rien Kaashoek was in Maryland on sabbatical leave as guest of David Lay and Seymour Goldberg. There he met Israel Gohberg who came to see Seymour. They were befriended with each other since Seymour's visit to the Soviet Union in 1964. Intensive discussions on topics of common interest took place, so I was told. And Rien thought that I should personally make acquaintance with Gohberg at the first reasonable possibility that presented itself. It was Gohberg's visit to Kaiserlautern, as mentioned above.

On my first day there, I met him briefly in one of the corridors of the math department. I asked him whether there might be a possibility to have an hour or so together in which I would be able to discuss mathematics with him. He suggested seeing each other that same evening, and at about eight o'clock we met in the office assigned to him. After exchanging some general remarks, I told him about the thing that I was working on, a problem on poles of relative inverses of meromorphic operator functions. He listened attentively and made some mildly appreciative remarks. What followed after that, however, left me confused and startled. In a friendly but at the same time engaged manner he effectively said that I should not go on in this direction. "It's the salad, not the meat" - that's what his remarks came down to. (Israel was not a vegetarian, as his friend Seymour Goldberg was.) That night I did not sleep very well, rethinking the conversation over and over again. It was to have a great influence on decisions that I made about one and half year later when I returned from a one year stay at the University of Maryland.

The stay in Maryland, made possible by a grant of the Netherlands Organization for the Advancement of Pure Research, took place in the academic year 1975/1976. I worked there with David Lay and Seymour Goldberg. The latter introduced me to the theory of semigroups of operators, and together we wrote a paper on almost periodic semigroups. Later, the familiarity with semigroups turned out to be helpful in the joint work with Gohberg and Kaashoek on exponentially dichotomous operators and bisemigroups.

During the year in Maryland, I met Israel Gohberg for the second time. I still remember the day he arrived. David Lay picked him up at the airport. After having brought him to his lodging address, David came to my office, clearly excited about something that Israel had told him on the way. Here is what aroused his enthusiasm. David, Rien and Israel had been working on linearization of analytic operator functions. This is a method where one associates to such a function an operator pencil (i.e., a linear operator function of degree one) which shares a good many important characteristics with the original function. The trio just mentioned had succeeded in choosing for the pencil the one familiar in spectral theory (where the leading coefficient is just the identity operator). The thing that David was excited about was that Israel Gohberg had found a surprising new way of constructing the 'spectral pencil' in question. As David said: "Harm, you will not recognize it." As so often, an area to which Israel Gohberg turned his attention changed forever.

In the fall of 1976, after returning from Maryland, I was offered to join Israel Gohberg and Rien Kaashoek in a project they were about to start. I could have gone on with certain problems in the theory of semigroups that had come up in the work with Seymour Goldberg. However, by then I had become deeply impressed by the uncanny sense of direction displayed by Israel. Also the conversation we had in Kaiserslautern was still on my mind. So I decided to accept the offer. I guessed that working in this line would be exciting and rewarding – correctly as became clear when things unfolded.

The project in question was about so called characteristic operator functions. These are functions associated to an operator having factorizations corresponding to invariant subspaces of the operator. The meeting where Israel laid out his plan is still vivid in my mind. With Brodskii's book "Triangular and Jordan representations of linear operators" as a guide, Israel explained the main ideas and results.

He also indicated that there was a connection with mathematical systems theory, and that he believed there was an overall picture behind the things that could be fruitfully explored.

The actual work started by bringing recently published material into the framework of characteristic operator functions. It was the factorization theory for monic operator polynomials that had just been developed by Israel Gohberg, Peter Lancaster and Leiba Rodman. After a couple of months we submitted a paper which was however rejected. The referee overlooked that the importance of the article was more in its point of view than in the complexity of the arguments. In his place, I might have made the same mistake. The work appeared as the opening paper in the first issue of the Journal Integral Equations and Operator Theory.

Israel's intuition had been right. The overall picture Israel had alluded to was there and it was fruitful, indeed. By the way, the connection with systems made itself felt in a quite remarkable way. On Monday 13 and Tuesday 14 February 1978 we had a workshop together with the group of Patrick Dewilde from Delft. My talk was on the second day when we were in Delft; the first day we had been in Amsterdam. The talk dealt with what we called the geometric factorization principle, a result which relates factorizations of operator functions (in realized form) with pairs of matching invariant subspaces. At the point where I came to this principle, Patrick Dewilde intervened and said to his Ph.D. student Paul Van Dooren who was to give his talk later that day: "Paul, don't you also have something like that?" The answer was: "yes, it's the same thing and I have nothing any more to talk about." The latter was not true because Paul's focus was to a large extent on numerical matters that we did not cover. The four of us decided not to make any point of priority, but to write a joint paper about the common discovery, with in the introduction an explanation about the origin of the result. I have always been happy that we did it this way, avoiding a non-decidable discussion on who had done things first.

Finding the geometric factorization principle referred to above did not go overnight. Guided by the theories of characteristic operator functions and by the then recent developments on factorization of monic operator polynomials, we had been looking for a correspondence between a factorization on the one hand and one single invariant subspace on the other. It took time before we realized that the single invariant subspace should be replaced by a pair of matching ones. I remember the intense discussions, the close reading of the literature pertinent to the subject, the counterexamples and, above all, the slow progress. It was Israel who kept us going with his steady conviction that there had to be something there and that we would find it, if not quick, then "little by little, step by step."

In line with such sayings, he would now and then cheer us up with an anecdote or a joke. In fact, he regularly used such verbal devices to make a rather serious point. One time when we were sitting together, a younger colleague came into the room telling us how mad he was about something that had happened. It was however clear that he actually took pleasure in the fact that he could be so justly aroused by something so unfair. This is the story that Israel told us after the person in question left. A man came to a rabbi and told him that he had to go to a country where it was impossible to have kosher meals. I will starve – so was his conclusion. But the rabbi disagreed. Life comes first, he said, and if you really cannot avoid it, I give you permission to have non-kosher food. The man left relieved. But just before he closed the door, the rabbi stopped him saying: but don't enjoy it!

The close cooperation with Rien and Israel that began in 1978 lasted for about ten years. Working with Israel was demanding, not only in matters of mathematical content, but also on the point of exposition when writing a joint paper or a book. Often when the work seemed to be ready, a new twist would make it necessary to go over the material again and make significant changes. I know of colleagues who found this difficult to cope with; what helped me carry on was that the final result almost always benefited from the process.

I have fond memories of the many research sessions we had during the period mentioned above. It was a pleasure to see Israel doing what he probably liked most: discussing mathematics. Mathematics seemed to be always on his mind. I do not know of him taking holidays in the normal sense of the word. During the conversations his expressive face would show his emotions: pleasure, dislike, impatience, astonishment. Sometimes it was not easy to convince him that a certain line of thought was not right. It needed almost more than a counterexample, especially when the thing in question did not fit in the general picture that he somehow had in his head. If in such a case we were finally able to sow the seeds of doubt his reaction was "now I have to think a little." After having done so, a new general picture, often more powerful, would emerge and suitably modified observations and conjectures were put forward.

For me two research sessions stand out in particular. One for mathematical reasons, the other because of a strong personal involvement. Let me begin with the first.

It must have been an afternoon in 1980 or 1981. Israel, Rien and I were talking mathematics. Suddenly Israel came back to a theorem in our book "Minimal factorization of matrix and operator functions" that had appeared at the end of 1979. He drew our attention to the fact that implicit in the proof of that theorem, there was an interesting observation on what could be called simultaneous reduction to complementary triangular forms of pairs of matrices. The remark led to a new chapter in Linear Algebra, ultimately bringing to light a surprising connection between factorization and the two machine flow shop problem from the theory combinatorial job scheduling. The project in question, which led to several papers, was carried out during my years in Rotterdam with Henk Hoogland, Philip Thijsse, Leo Kroon and my Ph.D. student Rob Zuidwijk.[1] Here was a case where one remark or one question sparked off a whole line of research. It was no exception.

---

[1]See also Part 3 of the recent book H. Bart, I. Gohberg, M.A. Kaashoek, and A.C.M. Ran, *Factorization of matrix and operator functions: the state space method*, Birkhäuser Verlag, Basel, 2008, which is entirely devoted to this topic.

Of the other research session whose memory is still strong in my mind, I know the exact date: February 10, 1978. At about three o'clock in the afternoon my wife Greetje called. She told me that I should not go on much longer because our third child was on its way to be born. After a few words with Israel and Rien, I went home and a couple of hours later our third daughter saw the light of life. We have a photograph, taken a few days later, of Israel Gohberg having the baby in his arms.

Israel not only became close as a colleague mathematician; familywise strong ties were built up too. Many times we were each other's guests, also sharing important family events. A highlight was Israel's presence at the 25-th anniversary of my wedding with Greetje. He spoke at that occasion, bringing out a toast and again using a story to make his point. An almost Biblical story, as he put it. The gist of the matter was that he wished us happiness, not just for ourselves, but especially for our three daughters Alet, Helen and Femke to whom Israel and his wonderful wife Bella had become like distant grandparents.

From the left to the right: Israel Gohberg with Greetje and Harm Bart and Bella Gohberg, at a party in Harm's garden in Bleiswijk, on the occasion of Israel's 75th birthday.

I consider myself privileged to have been able to work with Israel Gohberg as a mathematician and grateful for having become one of his friends.

Linear and One-dimensional

Albrecht Böttcher (Chemnitz)

The first time I encountered the name Gohberg was in 1978 when Bernd Silbermann told me about Gohberg's invertibility criterion for infinite Toeplitz matrices. It says that such a matrix is invertible if and only if the function whose Fourier coefficients are the entries of the first row and first column does not have zeros on the unit circle and has winding number zero about the origin, provided, of course, that this function is continuous. I was deeply impressed by such a fascinating interplay of operator theory, harmonic analysis, and topology, and now, 30 years later, I have to state that this theorem by Gohberg has actually determined my entire mathematical career. In the late 1970s the three books by Israel Gohberg with Israel Feldman, Naum Krupnik, and Mark Krein occupied the most prominent place on my desk and introduced me to a world full of excitement and profound beauty.

The function appearing in Gohberg's theorem is referred to as the symbol of the Toeplitz operator. One of the next things I learned was that if the symbol is no longer continuous but has jumps, then one has to fill in line segments between the endpoints of the jumps when working in ℓ^2, and circular arcs when considering the operator in ℓ^p. This was discovered by Israel Gohberg and Naum Krupnik, and independently and in a slightly different context, also by Harold Widom. It was that mysterious emergence of circular arcs that bewitched me more than anything else and caused me falling in lifelong love with Toeplitz operators generated by piecewise continuous symbols.

In the early 1980s, Israel Gohberg already became one of my great mathematical heroes. At that time the chance to meet him at all in person was nearly zero. Nevertheless, in 1983, I had the opportunity to visit Kishinev and to meet his friends and former colleagues there, including Naum Krupnik, Israel Feldman, Alexander Markus, Igor Verbitsky, and Georg Heinig. In those days it was known that if one considers certain approximation methods for Toeplitz operators, then one has to join the endpoints of the jumps by two circular arcs. The continuous analog of Toeplitz operators is Wiener-Hopf integral operators, and in my dissertation I had just proved that in the Wiener-Hopf case one must join the endpoints of the jumps by a lentiform domain in order to decide about the convergence of a certain projection method. The figure below is from my dissertation and shows lentiform domains between the jumps of a piecewise continuous function. I was very happy to give a talk on this fresh result in Kishinev and remember to this day the competent appreciation and kind reception of my talk by the listeners.

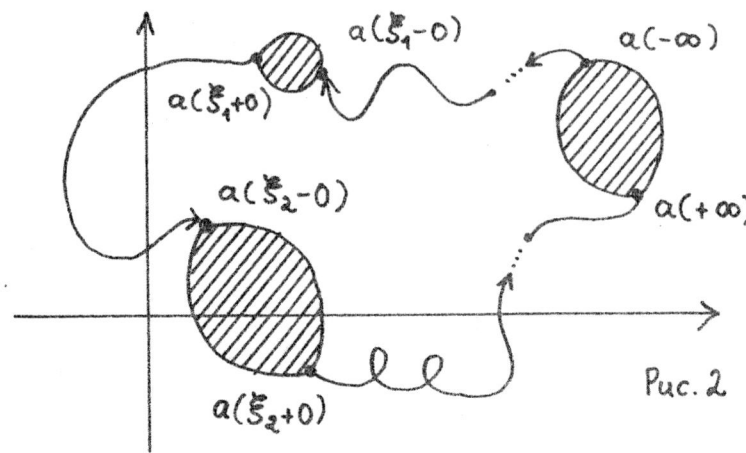

Puc. 2

Throughout the 1980s, Israel Gohberg remained a legend beyond my reach. It was in 1989, at an Oberwolfach conference organized by him, Rien Kaashoek, and Ehrhard Meister, that I met him in person for the first time. The moment we first faced each other is still in my mind and I perfectly remember that I turned to him in fluent Russian with the words "That means, you really exist."

Since then we have met regularly and each meeting was a source of inspiration for me. In 1993, he invited me to Tel Aviv. We then thought about organizing a German-Israeli workshop in Tel Aviv and, among other things, discussed several titles for the workshop. However, we didn't come to a decision. Another day, while we were strolling through olive groves, I incidentally told him that a recent application by myself for a position was turned down by the remark that research into linear and one-dimensional singular integral operators was no longer current. And Israel immediately replied that he now had a good title for the workshop: Linear and One-Dimensional Singular Integral Equations. The workshop itself took place under exactly this title in Tel Aviv in March 1995 and was a great success. My highlight at this workshop was the personal acquaintance with Uri Toeplitz, a son of Otto Toeplitz.

In fact, in the following years linear and one-dimensional singular integral operators have undergone a spectacular development. Ilya Spitkovsky discovered that Gohberg, Krupnik, and Widom's circular arcs become horns when considering the operators on spaces with general Muckenhoupt weights. Subsequently, Yuri Karlovich and I were able to treat the case of operators on general Ahlfors-David-Carleson curves with general Muckenhoupt weights and to show that there is an undreamt metamorphosis of the circular arcs and horns through logarithmic double spirals and logarithmically twisted horns up to so-called leaves with a halo.

Linear and one-dimensional Toeplitz, Wiener-Hopf, and singular integral operators have always been and still are among my top favorites. I believe Israel Gohberg shares that preference, because finding explicit solutions and effectively verifiable criteria has perpetually been one of his maxims. Linear and one-dimensional operators have given me permanent delight (and income) throughout my life and I don't miss any opportunity to repeat that my love for them originated just with such men as Bernd Silbermann and Israel Gohberg.

Rehovot Days Redux[2]

Harry Dym (Rehovot)

The story of how I met met Israel Gohberg and began to work with him is related earlier in the "Gohberg Micellanea" which were written for Israel's 60th birthday celebration and are reproduced in Part 3 of this book. To amplify just a little: In those days our department was concentrated in one wing of the Ziskind building with ten offices, five on each side of the corridor. The occupants of eight of them kept their doors closed. The departmental secretary (who had no choice in the matter) and Israel (who had a choice in the matter) kept their doors open.

Israel's first student at the Institute was Sonia Levin (daughter of the well known *refusenik* Alexander Lerner). If memory serves me right, I suggested to Sonia that she approach Israel to ask him if he would be willing to serve as her PhD supervisor. By a peculiar quirk of fate, some thirty years later, Sonia was to work with my son Jonathan (Z″L) designing algorithms for a new radar system that was being developed at the Elta Corporation in Ashdod, Israel.

Shortly after his arrival in the West, Israel founded the journal *Integral Equations and Operator Theory* (IEOT); the (OT) book series *Operator Theory: Advances and Applications* and initiated the biannual IWOTA (*International Workshop on Operator Theory and its Applications*). In the early days it was far from clear that any of these activities would be long lived. The journal IEOT started life as a quarterly in 1978 and has since grown to a monthly. Two other journals devoted to similar topics that started out at roughly the same time ceased to function. The OT book series is approaching the 200 volume mark and the IWOTA meetings have become annual events with groups from many different countries applying to host them.

Israel often pays tribute to M. G. Krein, his teacher, mentor and collaborator, and speaks fondly of M. G.'s penchant for rewriting. He relates the following conversation with Lev Sakhnovich that took place in Odessa many years ago:

Lev: *How is your book with Mark Gregorovic going?*

Israel: *Its about 85% done.*

Lev: *Thats wonderful. Why so sad?*

Israel: *If you would have asked me yesterday, I would have said 95%.*

Israel also tells the story of his Mother watching him throw away pages and pages of unfruitful calculations day after day. Finally one day she said, "don't you think that you should work on something easier".

It was a great piece of good fortune to have the opportunity to work intimately with Israel for many years and to become friendly with his wonderful family and many of his other collaborators. Apart from Israel's obvious mathematical talents, he is blessed with an open friendly manner, a cheerful optimistic approach to life and lots of energy that continues to propel him to new heights when he really could sit back, relax and view the many projects that he initiated with satisfaction.

I close with the hope that the rest of us will have the strength, health and vitality to help Israel celebrate his ninetieth birthday.

Israel Tsudikovich: My Older Colleague and Older Friend

Yuli Eidelman (Tel-Aviv)

My acquaintance with Izrail Tsudikovich began in my early childhood. My father and he were great friends. Uncle Iz'a was mentioned often in the family discussions. From about the age eleven, I remember his association with my father and the deep impression Izrail Tsudikovich made on me, as a large and kind man with much humor, who called me "druzhische (nice guy)".

In 1966 my family went on vacation in Odessa. Izrail Tsudikovich was also there and spent quite a lot of time with us; some of which was at Mark Grigor'evich Krein's "dacha" in Arkadia; I knew Krein's grandson. It was only many years later that I realized what a historical place I have visited, what mathematical masterpieces had been created their. I also understood how much Izrail Tsudikovich had in common with my father, their outlook of life and their fanatical passion for mathematics. The friendship between them, which began in '50s lasted until my father passed away in 2005.

In 1967 Izrail Tsudikovich attended the very first Voronezh Winter Mathematical School and visited our home during his trip. These schools became a notable event in the soviet mathematical life.

I did not cross paths with Izrail Tsudikovich until 27 years later, as a very new immigrant I came to look for work at Tel-Aviv University. Izrail Tsudicovich greeted me very warmly but in the same time explained the situation to me very clearly. He felt I had to change my direction of research, and suggested that I work in numerical methods for structured matrices. Such a topic seemed to surprise some of his colleagues as Izrail Tsudikovich was renown as an outstanding expert in pure mathematics. However, I was not the first to get involved in this direction of his work. On the moment of my appearance in Tel-Aviv my two predecessors in this field were Israel Koltracht and Vadim Olshevsky; they continued their careers in the USA. It was not easy for me to make the switch to numerical issues, but in time I got used to it and also my previous programming experience turned out to be useful in this.

My joint work with Izrail Tsudikovich is devoted to an interesting and important class of structured matrices. This class contains the band matrices, the semiseparable matrices, inverses of band and of semiseparable matrices, and other interesting examples. Various algorithms for such matrices are the subject of our interest till now. Not only our involvement in this direction has grown but also that of other researchers, which points to how topical the subject has become. In working so closely together, I have witnessed Israel's great intuition, how he follows "the scent" to use an expression employed by Iz'a Koltracht. One can absolutely rely on Israel's innate ability to understand what is worth paying attention to, what is worth making an effort to achieve and what is not worth pursuing. We have a series of papers on our results: together with Izrail Tsudikovich and later jointly with our colleagues Vadim Olshevsky, Tom Bella and Israel Koltracht from Storrs (Connecticut), Dario Bini and Luca Gemignani from Pisa and Eugene Tyrtyshnikov from Moscow. Now we are working on a book which will put our achievements in some order and present a clear picture.

Izrail Tsudikovich recently said the following about our long-standing acquaintance and my present age: "Now every boy is 50 years old."

Eight Friends

Robert L. Ellis (College Park)

I first met Israel Gohberg in a Mexican restaurant after he gave a talk at the University of Maryland, shortly after he emigrated from the Soviet Union. Since he spoke little English, I really did not interact with him at that time. Later he began to visit the University of Maryland for two months every year, his English improved, and I found his talks very interesting. Eventually we began working together and with David Lay on some projects, and our close collaboration continued for more than twenty years, not only at Maryland but also at conferences and other universities. The two-month annual visit always meant a period of maximal effort that excluded most other activity, the most notable exceptions being a few evenings of dinner and conversation at one of our homes. Even though our families always looked forward to Israel's visits, they experienced a certain amount of relief when he left, because life resumed a less torrid pace.

Israel's influence at Maryland extended far beyond the people with whom he worked directly. He spent more hours in his office than most of the resident faculty and was frequently visited by other faculty members who sought his advise on operator theory and their own research interests. There were interactions with people from other departments, especially electrical engineering, and more invitations from other institutions than he could accept.

Obviously he is a great mathematician, but anyone who knows Israel well cannot fail to be influenced by his devotion to his family and his religious convictions. Being Jewish made becoming an academic mathematician in the Soviet Union nearly impossible. The mere fact that Israel became such a successful mathematician and a corresponding member of the Moldavian Academy of Sciences already attests to his unusual ability in mathematics. In Tel-Aviv I had the opportunity to meet Grigori Freiman, who wrote the book "It Seems I Am A Jew", in which he describes the extreme difficulties of Jews just to survive in the Soviet academic system.

Israel likes to relate interesting stories and jokes about dealing with communist officials in the Soviet Union. I don't remember the exact details, but I recall him telling about someone being questioned about frequently changing nationalities.

The response was that he was always living in the same place; only the borders were changing.

Israel has unexpected talents and abilities. For example, he can easily determine whether a watermelon is overripe or ready to be eaten. This he learned as a youth working on a farm. Once we went to the beach at Ocean City, Maryland, and I wondered whether he would be able to swim in the ocean. But as soon as we laid our towels on the beach, he ran into the water and dove in headfirst. He stayed out for about 15 minutes.

Israel was frequently invited to colleagues' homes and he was usually the life of the party. So it was expected that whenever there was something to celebrate, someone would give a party. Once at lunch in the cafeteria at the Free University in Amsterdam, Rien Kaashoek mentioned that Israel had just received a nice raise from the University, and someone immediately suggested that we have a party. Israel naturally joined in and said, "Yes, let's have a party!" Then someone said, "And since Israel is the one who got the raise, he should pay for the food." It was the only time I saw him blush, and everyone, including Israel, had a good laugh. Here it should be mentioned that Israel was very generous in returning invitations, both in restaurants and in his own home.

Israel has many friends, but he always focuses on the ones at hand. I remember a remark he made at a dinner in a restaurant in Maryland with eight friends. He said, "One doesn't need a lot of friends; it is enough to have eight". Israel has eight friends everywhere he has lived.

Israel Gohberg and Engineering

A.E. Frazho (West Lafayette)

It is a great honor for me to share some of my experiences with my friend, mentor and colleague Israel Gohberg on the occasion of his 80-th birthday. My background is in engineering and I am not a professional mathematician by training. When I was a graduate student, it became clear to me that operator theory played a fundamental role in control theory and signal processing. For example, the projection theorem is the basic tool used derive the Kalman and Levinson filter. In fact, the first time I was introduced to some of Gohberg-Krein's work was as graduate student in engineering. In our engineering department they were viewed as giants. One of the things that I find most intriguing about Professor Gohberg's mathematical work is how his research has played such an important role in engineering, especially in control theory and signal processing. The Gohberg-Semencul formula for inverting Toeplitz matrices is now standard material in many courses on signal processing and can be found in various engineering text books. Professor Gohberg is a true genius at solving important mathematical problems that have immediate ramifications in engineering and applied sciences.

I first met Israel Gohberg in the early 1980's. I remember that he was giving a seminar in the mathematics department at Purdue university. At that time I just finished my lectures on Wiener filtering using standard engineering techniques in a graduate engineering course. Then I heard Professor Gohberg's seminar on Wiener-Hoff integral equations. I was very impressed with his talk and remember it like it was yesterday. The lecture was so clear that I did not even have to take notes to understand his solution. I could easily modify his methods to solve the Wiener filtering problem, and derive a computer algorithm to compute the Wiener filter. Immediately after the seminar I changed my lecture notes to derive the Wiener filter according to Professor Gohberg's methods. Then I presented his approach to my class, and have done so ever since. Several years later I invited Professor Gohberg to give a seminar in the Aerospace engineering department at Purdue University. I encouraged many engineering students to attend this seminar, and they were hesitant to hear a mathematician's talk. However, after his talk many students were very impressed and keep asking me questions about his seminar concerning the inversion of infinite matrices and their applications to engineering.

I have found that all of Professor Gohberg's seminars are a real learning experience, and I have always walked away with a clear understanding of the subject area. In fact, after his seminars, I find myself pondering the subject area for weeks and trying to find alternate proofs. Of course, I have never found a better proof for any one of his results.

Operator theory plays a fundamental role in control engineering, communications and signal processing. Since engineers do not have the background of most students in mathematics, it can be difficult to introduce this subject area to engineers. Moreover, some topics in operator theory are more relevant to engineering, than other aspects of operator theory. This makes it difficult for an engineering student to decide exactly topic what to study. However, Professor Gohberg's book: Basic Classes of Linear Operators (with S. Goldberg and M.A. Kaashoek) has solved this problem. In my opinion this is a marvellous text on fundamental operator theory that is readily accessible to engineering students, even undergraduate engineering students. I strongly encourage every student in control theory and signal processing to read this monograph and do the problems. Not only is this book beautifully written and mathematically elegant, it is also loaded with many important results that can be immediately applied to engineering problems. One can almost open up the book at random and find applications to engineering.

Professor Gohberg is the ultimate teacher. I have learned a great deal from him about mathematics and more importantly about life. He has taught me to pick my problems carefully and to work the problems to the end, including all the formulas and algorithms. On the other hand, he has helped me understand what kind of problems to avoid. He has also giving me advice on how to improve my seminars which is greatly appreciated. I always try hard to follow his advice. I am especially happy when he is listening to one of my seminars, because I know that he will always help me to improve my talks.

Finally, it has been a great pleasure and honor working with Israel Gohberg on joint projects, including a book, and learning from him throughout the years.

Professor at Amsterdam: Thank you Israel!

Rien Kaashoek

Two professors from the Vrije Universiteit at Amsterdam (VUA). Israel Gohberg is on the left, Rien Kaashoek on the right.

When I was a young Ph.D. student at Leiden University, with A.C. Zaanen as supervisor, the second mathematics paper that I read was the seminal Gohberg-Krein paper from 1957 about defect numbers and indices of Fredholm operators. At that time Fredholm theory was an exiting new area. This more than 100 pages paper, with its beautiful applications to integral operators of Wiener-Hopf type, made a big impression on me, and I talked so much about this paper that the names

Gohberg and Krein became well-known in my family. Unfortunately M.G. Krein I never met – we exchanged a few letters – but Gohberg became my greatest mathematical friend, and not only mathematical.

The first time we met was in September 1970 at an international conference on "Hilbert space operators and operator algebras" in Tihany (at Lake Balaton in Hungary). The conference was organized by Béla Szökevalvi-Nagy, and a large number of very distinguished mathematicians from the USA, Western-Europe and from the communist countries participated in the conference. In fact, at Tihany many operator theory people met for the first time their colleagues from the other side of the Iron Curtain. Also young "talents" like Bill Helton and Jürgen Leiterer attended this conference.

At that time Gohberg was the head of the functional analysis department of the Moldavian Academy of Science at Kishinev and he did not speak English very well. I was a young professor from the Vrije Universiteit at Amsterdam and did not speak Russian at all. Communication was slow.

At Tihany, from the left to the right: Joe Pincus, Israel Gohberg, Lew Coburn and the late P. Masani.

Unexpectedly we met again five years later in the USA, in College Park. This second meeting would be the beginning of a highly stimulating and productive co-operation[1] for which I am extremely grateful. It changed my life, and it influenced that of Israel. He became a professor at VUA.

---

[1] See the article "The Dutch connection" which I wrote for the "Gohberg Missellania" in *Operator Theory: Advances and Applications* OTAA **40**, and which is reproduced in Part III of the present book.

A simple question

One of the characteristics of working with Israel is connected with a simple question he is asking many times in different versions: *Did you see this - - -? Did you observe that - - -? Did you notice how - - -? Did you pay attention too - - -?* Often I had not seen, observed, noticed, payed attention, but Israel did. Sometimes the questions refer to such mundane things as new shoes of a young girl, the fruits on a tree, or the geometry of a bush on the Tel-Aviv campus. But when talking mathematics, Israel's simple questions are eye-openers, they lead to new inventions, new plans, new results.

Although we are now both retired, I am happy to say that we still collaborate. After his retirement as professor at the Vrije Universiteit in 1998 Israel has been in Amsterdam almost each year for at least a month or so, I have visited him in Tel-Aviv, and we continue to work on joint projects. In fact, last year we sent to Birkhäuser Verlag a new book, written jointly with Harm Bart and André Ran, and the four of us are working on a second book. Together with Leonid Lerer from the Technion in Haifa we completed a series of four papers on continuous operator analogs of Bezout and resultant matrices, which resulted in the solution of the inverse problem of Krein orthogonal functions for the matrix case. These are all topics that have been close to Israel's interests for more than 35 years. More is to come.

Israel: a man of principles.

When Harm Bart and I started to work with Israel on joint projects, he told us some rules for successful collaboration. "In joint work," he said, "one does not make claims about who did what or contributed what. Joint work is based on discussions, on sharing ideas, on combined efforts. Sometimes person A may come up with a suggestion, person B will change it into a conjecture, then person A sees a counterexample, person C presents a positive result, and suddenly person B sees the complete picture and states a theorem, which is then proved in detail by person A. So to whom to contribute the result, who can claim it to be his? Also, the order of the authors should always be alphabetically." We protested: it seemed the rule did not do justice to Israel's leading role. But he convinced us. We agreed to follow it: never to claim that section 1 in a joint paper is from A and section 4 from B, and so on. Israel told us that he learned this rule from M.G. Krein, and at that time we knew already that when Israel tells that he learned something from M.G. Krein, you better take it seriously. The rule has worked greatly, none of us worried about priorities, about competitive claims, and as a result the working atmosphere was relaxed and never tense. I have passed on this rule to other groups I have been working with, and it is always working fine.

Sunday afternoons in Haarlem

When Israel was in the Netherlands he was always invited to visit me and my family on Sunday afternoons in Haarlem. In Summer and Autumn he would usually come on the bicycle (a trip of about 20 km, one way). In the course of the years a fixed pattern developed. First a telephone conversation at noon, at 12 o'clock sharp, to discuss the weather and what to bring. Is the wind not too strong (too anti-semitic) to come on the bicycle, and what about the rain? When he would come on the bicycle, I would often meet him half-way, and we would make little detours into the Haarlemmermeer polder or to the beach. At home we first talked about the families. How is Bella, how is your mother, how are the children, how is your sister? He had stories about what was happening in Israel; we talked about politics, and about what Israel had seen during his bicycle ride: people whom he had met on the way or what on a hot Summer day the Dutch people were doing in the Amsterdam park he had passed through. Next we went to work in my study. We reviewed the plans for the coming weeks, talked about the operator theory seminar (who is going to talk, whom to invite?), discussed the ongoing research, wrote drafts of papers. During these relaxed sessions many new ideas emerged. After the work came the dinner. Israel entertained my children, told jokes and stories. Also Jewish customs were introduced. He gave my three children "Chanukah geld" (in real money, not as chocolate coins), and promised them that they would get next year the same amount increased according to inflation, however provided they would make the right calculations. And, indeed, next year my children presented him their calculations and got considerable raises. It was at the time that inflation was about 10% per year. Sometimes the plans for the dinner changed at the spot. On the way Israel had picked mushrooms and together with Wies, my wife, he cooked a wonderful three course mushroom diner.

Israel Gohberg at IWOTA 2005 with Henri Landau (on the left) and Rien Kaashoek.

The IWOTA conferences

In 1983 we were both at an international operator theory workshop in Santa Monica, CA, USA. Hotels in Santa Monica are not cheap and we shared a room. The day the conference ended I departed for San Diego. Israel was going to stay a night longer and would come to San Diego later. Not ready to pay the full room price, Israel went down to the desk and said to the clerk: "My friend left me today, could you send another gentleman to my room." The clerk stiffened, straightened his back and replied: "Sir, we don't do such things in California." Israel came to San Diego the same day, and the first thing he told us was this event and how it had amused him.

This conference in Santa Monica was an idea of Israel and Bill Helton, and they started it as the first in a series of international workshops on operator theory and its applications (IWOTA). At the beginning these workshops were biannual and meant as satellite conferences to the engineering conferences on Mathematical Theory of Networks and Systems. More recently they are organized annually, and the series is truly international. Up to now there have been 18 IWOTA workshops, held in America, Europe, the Middle East, Africa, and Asia, with three in the Netherlands. Israel is the president of the IWOTA Steering Committee and he introduced two simple and effective rules for the work of the Steering Committee: (1) the committee chooses the site for the next meeting and elects the chief local organizer(s); (2) the sub-themes of an IWOTA workshop and the lecturers are chosen by the local organizing committee.

Thank you, Israel! For your mathematics, your "simple" questions, your good rules, your suggestions, your stories, your friendship. For sharing with me your family and friends.

IG and Me

Peter Lancaster (Calgary)

I was born in 1929 in Appleby, a small county-town in the north of England. I lived with my family for the first 21 years of my life, and we moved around the north-west of England as my father's work in an insurance company dictated. I attended several different primary and secondary schools, but completed secondary education at the Liverpool Collegiate School in 1948. At that time my favourite subjects were art and mathematics. I attended the University of Liverpool for four years, studying architecture for one year before starting in the honours mathematics programme. It became clear that I really enjoyed mathematics and some of its applications and I developed the ambition to teach and do mathematics.

On graduating in 1952 I chose to take a job in the aircraft industry rather than go into the armed forces (national service was compulsory at that time). I worked on interesting problems involving both vibrations and aerodynamics – transonic flight was still in its infancy. As a result, when I was able to change occupations in 1957, and find a university teaching/research position, I had a source of research problems that has influenced my research to this day.

This first university position was at (what is now) the University of Singapore and it was in this period (1957-1962) that I first heard of the research work of the Krein/Gohberg group. I learned that we had problem areas in common, although they brought to bear (for me) a new depth of mathematical technique. In 1962 I moved with my family to what is now the University of Calgary, Canada.

Through the 1960's my contact with the Krein school of mathematics developed further through Heinz Langer, who was a student of M.G.Krein (and therefore well-known to IG), and who spent the 1966-'67 year in Canada. I also learned more of their work through the publication of the 1969 English edition of the famous Gohberg-Krein book on "nonselfadjoint operators".

IG emigrated from Moldavia to Israel in 1974. Then in 1975 he made his first visit to Canada, and I was able to negotiate a stay in Calgary. A fond relationship with me and my extended family evolved quickly. This first visit was in a cold February and IG obviously thoroughly enjoyed his time in Calgary, including the visits we were able to make to the mountains and our cottage at Lac des Arcs. My family also fondly recalls my daughter Jill's fifteenth birthday which we celebrated at one of the best restaurants in town, and which IG shared with us – somewhat startled at the western opulence.

We met again in Europe and in Israel in 1976 and collaborative work got underway. Then I met IG's student Leiba Rodman and an extraordinary mathematically productive period covering several years began. During this time IG made some extended visits to the University of Calgary and, among many other things, we learned of his skill in collecting mushrooms. He had an uncanny ability to find them – even when covered in snow. We would return to the cottage with rucksacks full of mushrooms, spend some time cleaning them, and then IG would usurp my wife, Edna, in the kitchen and prepare the mushroom supper.

These visits were reciprocated in the form of several visits for Edna and me to Israel, where we enjoyed getting to know IG's family – and the country of Israel itself. I learned to enjoy walking and hiking in Israel, especially in Judea and the Negev. Through my friendship with IG I came to build a network of friends in Israel. These included some of IG's colleagues from Kishinev, notably Alek Markus of Be'er Sheva, with whom we enjoyed friendship and some fine mathematics.

It was a great pleasure to take the lead in organising a meeting in honour of IG's sixtieth birthday. This was held in Calgary in 1988 and attracted an extraordinary mix of mathematicians from around the world. I believe they included all of IG's many collaborators in the western world, and it was a great experience for me to get to know more of them. The "iron curtain" was still in place, and we were particularly interested in a group of participants who got permission to come from the Soviet Union; some with invitations and some without. And to top it all, IG's family was also able to join us!

I also like to recall a time in 1984 when I was visiting (appropriately enough) the University of Lancaster – in England. IG attended a conference held at the university which was memorable for poor food but, more importantly, it gave me the opportunity to take IG to my birthplace – Appleby – to meet my octogenarian parents.

To sum up, knowing IG, and being able to join in mathematics projects with him has been, and continues to be, a privilege and a very great pleasure; among my life's most pleasant surprises.

To Israel

Henry J. Landau (Murray Hill, NJ)

When trying to write about Israel Gohberg, his prodigious contributions to mathematics, his inexhaustible optimism, humor, energy, his triumphs over every kind of adversity, where is one even to begin? "Begin at the beginning," I can hear him say, and so I will.

I first met him in the early 1980's, when he visited the Bell Labs. I remember his coming into my office and looking (approvingly) at his book with Feldman which was on my desk, and I remember how quickly we discovered a common interest in entropy. It happened in the same way as he reads a paper, leafing through the pages quickly, at the end of which he knows exactly what was done, what parts (if any) were new or good, and which way the question could be extended or deepened. As for so many of us, it led to wonderful collaborations and personal closeness, first with him and his family but also with so many of his friends.

Of course everybody knew about his brilliant mathematics, but it was a while before I learned of the relentless hardships of his earlier years: his father taken away by the KGB on Israel's 12th birthday, never to be seen again; the family's hairbreadth escapes and displacements during the war; his struggle for an education in Kazakhstan and later, alone, in Kishinev; and throughout, corroding everything, the official Soviet antisemitism. Characteristically, even without information about the West, which was unremittingly vilified by all the Communist media, he understood the true situation. Convinced that their daughters would have no life under the Soviet system, he applied to emigrate to Israel. As he foresaw, this unleashed a huge retaliation and immediately put the entire family into misery and danger. They lost their jobs, he was expelled from the Moldavian Academy, they and their children were pilloried in public meetings, erstwhile friends were afraid to be seen with them, and they were menaced by informers and false rumors. Finally, after a year of such persecution, they prevailed. Of course they then found themselves in a completely new world where they didn't know even the languages, so that Israel, when invited to teach at Stony Brook, had to memorize his lectures phonetically. Still it was not long before his influence took wing. Thanks to him we now have over 500 papers and books, each with some new idea, or new application, or excursion into a new field; problems which themselves seem to generate energy;

journals he founded and edits; centers of research that flourish wherever he goes; and, adding to all this, the warm atmosphere of friendship. One can say simply that by tending to both roots and branches he makes operator theory blossom everywhere.

The role of the past took an ironic turn when the Academy in Kishinev reinstated Israel as a member, and the Universities of Kishinev and of Beltsy, where he began his career, invited him in 2003 to receive Honorary Doctorates. It was not easy for him to agree, but he did, and he, Bella, Zvia and Yanina all returned to Kishinev for the first time in 30 years, to a red-carpet reception. Of course they had not forgotten. Israel accepted the formal presentation in Moldavian, finishing in Hebrew. But the most moving tribute came afterward, when many people joined him informally in a lecture room. Israel acknowledged reality from the outset, saying that people had not always acted well but that he understood the pressures. There followed an outpouring of emotion. One after another, speakers told how clearly they remembered him, and how much his contagious joy in mathematics had meant to them. All seemed to feel a personal loss.

I remember that once at the end of a Toeplitz Conference everybody gathered at Israel and Bella's. There was a large crowd of mathematicians from all over the world, friends, some family and pictures of others, beautiful things in the beautiful house, a wonderfully festive mood, and Israel looked around, seeming suddenly aware of it all, and said, "And I did this all by myself..." You certainly did!

There is a Russian proverb that says, "Don't have a hundred rubles, have a hundred friends" (it sounds better in the original). I propose to improve it: "Don't have a hundred friends, have only one – just make sure it's Iz'ia!"

From a Young Student to an Older One

Jürgen Leiterer (Berlin)

In Spring 1968, at the invitation of Albrecht Pietsch, Israel Gohberg visited and lectured at the University of Jena. At that time I was a young student in Jena, just finishing my diploma thesis with Albrecht Pietsch. From the talk given by Israel, I did not understand much. But still today I see him acting in front of the blackboard, and I remember how I was impressed by the colossal energy and optimism communicated by him, accompanied by kindness and humor.

Then Albrecht Pietsch had the good idea to send me to him as a Ph.D. student. I immediately agreed and, fortunately, so did Israel. In September 1968, I arrived in Kishinev (now Cisinau), and joined the group of colleagues and students headed by Israel. The atmosphere there was very pleasant and I felt happy. A problem was my Russian. I needed a year to learn it to some extent. Fortunately, Israel's German is quite good, and in the beginning he spoke German to me.

In Kishinev I became acquainted with Israel's outstanding capability to fill students and colleagues with enthusiasm. Israel was not only my 'supervisor', he was my 'Doktorvater' in the true sense of the German word for 'supervisor'. I told him that I did not like singular integrals and, in fact, was somewhat afraid of them. He accepted this, and proposed a problem on Wiener-Hopf operators. But, as good fathers sometimes do, he did not tell me the full truth, namely, that this is simply the same. So, at the end, I wrote a thesis on singular integral operators, and I was happy and proud to succeed in a field which I found too difficult before.

In Israel's home I found two 'doctor mothers'. His mother Clara and his wife Bella. They cooked for us when I visited Israel at home to work with him. The meals were always very tasty. On holidays we were often invited to have dinner at Israel's apartment. It was always a big pleasure for me to be with this wonderful family. Sometimes it was not so easy though to satisfy the high demands of Israel's mother, who did not accept empty spots on a dinner plate.

After my Ph.D. thesis was finished, Israel introduced me to problems on factorization of matrix and operator functions. In the beginning of the seventies this led to an intensive collaboration and our first joint papers appeared. Working in this area, we realized that methods from complex analysis of several variables are useful in obtaining results on factorization of operator functions.

That time, at Israel's invitation, Mikhael Shubin from Moscow University visited Kishinev and gave two talks about applications of such methods from complex analysis of several variables. This visit had an important influence on us. Very soon we wrote a series of papers on operator valued cocycles in the case of one variable with new direct proofs and also with new results and applications to operator functions.

After my return to the University of Jena in 1972, this collaboration continued. So in 1973 I visited Israel again in Kishinjev. This was a very important period in my life; I learned a lot from Israel. That time also the idea came up to write a joint book on holomorphic operator functions. Also, encouraged by Israel, I tried to understand more of complex analysis of several variables.

Things changed when Israel emigrated to the State of Israel. Now the Iron Curtain was between us. Rien Kaashoek made an attempt to bring us together in Amsterdam, inviting Israel and me for the same period. But this was made impossible by the East German authorities and after some time our projects were almost forgotten.

After Israel's emigration I became more and more involved in complex analysis of several variables and, finally, I completely switched to this area.

Twenty years later, the Iron Curtain had disappeared and I met Israel again. He proposed to return to our old project and to write the book on holomorphic operator functions. In the beginning of 1996, we wanted to start. For that I planned to visit Tel Aviv for a month. The schedule was already fixed. Then I suddenly got some difficult problems in my private life, and I could not go to Tel Aviv. Although the private problems then were solved more or less, this was again the beginning of a longer interruption of our contacts. Only in autumn 2003, there was a phone call by Israel. He was in Germany and reminded me about the book. I invited him to Berlin, and we made concrete plans once again. This time we were successful. Israel was in Berlin at quite a few occasions, and I visited Tel Aviv several times for longer periods. Now we have February 2008, and there is a realistic hope to finish the book by Israel's 80th birthday.

During the work on the book, I became again involved in this interesting area of mathematics. Once more Israel filled me with enthusiasm and again I learned a lot from him. There is only one difference: I became older.

What About Stability?

André C.M. Ran (Amsterdam)

As a student in 1978 I followed a course on Operator Theory at the Vrije Universiteit in Amsterdam. The professor who gave the course was Rien Kaashoek (although we as students did not dare to address him as Rien, nor even spoke about him as Rien). At some point during the course professor Kaashoek announced that the following week, and the weeks after that, the lectures would be given by a professor Gohberg, who was visiting. He would lecture on Toeplitz operators. The combination of these two teachers hooked me on operator theory, and I decided to try, if I could, to do a PhD under their supervision.

The following year I started as a Ph.D. student on a topic suggested by Israel Gohberg. It was a hard problem, and only much later did we make any progress. Nevertheless, this was good experience. After about a year the topic was changed, and on the new topic rapid progress was made. At about that time also Leiba Rodman was visiting the Vrije Universiteit and we shared an office for a while. It was a question of Israel after a lecture of Leiba which started the collaboration between Leiba and myself. Leiba lectured on invariant neutral subspaces for selfadjoint matrices in spaces with an indefinite inner product. After the lecture, Israel asked: "What about stability?" At that time, stability of invariant subspaces for general matrices had just been settled, in work by Bart, Gohberg and Kaashoek, and by Campbell and Daughtry. Israel's question about stability of invariant subspaces of special classes of matrices turned out to be an important one, and is something that Leiba and I are still working on after more than 25 years (and certainly as many papers).

The first topic suggested by Israel for my Ph.D. work led only much later to some results. The topic was concerned with the result by Markus and Macaev stating that the circle is characterized by the fact that it is the only contour with the property that rational matrix functions of the form identity plus a contractive function always admit both left and right canonical factorization. Israel asked me to study this result in terms of state space realizations. As stated before, I made little progress initially. In 1993, Israel and I co-authored a paper, entitled: *On pseudo-canonical factorization of rational matrix functions*. In it we also discuss the canonical factorizations of rational matrix functions of the form identity plus a contraction. It turns out that dissipative matrices in an indefinite inner product space, and their invariant subspaces play an important role in the state space description of such a factorization. Our result makes it easy to construct examples of non-circle contours and a matrix function of the form identity plus a contraction for which there is either no left or no right canonical factorization (or both). On the other hand, the problem to prove the result of Markus and Macaev via state space methods is still open.

The paper appeared in the Indagationes Mathematicae. As usual, we got a bunch of reprints, with authentic looking covers, which turned our to be useful later. In 1996 I went on a visit to Tel-Aviv, flying via Vienna. On the airport in Vienna, at the security check I was asked numerous questions concerning my intended stay, first by a young lady, and later she was joined by a young man. Obviously, I was a suspect. What saved me in the end, and what got me on the plane, was showing them a reprint of this paper, which fortunately I had with me. After inspecting this they finally believed that I was really a mathematician who wanted to spend a month at the University of Tel-Aviv to work together with Israel Gohberg.

Israel Gohberg and his Influence on my Mathematical Life

Leiba Rodman (Williamsburg)

On August 22, 2008, Professor Israel Gohberg will be 80 years old. I have been in close association with Gohberg for many years in several capacities, as a graduate student, collaborator, junior colleague. On this occasion it seems appropriate to share with others my reflections on Israel Gohberg and on how his mathematics and working with him shaped my professional life.

First, a little background. I was born and raised in Riga, Latvia, at that time part of the Soviet Union. I attended high school with specialization in mathematics and physics, participated in Mathematical Olympiads, and it was clear to me already during teenage years that mathematics or physics was my calling. I finally chose mathematics at the age of 17, enrolling as a math major at the Latvian State University in Riga. After graduation, I emigrated to Israel with my family. I did my graduate studies at Tel Aviv University, eventually obtaining Ph. D. in mathematics under the supervision of Israel Gohberg.

In the fall of 1966 I was a freshman mathematics major. Also, in the same year, the International Mathematical Congress took place in Moscow. This was a very important event in the mathematical life of the Soviet Union, and prominent mathematicians were written up in popular press, including newspapers. As an eager student, I took interest in these publications, and it was then that I learned about the famous mathematician GoKr[1] and his remarkable work in Operator Theory. This was the first time I have encountered Gohberg's name, although in disguised form.

My first meeting with Israel Gohberg in person was inconsequential. I saw him when he came for a short visit at the Latvian State University some time during the late 1960's. Our first consequential meeting took place at Tel Aviv University in the Spring of 1975. At that time I was a graduate student at the Tel Aviv University, enrolled in both M. S. program in Statistics and Ph. D. program in Mathematics, and I was eager to meet Israel Gohberg for the purpose of discussing

---

[1] A mathematician of this name appeared for the first time at the Mathematical Congress held in Moscow in 1966; see Gohberg's Mathematical Tales in Part 1 of this book.

with him possibilities to do my dissertation research under his supervision. My mathematical background was heavily influenced by the school in Abstract Algebra, led by Professor Boris Plotkin, at the Latvian State University. However, my efforts to start doctoral research in this area under algebraists at Tel Aviv University were not productive. One of them suggested that I should contact Israel Gohberg. Since at the time he was traveling abroad, I had to wait for Gohberg's return home, checking frequently if he has come back yet. And quite remarkably, Israel Gohberg decided during our very first meeting, to give it a try. Finding a suitable topic was not easy, as I later understood, given my strong background in Abstract Algebra and then relatively weak background in Operator Theory. Gohberg settled on Matrix Polynomials, a topic that he worked on at that time together with Georg Heinig, Leonid Lerer, and others. This choice turned out to be very successful. Later, Professor Peter Lancaster of University of Calgary came to visit, and the three of us started our long term collaboration that resulted so far in 4 books (the first one, "Matrix Polynomials" was published in 1982) and numerous papers.

Israel Gohberg's influence on my professional life was decisive. I consider him my mathematical father, in many respects. Many of my joint works (books and papers) were done with Gohberg, inspired by his ideas, several of them long after I completed my doctoral studies under his supervision. And much of my subsequent work was inspired by Gohberg as well.

Here, I would like to emphasize just one aspect of mathematical work, namely, writing of mathematical texts, especially original research articles. It is commonly known that writing up mathematics well is notoriously difficult, not the least because one has to explain precise mathematical constructions in terms of imprecise human language. Many mathematicians never fully develop the art of their mathematical writing, to the detriment of mathematical community. Well, this could not have happened to me. From the very beginning of my work under Israel Gohberg, he strongly emphasized quality of writing, in addition to high quality of mathematics (which was understood, of course). It was difficult for me at times to accept that, but I accepted nevertheless. What, I have to rewrite the whole thing just to improve a little one particular paragraph? (This was in the pre-computer days, and the copy and paste option, as well as labels, did not exist, except by scissors and tape.) Only later did I realize what an excellent, valuable school it was. Colleagues often tell me that I write mathematics well; each time I am thankful anew to Gohberg and his schooling.

It was the greatest privilege of my professional life to study and do research under Israel Gohberg's mentorship.

The First Visits of Israel Gohberg to Karl-Marx-Stadt (Chemnitz)

Bernd Silbermann (Chemnitz)

After getting the diploma from Lomonosov University Moscow in 1967, I went to Siegfried Prößdorf who was at that time an associate professor at the Technische Hochschule Karl-Marx-Stadt (Chemnitz was renamed from 1953 until 1990 as Karl-Marx-Stadt). Siegfried Prößdorf was interested in so-called degenerate one-dimensional singular integral operators with continuous coefficients. The degeneracy means here that the symbol of the operator vanishes at a finite number of points. Such an operator is Fredholm if and only if its symbol is invertible. The necessity of this claim was proved by Israel Gohberg in a remarkable paper from 1952 using Banach algebra techniques. Because Siegfried Prößdorf proposed to me a problem concerning degenerate operators, I came across Israel's paper in the very beginning of my career, and it showed me that Banach algebra techniques proved to be a powerful tool in operator theory. So Israel Gohberg entered into my life and he stayed there until now.

Already in my first years with Siegfried Prößdorf the mathematical research of Israel was exceedingly important for us. Especially his investigations about singular integral operators with piecewise continuous coefficients and those about projection methods attracted our attention.

It is one thing to know that there is a person named Israel Gohberg, but it is a quite different thing to see a person in real life. The latter happened when Israel visited Siegfried in the year 1969 in Karl-Marx-Stadt. I do not know exactly but I guess that Salomon Michlin, the teacher of Siegfried, introduced him to Israel. Anyway, in the end of the sixties there were already close relations between Israel and Siegfried. Gohberg gave two talks, one for the staff (on singular integral operators with piecewise continuous coefficients) and one for students. I was surprised to see that Israel talked to the students about a completely different topic namely about the illumination problem from Combinatorial Geometry. I also remember the discussions with Israel about a number of mathematical problems. Afterwards I understood much better what Siegried had in mind when he said that Israel Gohberg is a "General of Science". The visit of Gohberg was also used to introduce

Georg Heinig, a student of Siegfried, to him and later Georg wrote his PhD under
Gohberg's supervision.

The second stay of Israel Gohberg in Karl-Marx-Stadt took place in 1972. He
talked about the stability problem for the finite sections of Toeplitz operators on
the spaces $l_p, 1 < p < \infty$, having piecewise smooth generating functions with
exactly one jump. Israel posed the question whether this result is valid for more
then one jump. Further he mentioned that maybe a suitable local principle will
be of use. Notice that for $p = 2$, Israel already proved that the finite sections of a
Toeplitz operator with piecewise continuous generating function are stable if and
only if the operator itself is invertible, that is a further extra condition does not
occur contrary to the case p not equal to 2. It turned out that he was right; an
answer was found years later, but this is a different story.

What I wrote indicates that Gohberg's spirit was of decisive importance for the
work done by my colleagues and myself in Chemnitz. We all were deeply impressed
by his personality, his humor and friendship.

Siegfried then spent one year with Israel in Kishinjew and wrote there his first
book. Coming back in the summer of 1973 he told me that Israel will leave the
Soviet Union. This was a shock for me because I was afraid that I never will see him
again. I was too pessimistic and met him again by the end of 1989 in Oberwolfach.
All the time, beginning with the seventies, I was aware of his support, help and
friendship. This fact is one of the most exciting experiences of my life. Thank you
very much, Israel.

My Gohberg Encounters

Ilya Spitkovsky (Williamsburg)

I first met Professor Gohberg in person only twenty years ago (the word "only" is appropriate here, considering that we are about to celebrate his 80th birthday) – his presence in my life, however, long predates that initial meeting.

When I was asked to contribute to this issue, I found myself reflecting on the role Israel Tsudikovich has played in shaping not only my mathematical career, but also my life.

The first significant mathematical problem I wrestled with as a young man was a completion problem, presented by Mark Grigorievich Krein, in the first reading stated for unitary matrices, then for unitary operators on Hilbert spaces, and finally in the setting of J-unitary operators on what is now known as Krein spaces. Incidentally, as I was informed in due course, the problem as stated for unitary matrices had already been solved much earlier by Gantmacher and Krein, and had been presented to me only as a test of my abilities.

This evolution lasted for two years, and although I was very much taken with the never-ending process of mathematical discovery, I could not help wondering whether a publication would ever result from the work - though I was far too shy to ever ask Mark Grigorievich directly. In a fortunate turn of events, Israel Gohberg soon paid a visit to Krein, who mentioned my recent work to him. To my delight, Professor Gohberg not only judged my work worthy of publication, but offered to consider it for "Matematicheskie Issledovaniya" (Mathematical Research) — the journal he founded and continued to edit in Kishinev. This marked my first "Gohberg encounter."

My second mathematical result stemmed from an unexpected coincidence. In the summer of 1973, I was studying two seemingly unrelated questions: partial indices of the Wiener-Hopf factorization — calculation of which is still an intriguing open problem — and a classical notion of the numerical range, examined as part of an enthusiastic study group following the famous *Finite-Dimensional Linear Analysis in Problems*, by Glazman-Lyubich. In a moment of clarity, the two things clicked, resulting in a factorability statement for matrix function with the numerical range not containing the origin.

I was pleased with the result – in fact, I still am – and proudly showed it to Mark Grigorievich, who then discussed it with Israel Tsudikovich. Upon hearing Krein declare, "If Israel and I had known this result when we were working on our joint paper for Uspekhi thirty years ago, we would have included it there," I was inspired to submit my work for publication. This marked my second "Gohberg encounter."

In 1974, Professor Gohberg emigrated from the Soviet Union to Israel. His influence remained in the form of articles and books he had written (and continued to write, with even greater efficiency as the years went by), but personal contact became virtually impossible for the next fifteen years. With Perestroika, however, things began to change, and in1989 I was able to secure permission to attend an international conference in Calgary, Canada, in honor of Israel Tsudikovich's 60-th birthday. That is where I finally met Professor Gohberg in person.

Needless to say, the Calagary conference was in many ways an eye-opening experience – this being my first trip beyond the Iron Curtain, at the age of 36. My most striking memories from that conference, however, were my conversations with Israel Tsudikovich and members of his family. Professor Gohberg's viewpoints, non-intrusive but firm, planted a seed that influenced my decision to follow his example of 15 years prior. This marked my third, and decidedly most important, "Gohberg encounter."

The logistics of emigrating from the Soviet Union took time, of course, the primary issue being the need to secure an invitation. My wife and I made several attempts, turning for help to friends and mere acquaintances alike. As the saying goes, many promised, but only one delivered. It is to Professor Gohberg that I owe the great fortune that my family was able to immigrate to the United States where we have lived for the last eighteen years. It is with great pleasure that, twenty years since our initial meeting, I am once again honored to take part in celebrating another jubilee in honor of Professor Israel Gohberg.

Becoming a Co-author

Freek van Schagen (Amsterdam)

In 1975 the chairman of the Mathematics Department of the Vrije Universiteit in Amsterdam advised the Board of the Department to accept a proposal of Rien Kaashoek to invite Israel Gohberg for a visit in the fall of 1976. At that time I was a junior member of the Board. I was happy to agree, not knowing how much this visit of Israel and the many visits to follow would change my own life.

In the first half of 1976 a group of mathematicians at the Vrije Universiteit studied a paper of Gohberg and Leiterer. Although not an analyst by training – my Ph.D had been in Algebraic Geometry – I was invited to join the group and made my algebraic background useful. In the fall of that year Israel Gohberg gave an inspiring series of lectures on singular integral equations and Toeplitz operators, and a seminar on the newly developed spectral theory of matrix polynomials. This seminar was to be the starting point of my research in Operator Theory. The first joint paper with Israel and Rien was on common multiples of operator polynomials with coefficients that were analytic in an extra parameter. Later the attention shifted to completion problems and canonical forms of partially given objects, and this topic became a true project from which a series of papers evolved. The results were compiled and extended further in our joint book "Partially Specified Matrices and Operators: Classification, Completion, Applications," OT 79.

Needless to say that listening to and discussing with Israel Gohberg is a very pleasant and efficient way to learn mathematics. But it was not only mathematics that I learned. For example, we, that is the operator theory group in Amsterdam, had the habit to have test lectures for conference presentations on joint work. Israel, but also Rien Kaashoek, helped me improving my lecturing greatly. They identified my weak points. For instance, they gave me the wise advice to finish a sentence before starting a new one, even when I already saw that the sentence I had started was not the optimal one. Also you should not try to expand 'bright' ideas on the presentation on the spot and during your lecture. Furthermore, be precise in your statements in order not to loose precious time in clarifying vagueness.

Currently I am director of the bachelors and masters programme in mathematics at the Vrije Universiteit. While teaching and managing these programmes, I am aware that I pass on ideas that Israel taught me. I mention a few of these things.

Leave things for others to do, even if you think that you yourself might do them better. Also, students should be stimulated to migrate to a research environment where their particular talents are developed best. So they should go to places where the right questions are posed. Further, not every mathematical problem is worth the effort it costs to solve it.

For my own research the guiding Israel Gohberg offered me, and which I think I almost always accepted, has led to joint papers, many of which I am still proud of. Often these papers and this is also true for our book, had many versions. The weak and strong points in preliminary contents and presentation had to be treated accordingly in order to end up with a result that satisfied the standard Israel required. I must admit that I was often happy with the result before he was.

I am sure many others had similar experiences with Israel and are equally grateful to him. To conclude: thank you very much Israel!

International Congress of Mathematicians, Moscow, 1966: Israel Gohberg in the second row; on the left of him are Professor A.R. Holves from Tbilisi (who is now in Israel), Israel Fel'dman and Alek Markus.

srael Gohberg: the picture was aken in 1967 soon after he became rofessor.

April 1959: Israel Gohberg and Chandler Davis in Kishinev, on the left Israel Fel'dman, on the right Alek Markus.

Israel Gohberg visiting Bela Sz-Nagy at the University of Szeged in 1969; picture taken in front of the building of the Department of Mathematics with staff members of this department. On the left of Sz-Nagy is Dumitru Gaspar from Timişoara.

Israel Gohberg with Thomas Hempfling, the present Birkhäuser Mathematics Editor.

At IWOTA 2006 in Seoul, Korea. From the left to the right: Woo Young Lee (Professor, Seoul National University, Korea), Jun Ik Lee (Professor, Sang Myoung University, Korea), Israel Gohberg, Rien Kaashoek, Bella Gohberg, and An Hyoun Kim (Professor, Changwon National University, Korea).

Israel Gohberg and Seymour Goldberg.

Israel Gohberg in his office at the University of Tel Aviv.

Israel and Bella Gohberg in a relaxed mood, back in the Netherlands (Haarlem) from Timişoara, after having received the honorary doctorate.

Pictures taken at a party in Harm Bart's garden in Bleiswijk on the occasion of Israel's 75th birthday

Israel Gohberg with Heinz Langer (left), Rien Kaashoek (first right) and André Ran (second right).

Israel Gohberg with Bernhard Gramsch (left) and Reinhard Mennicken (middle).

Israel and Bella Gohberg.

Tête-a-tête between Israel Gohberg and Harm Bart.